WHAT OTI

PARALLELS —
Ancient Insights into Modern UFO Phenomena

by Richard L. Thompson, Ph.D.

"Carefully researched and well written . . . [Richard L. Thompson's] approach to the UFO problem is refreshingly innovative. . . . *Parallels* is a major addition to ufological literature, too much of which in recent years has come to seem overspecialized and repetitive."

— Roger W. Wescott, Professor of linguistics and anthropology, Drew University

"*Parallels* is stimulating, well-organized, and an encyclopedic smorgasbord of UFO data and ancient Indian counterparts. . . . What a refreshing treat. Richard L. Thompson is obviously a scholar who knows his subject and has read and integrated a wealth of literature. . . . He demonstrates considerable skill, knowledge and penetrating insight with his divergent hypotheses for the disparate pieces of information. . . .

Parallels: Ancient Insights into Modern UFO Phenomena, will not be read on one sitting, nor will it be easily set down. Beware! For like an excellent detective story, the book can become cognitively addicting. . . . *Parallels* stands to become a classic."

— Berthold Schwarz, M.D., psychiatrist and author of *UFO Dynamics*

"[*Parallels*] will contribute a valuable insight into the possible origin of UFOs. Bravo! *Parallels* is the much searched for Rosetta Stone of UFO literature."

— Clark C. McClelland, pioneer NASA aerospace engineer, United States Space Program, Kennedy Space Center, Florida

"Though a few books and articles have briefly flown over the common ground of UFO sightings and contact, and the descriptions of aerospace craft and nuclear-like weapons in Hindu scripture, Richard L. Thompson's *Parallels: Ancient Insights into Modern UFO Phenomena* turns a scholar's satellite lens on this fascinating and mysterious subject. . . . *Parallels* is a mind stretcher that goes as far as the evidence can allow for UFO-type phenomena in ancient India."

— *Hinduism Today*

"As an ordained Tibetan Buddhist monastic, I refer to texts from India as the ultimate source material by which to answer questions. Both oral and written teachings which I have given throughout the years have focused on comparing the descriptions of "vimanas" in the Vedic texts to modern Western UFO investigations. Having been involved in this work for decades, I highly recommend Dr. Thompson's book as one which will help Westerners understand the Vedic knowledge pertaining to so-called UFO phenomena."

— Reverend Mary Teal Coleman, Dharma Institute

"*Parallels* provides the reader with a perspective that not only compares UFO events from different cultures but also from different ages. . . . [Richard L. Thompson] offers a balanced approach for the psychological evaluation of UFO Experiences, but he shows courage and scholarship in addressing all levels of evidence: physical, biological, psychosocial and spiritual levels of experiences. Physical traces, abduction reports, the role of the government, and UFO communications, all are grist for the author's mill. . . . Part One is an excellent summary of the problems and findings of modern UFO research."

— R. Leo Sprinkle, Ph.D., Counseling psychologist, Wyoming, U.S.A.

"Like Jacques Vallee's *Passport to Magonia, Parallels* is absolutely seminal, a real ground breaker. It looks at this peculiar phenomenon with marvelous insight and imagination. A "must-read" addition to the field."

— Whitley Strieber, author of *Communion*

"Brilliant! *Parallels* is a remarkable comparison of some of the best and least-studied information we have on UFOs. . . . Thoroughly researched, thought-provoking and eminently readable, I could not put it down!"

— Tricia McCannon, founder UFO Forum, Atlanta, U.S.A.

"Excellent research material for scholarly UFO students . . . for advanced students and beginners alike."

— *Ūnicŭs* magazine

"A *Summa Ufologica* of great importance. . . . Thompon has written perhaps the most honest UFO book extant in the English language."

— *Gnosis* magazine

"*Parallels* is brilliant. Thompson's work is both universal and detailed, yet put over in a very readable layman's way. . . . I highly recommend this book."

— Ken Phillips, editor of *BUFORA* magazine, U.K.

"A new and important book now joins the expanding literature on UFOs. But this book is refreshingly different in its approach. Richard L. Thompson has used a much larger canvas to present his logical and scholarly work. . . . The author has meticulously researched material from all over the world in writing this book. [*Parallels*] is a veritable encyclopaedia of information on this subject. It is likely to become a classic in the world of UFOlogy."

— Joan Wilder, *Flying Saucer Review,* London, England

"I enjoyed the book, and found the parallels Dr. Thompson points out in Vedic literature to modern UFO lore to be most interesting. His review of UFOlogy is also well done. He strikes a fine balance between the physical and psychic aspects of UFO encounters. The book stands well on its own, and should be an essential addition to the library of anyone interested in a transcultural and historical analysis of the UFO phenomenon. . . . I can (and will) recommend it highly to my colleagues, to experiencers, and to anyone interested in the UFO phenomena."

— David Gotlib, M.D., Toronto, Canada,
Editor of *Bulletin of Anomalous Experience*

PARALLELS

ANCIENT INSIGHTS INTO MODERN UFO PHENOMENA

PARALLELS

ANCIENT INSIGHTS INTO MODERN UFO PHENOMENA

RICHARD L. THOMPSON, PH.D.

INSTITUTE FOR VAISHNAVA STUDIES
Gainesville, Florida

First edition, first printing: 1993
First edition, second printing: 1994
Revised second edition, first printing: 1995
Third edition, first printing: 2017

Published by Institute for Vaishnava Studies (IVS)

Contact information:
Web: www.richardlthompson.com
Email: rlthompsonarchives@ivs.edu
Write to: Richard L. Thompson Archives, 21834 NW 151 Road, Alachua, FL 32616

Cover images:
UFO: ©mik38 | iStockphoto.com
Background: ©suratoho | iStockphoto.com

Cover and book design: Robert Wintermute

Cataloging-in-Publication Data

Thompson, Richard L.
Parallels: ancient insights into modern UFO phenomena /
Richard L. Thompson.
p. cm.
Includes bibliographical references and index.
Preassigned LLCCN: 92-76167
ISBN 978-0-9981871-4-3

1. Unidentified flying objects—Sightings and encounters.
2. Unidentified flying objects—Sightings and encounters—India.
3. Civilization—Extraterrestrial influences. 4. Civilization, Hindu—
Extraterrestrial influences. 5. Unidentified flying object literature.
I. Title.

CB156.T46 1993 001.942'0934
QBI93-720

PUBLISHER'S NOTE: This book was originally published as *Alien Identities: Ancient Insights into Modern UFO Phenomena,* in two editions. This third edition is a reprint of the second edition of *Alien Identities*, minus the Foreword. The new title, *Parallels*, was the author's original choice.

Dedicated to

His Divine Grace
A. C. Bhaktivedanta Swami Prabhupāda

who wrote *Easy Journey to Other Planets*

Other books by Richard L. Thompson

MECHANISTIC AND NONMECHANISTIC SCIENCE
An Investigation into the Nature of Consciousness and Form

MYSTERIES OF THE SACRED UNIVERSE
The Cosmology of the *Bhāgavata Purāṇa*

VEDIC COSMOGRAPHY AND ASTRONOMY

MAYA
The World as Virtual Reality

GOD AND SCIENCE
Divine Causation and the Laws of Nature

FORBIDDEN ARCHEOLOGY
The Hidden History of the Human Race
(Co-authored with Michael A. Cremo)

THE HIDDEN HISTORY OF THE HUMAN RACE
(Abridged version of *Forbidden Archeology*)

CONSCIOUSNESS: THE MISSING LINK
by His Divine Grace A. C. Bhaktivedanta Swami Prabhupāda,
Dr. T. D. Singh and Richard L. Thompson

Contents

Acknowledgements

Progress in the study of human experience inevitably depends on the accumulated efforts of many people. I am indebted to the following UFO authors, investigators, and witnesses for providing important information used in this book:

George Adamski, Maury Albertson, Orfeo M. Angelucci, Thomas Bearden, Charles Berlitz, Michael Bershad, Ted Bloecher, Howard Blum, Charles Bowen, Lyle G. Boyd, Richard and Lee Boylan, Thomas E. Bullard, John Carpenter, Martin Cannon, Bill Chalker, Aphrodite Clamar, Edward U. Condon, Ed Conroy, Gordon Creighton, William Curtis, Terence Dickinson, Paul Dong, Barry H. Downing, Ann Druffel, George Earley, Don Elkins, Lawrence Fawcett, Edith Fiore, Raymond Fowler, Stanton Friedman, John G. Fuller, Paul Fuller, Encarnacion Garcia, Timothy Good, Elmer Green, Barry J. Greenwood, Bryan Gresh, Richard F. Haines, Leah Haley, Richard H. Hall, James Harder, Richard C. Henry, William J. Herrmann, Charles Hickson, Cynthia Hind, Budd Hopkins, Linda Moulton Howe, Antonio Huneeus, J. Allen Hynek, Philip J. Imbrogno, David M. Jacobs, Donald A. Johnson, Carl G. Jung, Parli R. Kannan, Alvin H. Lawson, Meade Layne, Desmond Leslie, Coral Lorenzen, Jim Lorenzen, Bruce S. Maccabee, John Mack, Victor Marchetti, William Markowitz, James A. McCarty, James E. McDonald, William Mendez, Donald H. Menzel, William L. Moore, Kanishk Nathan, Thornton Page, Ted R. Phillips, William T. Powers, Bob Pratt, Kevin D. Randle, Jenny Randles, Antonio Ribera, Kenneth Ring, Franklin Roach, August Roberts, D. Scott Rogo, Carla Rueckert, Carl Sagan, John R. Salter, Virgilio Sanchez-Ocejo, Ivan Sanderson, David R. Saunders, Larry Savadove, Donald R. Schmitt, John F. Schuessler, Berthold E. Schwarz, Frank Scully, Michael Seligman, Margaret Shaw, Elizabeth Slater, Nicholas Spanos, William S. Steinman, Wendelle C. Stevens, Ronald Story, Whitley Strieber, Barry Taff, Rolf Telano, Karla Turner, Denise and Bert Twiggs, Jacques Vallee, Jean-Jacques Velasco, Ed and Francis Walters, Travis Walton, David F. Webb, Walter N. Webb, Roger Wescott, Dan Wright, Jennie Zeidman, and Lou Zinsstag.

I am also indebted to many other authors, including the following:

D. P. Agrawal, S. Maqbul Ahmad, Vaman S. Apte, St. Augustine, Alice A. Bailey, Prithivi Bamzai, Raja Bano, John Bentley, Bhaktisiddhānta Sarasvatī Goswami Ṭhākura, His Divine Grace A. C. Bhaktivedanta Swami Prabhupāda, Stephen E. Braude, David H. Childress, William R. Corliss, Lawrence Daly, Ramachandra Dikshitar, Theodosius Dobhzhansky, H. M. Elliot, Walter Y. Evans-Wentz, Roland Mushat Frye, Kisari Mohan Ganguli, Robin Green, Edwin S. Hartland, Hudson Hoagland, Hridayānanda Goswami, Francis Johnston, G. R. Josyer, Walter Kafton-Minkel, Dileep Kumar Kanjilal, Kenneth Lanning, Terence Meaden, Janardan Misra, F. W. H. Myers, Christian O'Brien, Richard Ofshe, Gustav Oppert, Satwant Pasricha, J. S. Phillimore, Marmaduke Pickthall, Pope Pius XII, C. S. R. Prabhu, V. Raghavan, Louisa Rhine, Robert Rickard, William Roll, Steven Rosen, Michael Sabom, Sanātana Goswami, Bapu Deva Sastrin, Satsvarūpa Dāsa Goswami, Hari Prasad Shastri, George Gaylord Simpson, Zecharia Sitchin, M. A. Stein, Ian Stevenson, J. A. B. Van Buitenen, Leonid L. Vasiliev, John A. Wheeler, and H. H. Wilson.

I would like to thank Raymond Fowler and Jay Israel for useful criticisms of the manuscript and Michael Cremo for many helpful discussions. I also thank Thomas Doliner, Gary Loscocco, and Jeffrey Soloman for proofreading, Christopher Beetle and Dave Smith for proofreading and typesetting, and Robert Wintermute for layout and design. Others who contributed in various ways are Michael Best, Sigalit Binyaminy, Austin Gordon, Tricia McCannon, and Scott Wolfe. Special thanks to James McDonough and Hawaii Vedic College for investing in the printing of the book.

Introduction

Today, if we scan the shelves of a university bookstore, we find many books describing the triumphs of science. These books tell us that physicists have understood the laws governing ordinary matter and molecular biologists have explained life in terms of these laws. Although some scientists still find consciousness perplexing, we are told this problem will soon be solved by studying the brain. The books assure us that modern science has understood the evolution of species, the origin of life from the primordial soup, and the processes that formed planets, stars, and galaxies. The frontier of fundamental physics has been pushed back to the beginning of the Big Bang, and we are on the verge of the final advance that will give us the Theory of Everything.

Yet these same shelves occasionally display books about anomalous evidence that contradicts accepted scientific theories. There are several categories of anomalous evidence, including psychical phenomena, out-of-body experiences, past-life memories, cryptozoology (e.g., Bigfoot), and archeological anomalies.

In recent years, however, the most prominent anomalous category has been UFOs—Unidentified Flying Objects. Covers proclaim that thousands of responsible people have seen something inexplicable flying in the sky. Other covers announce visitations by strange-looking aliens, and still others hint darkly at sinister conspiracies and cover-ups. These books say that for decades people have been observing unknown flying objects that drastically violate the known laws of physics. They also declare that people have encountered humanlike beings that pilot these strange craft and exhibit powers contrary to both science and common sense.

For years I have been interested in the relationship between modern science and the ancient Vedic world view of India. I have been particularly concerned with the contrast between the mechanistic model of life developed by modern science and the animistic (or soul-based) conception of life that lies at the foundation of Vedic philosophy.[1] The scientific model of life is based on experiments and careful reasoning, but by reducing life to a combination of atoms, it deprives it of all higher purpose and meaning. It reduces human values to behavioral patterns produced by cultural and physical evolution. These behavior patterns are contingent on accidental historical circumstances, and they have little to do with the fundamental nature of things.

In contrast, the Vedic philosophy gives meaning to life by linking it with a transcendental level of reality, but in the process it brings in phenomena and categories of being that have no place in the theoretical picture of modern science. This naturally gives rise to the question of where the truth lies. Has modern science given a complete outline of life's fundamental principles, or has it perhaps given only a detailed but narrow account of certain limited aspects of life?

With these interests, I was naturally somewhat intrigued when books on UFOs (such as *Intruders*, by Budd Hopkins) first began to appear prominently in the science sections of university bookstores. These books seemed as though they might shed some light on the nature of life, since they reported encounters between human beings and other intelligent life forms. But were any of the UFO reports credible?

Like many people, I had always regarded the topic of UFOs as disreputable. But on reading some of these books more carefully, I saw that they seemed to contain substantial, though anecdotal, evidence for some very unusual occurrences. In particular, they seemed to be giving contemporary eyewitness accounts of many different life-related phenomena that are described in ancient Vedic texts. This led me to investigate the UFO phenomenon in greater depth and finally to write this book.

This book is a comparative study of UFO literature and the Vedic literature of India. In the first five chapters, I give a broad survey of the material written on UFOs from the late 1940s up to the present. I have included this survey in an effort to give the reader an overview of the reported UFO phenomena.

The remaining six chapters introduce the Vedic literature and present detailed comparisons between phenomena reported in Vedic accounts and corresponding phenomena mentioned in UFO reports. The Vedic material is taken mainly from the *Bhāgavata Purāṇa* and the *Mahābhārata*. I also draw on the *Rāmāyaṇa* and various late medieval texts that follow Vedic tradition.

For Indologists I should point out that the *Purāṇas*, the *Mahābhārata*, and the *Rāmāyaṇa* are called the fifth *Veda* in text 1.4.20 of the *Bhāgavata Purāṇa*. Therefore, I will freely use the term "Vedic" to refer to them, even though some scholars insist that this term can be properly applied only to the *Ṛg Veda*.

The idea of comparing UFO accounts with Vedic literature is not new, but generally this has not been done in a scholarly way. The first attempt that I am aware of is the 1953 book entitled *The Flying Saucers Have Landed,* by Desmond Leslie and the famous contactee George Adamski. In the first part of this book, Leslie quoted a number of passages from the *Rāmāyaṇa* and the *Mahābhārata* describing *vimānas,* or Vedic flying machines,[2] and a number of passages describing remarkable weapons that were used in Vedic times.[3] Unfortunately, many of these passages are badly mistranslated, and Leslie's account is practically worthless.

Similar mistranslated passages from the *Mahābhārata* have appeared in a number of popular books that follow in Leslie's footsteps. Here is an example showing how misleading these bad translations can be. Leslie quotes the following passage from the *Karṇa Parva* in Pratap Chandra Roy's edition of the *Mahābhārata*:

> Karna took up that terrible weapon, the tongue of the Destroyer, the Sister of Death, a terrible and effulgent weapon. When the Rakshasas saw that excellent and blazing weapon pointed up at them they were afraid. . . . The resplendent missile soared aloft into the night sky and entered the starlike formation . . . and reduced to ashes the Rakshasa's vimana. The enemy craft fell from the sky with a terrible noise.[4]

This passage appears in the *Droṇa Parva* of the *Mahābhārata,* not the *Karṇa Parva*, and here is what it actually says in the edition of Pratap Roy:

> . . . that fierce weapon which looked like the very tongue of the Destroyer or the sister of Death himself, that terrible and effulgent dart, Naikartana, was now hurled at the Rakshasa. Beholding that excellent and blazing weapon capable of piercing the body of every foe, in the hands of the Suta's son, the Rakshasa began to fly away in fear. . . . Destroying that blazing illusion of Ghatotkacha and piercing right through his breast that resplendent dart soared aloft in the night and entered a starry constellation in the firmament. Having fought . . . with many heroic Rakshasa and human warriors, Ghatotkacha, then uttering diverse terrible roars, fell, deprived of life with that dart of Sakra.[5]

Instead of reducing a *vimāna* to ashes, the weapon killed the Rākṣasa Ghaṭotkaca, and instead of an enemy craft falling with great noise, the Rākṣasa fell while uttering terrible roars. I don't know how Leslie came up with his mistranslation, but it is typical of his book and others of its genre.

Nonetheless, there is a great deal of material in Vedic literature about flying machines, called *vimānas,* that show striking resemblances to UFOs. Even more important are Vedic accounts of the behavior and powers of humanlike races that use these flying machines. There are many parallels between specific details in these accounts and corresponding details in UFO close-encounter cases. These parallels provide my main impetus for writing this book.

Parallels between UFO cases and old Celtic and Germanic folklore have been explored by Jacques Vallee in two books, *Passport to Magonia* and *Dimensions*. In a sense, this book is an extension of Vallee's comparative method to the domain of ancient Indian culture.

Epistemology

Many people regard UFO research as something that is not intellectually respectable, and this includes many who are interested in Vedic thought either from a scholarly or a traditional religious point of view. At the same time, many serious UFO researchers feel that to bring old mythology into a discussion of UFOs is unscientific and can only lead to useless mystical speculation. It is therefore important for me to give some justification for writing on the theme of UFOs and Vedic literature. I will begin by making a few observations on the shortcomings of UFO evidence, and then I will point out how the study presented in this book might help overcome them.

One notable weakness in this body of evidence is that there appears to be no way of performing a reproducible experiment that will give us reliable information about UFOs. The problem is that UFOs are connected in many reports with humanlike beings that seem to have superhuman technological powers. If this is true, then we can study these beings only to the degree that they are willing to reveal themselves to us. But if "they" exist, they have shown little willingness to cooperate with human investigators. There is even evidence suggesting that they may try to deliberately keep people in the dark about their activities and their real nature. Thus UFO phenomena may be inherently difficult to study by standard scientific methods.

Yet even if a phenomenon is completely unpredictable and uncontrollable, it might still be expected to leave some "hard" evidence that can be scientifically evaluated. Where are the photographs and instrument readings that record the flight of UFOs? Where can we find UFO hardware or tangible physical evidence of UFO landings and other activities?

Interestingly enough, there are many reports of hard UFO evidence in the form of ground traces of landings (pages 66–69), photographic records (pages 73–77), and physical injuries suffered by UFO witnesses (pages 129–30 and 346–48). In addition, books have been written arguing that government authorities have collected a great deal of high-quality UFO evidence, which they keep hidden on the grounds of military secrecy. I myself have been told by an engineer involved in military weapons testing that high-quality UFO photos were regularly taken in the 1950s by technical personnel known to him, but that these photos have never been revealed to the public (pages 29–30). I discuss this controversial topic in Chapters 1 and 3.

Although hard evidence for UFOs does exist, the evidence readily available to the public says very little by itself. It can only take on significance in the context of a complete story involving witnesses whose honesty and competence can be evaluated.

This is illustrated by the following example. In 1987, a building contractor named Ed Walters claimed to have videotaped a UFO flying near his backyard in Gulf Breeze, Florida. Dr. Bruce Maccabee, a physicist and UFO investigator, made the following evaluation of this videotape:

> If the only information about this sighting were Ed's testimony and the pictorial information contained within the videotape itself, . . . I would seriously consider the hoax hypothesis in spite of the demonstrated difficulty in duplicating the videotape. However, taken in the complete context of the sighting, with the other witnesses testifying that they saw Ed videotape the UFO, I conclude that Ed did not produce his videotape using a model. Rather, I conclude that he videotaped a True UFO.[6]

Maccabee was able to rule out the hoax hypothesis only by interviewing the people involved in making the tape and evaluating their character and motives. If his judgement was right, then the tape provides useful information about the appearance and flight of UFOs. But whether he was right or wrong, most people must depend on his report in order to assess the validity of the tape. The only alternative would be to go to Gulf Breeze and conduct one's own investigation. But as time passes this option becomes less and less feasible.

We can conclude that most readily available UFO evidence takes the form of reports in which the testimony of witnesses and investigators is of crucial importance. Since one cannot arrange to see UFO phenomena at will, and hard evidence is meaningless without accompanying testimony, one has no alternative but to rely on such reports or make one's own investigations. And one's own investigations simply result in more reports for people to read.

In this book, I will take UFO evidence mainly from a wide variety of available reports, but unfortunately I will not be able to prove the truth of any of the reports that I cite. Proof, in so far as it is possible, can come only from in-depth investigation of particular cases. Of course, some of the cases I present have been extensively investigated, and the investigators have concluded that the cases are genuine. However, I am not in a position to prove that they are right or show that their investigations were really carried out properly.

Since my survey of UFO material is intended to be comprehensive, it inevitably combines material that is relatively well attested with material that seems particularly dubious. I have included some dubious material because suppressing it would result in a false picture of the UFO evidence. An artificially sanitized picture of the UFO scene would not be realistic, and I should warn the reader that when I present the evidence for certain claims, that does not necessarily mean that I regard those claims as valid. In some cases, my aim is to alert the reader to the kind of false material that can be found in the UFO field.

Some doubtful-looking UFO-related material is certainly phony. Yet we should be cautious about superficially rejecting things as false simply because they seem absurd. Information that initially seems absurd or meaningless may prove to be highly significant when seen later on in a broader context. Absurdity is defined only in relation to an accepted theoretical view, and as theoretical insight develops, the status of evidence as absurd or anomalous may also change.

It has been said that to listen sympathetically to "absurd nonsense" compromises one's credibility as an objective, scientific thinker. Perhaps this is so, but I would suggest that sympathetic attention to evidence is essential if real advances in scientific knowledge are to be made. As science advances, ideas once regarded as absurd may later become orthodox. Examples would be the idea that continents drift across the surface of the globe, or the idea that electrons tunnel through energy barriers. Of course, other ideas may prove to be actually invalid, including some that are accepted by mainstream scientists.

To keep this book from becoming too large, I have inevitably given emphasis to some UFO cases and neglected others. The same is true of my treatment of Vedic material. Hopefully, the sample of cases I have chosen is representative, and the same points could be illustrated using a different representative set of cases. Although certain cases are repeatedly mentioned, this does not mean that I regard them as being uniquely significant.

Since I am concerned with tracing patterns in collections of modern and ancient reports, this book can be regarded as a comparative study of folklore. It is certainly legitimate to study folklore, and for many readers this may be the best way to make an initial approach to the subject matter of this book. However, in the background of any study of folklore, there is always the question of how the folklore really originated. Is it simply a product of imaginative storytelling motivated by psychological factors, or does it have a basis in objective reality? In the next section, I will make a few preliminary observations on this topic.

Explanations for the Origin of UFO Lore

There is a saying that "amazing claims require amazing proof." This commonsense idea creates a problem in the interpretation of UFO evidence: a report of amazing proof is itself an amazing claim which, in turn, requires more amazing proof. The ironic result of this is that a case with elaborate proof may seem less credible than a case with relatively little in the way of proof.

For example, suppose that someone claims to have seen a flying object unlike any known manmade vehicle. This is an amazing claim. But if he offers a photograph of the object as proof, then that photograph constitutes another amazing claim. We can always suppose that the photograph might have been a hoax.

If he offers a large number of high-quality photographs as proof, then his claim becomes even more amazing, and our suspicions of fraud may become even greater. For example, Ed Walters of Gulf Breeze, Florida, published a book containing many remarkable UFO photos that he claimed to have taken with a Polaroid camera.[7] These photos were judged to be genuine by an optical physicist (Bruce Maccabee). But many readers reacted by saying that because of their very quality, they must have been faked. As one reviewer put it, "I am reminded of the warning frequently given by bunco squad detectives: "If it sounds too good to be true, it probably is.' Gulf Breeze sounds too good. . . ."[8]

To make things worse, there is evidence suggesting that there have been massive UFO hoaxes on a scale requiring considerable amounts of manpower and money. A possible example is the case of the Swiss contactee Eduard Meier, which is supported by—among other things—high-quality photos and movies, eyewitness testimony, photos made by eyewitnesses, UFO sound recordings, UFO ground traces, and professional analysis of UFO samples by a prominent IBM engineer.[9, 10] As far as I am aware, no one has actually proven that this case is a hoax, but it is a definite possibility. Such cases simply add more weight to the idea that large amounts of high-quality evidence are a cause for doubt rather than confidence.

If UFO reports become more doubtful the more evidence there is to support them, then why should any reports of this kind be taken seriously? The reason is that there are large numbers of apparently independent UFO reports from around the world, and these tend to be very similar in content. These reports are often made by respectable people who have no obvious motive for making up a bizarre story and exposing themselves to ridicule. Roughly speaking, five possible explanations are generally advanced to account for this:

1. UFO reports result from natural illusions or misperceptions. For eample, people may mistake stars, planets, or weather balloons for UFOs.
2. There is a mental aberration that causes large numbers of people to report UFO experiences, even though these experiences are not veridical. The content of these people's UFO stories is derived either from information transmitted by normal means (such as news media) or from delusory mental processes.
3. There are substantial numbers of people who sometimes lapse into dishonesty, even though they have reputations for being honest. During these lapses they sometimes make up UFO

stories, drawing on normal sources of information for guidance.
4. There is a massive hoax organized on a worldwide scale. The perpetrators of this hoax induce people to report UFO experiences, using methods ranging from bribes to the skillful use of Hollywood special effects and mind control techniques.
5. Although liars, frauds, and lunatics do exist, many people who report UFO encounters are experiencing real phenomena, which are worthy of careful observation and analysis.

It is widely acknowledged that explanation (1) applies to many (but not all) UFO reports involving objects or lights seen at a great distance in the sky. However, it cannot apply to close-encounter reports, in which people have a close view of strange craft and even stranger beings. If these reports are not lies, the only conventional explanation for them is that they involve highly abnormal mental states.

Explanation (2) suffers from the drawback that many UFO reports, including those of close encounters, are made by normal people who are considered to be sane and responsible by their peers. In many cases, people reporting extremely bizarre UFO encounters have been tested by psychologists or psychiatrists, who judged them to be free of mental illness (pages 61–63 and 152–58). To me, this is one of the strongest arguments for the reality of UFO phenomena. Large numbers of simple, direct statements by normal, levelheaded persons carry much more weight than a few sets of fancy photographs.

One skeptical reply to this involves the false memory syndrome (FMS). In recent years there has been great controversy over techniques of psychotherapy which supposedly enable people to recover repressed memories of childhood sexual abuse. Adults confronted with accusations of such abuse have countered that these accusations are based on false memories generated in the accusers' minds by the therapeutic process.

One can argue that UFO close encounter stories are based on false memories generated in people's minds by zealous UFO investigators and by UFO stories circulated through books and the media. My own impression is that there is some truth to this hypothesis, but I do not think that it can explain all UFO close encounter reports. After all, even the most vigorous proponents of FMS do not claim that *all* recovered memories of childhood abuse are false. I discuss this topic in greater detail in Chapter 2 (pages 50–53).

There are many real or alleged quirks of the mind which can be invoked to explain UFO phenomena. These include fantasy proneness, temporal lobe lability, dissociation induced by abuse in childhood, and ambulatory schizophrenia. Although it is legitimate to consider that such mental conditions may give rise to UFO reports, it is important for researchers to demonstrate the existence of a real causal link between the hypothesized conditions and the reported experiences. Otherwise, there is the danger that scientific progress can be hampered by allowing a false, blanket explanation to block real inquiry and understanding.

The way is also opened for people in positions of authority to use charges of mental malfunction to persecute persons holding unwanted beliefs. For example, consider ambulatory or "sluggish" schizophrenia. In the Soviet Union this alleged disorder was used to imprison political dissenters:

> The principles established by the Serbsky Institute of Forensic Psychiatry in Moscow have an important place in Soviet psychiatric method. Particularly relevant to psychiatric abuse are the theories of Dr. A. V. Snezhnevsky, a leading psychiatrist at the Institute and a member of the Academy of Science of the USSR. Dr. Snezhnevsky's concept of "sluggish schizophrenia"—a mental illness with no visible symptoms—has been used in psychiatric diagnoses that have secured the compulsory confinement of scores of known dissenters since the 1960's.[11]

The key point here is that sluggish or ambulatory schizophrenia has no symptoms other than the unwanted thinking that it is invoked to suppress. One could similarly invalidate UFO testimony by interpreting it as symptomatic of a disorder which discredits both the testimony and the persons making it. But we should avoid this, since it is both unscientific and unjust.

We should also avoid the temptation to label someone as a fraud or a liar simply on the grounds that they have made a claim that strikes us as implausible or absurd. In the course of this book, I will introduce a great deal of testimony that will strike many people (often including myself) as absurd. Any of this testimony *could* be fraudulent, but I will only argue that this is so in cases where I am aware of definite evidence of fraud.

There is not a single case cited in this book in which I can *prove* that the testimony was not fraudulent. This is inevitable, given the fact that I am simply reviewing reports written by others. I presume that a certain percentage of the material that I cite is phony, but I cannot say what that percentage might be. I can say only that I have not seen enough evidence of fraud to justify explanation (3), which maintains that UFO reports from ostensibly responsible people are generally (or always) lies.

Jacques Vallee has advocated explanation (4)—the worldwide hoax theory—for many UFO incidents, although he thinks some UFO cases involve genuine paranormal phenomena.[12] In recent years there has been an increase in testimony linking the U.S. military and intelligence establishments with the UFO abduction phenomenon (page 47). There is also reason to think that some purported UFO documents connected with these establishments may be part of an organized campaign of disinformation (pages 110–15). However, I have not yet seen evidence indicating that UFO reports in general are due to a hoax organized on a global scale.

This leaves us with explanation (5), the hypothesis that many UFO experiences are due to a real but unknown phenomenon. As I have already mentioned, one of the most compelling reasons for adopting this explanation is that many ostensibly sane people from widely separated parts of the world have made UFO reports. Even though these reports seem to originate independently, they are linked together by standard features that appear repeatedly in report after report.

Of course, one can argue that reports made within recent years cannot be proven to be independent, because there are many ways in which UFO information can spread from one person to another. It is here that comparisons between UFO accounts and Vedic literature become useful.

It turns out that there are many detailed parallels between typical UFO close-encounter reports and certain accounts from Vedic texts. Most of the UFO encounters I will be discussing took place in Western countries, where most people are almost completely unacquainted with Vedic ideas. Thus we can rule out the possibility that most UFO accounts have been influenced to any significant degree by Vedic literature. Likewise, the Vedic literature was written long before the modern period of UFO reports, and could not have been influenced by this material.

Vedic Literature and Culture Shock

Thus far I have defended the study of the UFO phenomenon, but I have said little to justify the introduction of Vedic literature into such studies. I will now offer a few suggestions as to how Vedic material could be approached by readers of this book. I also make some additional points on the interpretation of Vedic texts in Appendix 1.

People have many different perspectives on Vedic literature, but since this book has been written in America, we should discuss the typical American or European response to the Vedic world view. To put it briefly, this response is often one of culture shock. This is a consequence of the overwhelming strangeness of Vedic thinking to the Western mind, combined with specific objections arising from religious, ethnic, political, and scientific considerations.

The religious and ethnic objections are unfortunately based on exclusivism and charges of exclusivism. In response, I can only recommend an open-minded approach to other people's religious and ethnic ideas. Perhaps the study of the UFO phenomenon will help us overcome barriers based on cultural differences within human society, since these differences may be dwarfed by the differences between human societies and those of nonhuman intelligent beings.

Perhaps the best way to overcome misunderstandings based on ethnic and religious differences is to openly discuss all aspects of the world views of different peoples. To do this effectively would require a massive cross-cultural study. My own conviction is that such a study would result in a unified picture of human cultures that attributes much greater reality to each culture's world view than modern science allows.

Such a study lies far beyond the scope of this book. But as a starting point, I can ask the reader to compare the ideas presented here with those of Barry Downing, a Christian minister with a Ph.D. in science and religion, who has written extensively about UFOs and the Bible.[13] One point made by Downing is that UFOs may provide evidence for the reality of Biblical phenomena, such as visits by angels, that seem mythological from our modern perspective.

A similar point can be made about the Vedic literature. According to Vedic accounts, ancient peoples used to be in regular contact with advanced beings from other worlds. If this is true, and if contemporary UFO reports seem strange to us, then shouldn't we expect the

Vedic world view to also seem strange? The strangeness of the Vedic world view should not be taken as an immediate reason for dismissing it as mythology.

This brings us to the scientific objections to the Vedic world view. These come from several scientific fields, including physics, biology, archeology, and cosmology. In this book I cannot discuss all of these objections, but I note that some scientific objections to the Vedic world view also apply to UFO reports. These are the objections to the "physically impossible" actions of both UFOs and their reported occupants. It turns out that many of these actions are paralleled by corresponding impossible actions described in Vedic accounts.

These remarks are in defense of the reality of the Vedic world view. But just as the reader can approach UFO reports as folklore, he can also view the Vedic literature as folklore. The parallels pointed out in this book can be studied from a strictly literary viewpoint. However, it is natural to ask if something real may underlie these parallels. I would suggest that just as UFOs may be more real than our scientific and cultural conditioning may have allowed us to believe, the same may be true of the world view presented in the Vedic literature.

PART 1

A Survey of the UFO Literature

1

Science and the Unidentified

In September of 1967, Dr. John Henry Altshuler was working as a pathologist at the Rose Medical Center in Denver, Colorado. He had heard about UFO sightings in that state's San Luis Valley. So one day, out of curiosity, he stayed overnight in the Great Sand Dunes National Monument park to see if he could observe anything.

> About 2:00–3:00 a.m., I saw three very bright, white lights moving together slowly below the Sangre de Cristo mountain tops. I knew there were no roads up on those rugged mountains, so the lights could not be cars. They were definitely not the illusion of stars moving. Those lights were below the tops of the mountain range and moved at a slow, steady pace. At one point, I thought they were coming toward me because the lights got bigger. Then suddenly, they shot upward and disappeared.[1]

Altshuler was caught by police in the park, and when they learned that he was a medical hematologist, they took him to see a strangely mutilated horse that had been found ten days earlier not far away. After helping them out in their investigation of the horse, he took leave of the police officers in a state of great anxiety.

> I begged everyone not to reveal my name or where I was from. I was unbelievably frightened. I couldn't eat. I couldn't sleep. I was so afraid I would be discovered, discredited, fired, no longer would have credibility in the medical community. My experience in 1967 was so overwhelming to me, I denied the experience to everyone, even to myself. It was a matter of self-preservation, trying to give myself an insurance policy in the medical profession.[2]

Unexpectedly, Dr. Altshuler had fallen into the danger of being publicly connected with socially condemned subject matter. His reaction may seem extreme. But ridicule and ostracism are very effective punishments, and everyone knows how willing people are to use them. Altshuler was visualizing the imminent destruction of the medical career that he had worked for years to achieve. The same fear, operating in various scientific and academic professions, may have a strong effect on the publication and study of all types of anomalous observational data.

Stephen Braude, a professor of philosophy at the University of Maryland, pointed out how the fear of negative social labeling affects the study of psychical phenomena. He observed that parapsychologists tend to avoid the study of large-scale psychokinesis (PK), in which heavy objects such as sofas and tables are reported to move and levitate. After listing some theoretical and ideological reasons for this avoidance, he added:

> Others, I believe, are simply embarrassed by the extreme nature of many of the reported phenomena, and fear that their interest in them will be judged to be uncritical, weak-minded, or unscientific. And that fear is not without foundation. Historically, as a matter of fact, serious investigators of large-scale PK have been treated very badly by fellow scientists.[3]

People naturally tend to ridicule things that don't fit into their familiar systems of thought. But unfortunately, one effect of ridicule is to reinforce the limits such systems impose. By discouraging the careful study of forbidden topics, ridicule restricts people's opportunity to learn about these topics. For example, large-scale PK may be reality or nonsense, but as long as people are afraid to carefully investigate it, it will remain a disreputable unknown.

Another effect of ridicule is that it allows absurd or irresponsible versions of a subject to flourish. There are always unscrupulous people who are willing to distort the truth to fool others or to make a fast buck. Unlike scholars with good reputations to maintain, such people do not tend to be discouraged by ridicule. Thus ridicule has the perverse effect of encouraging ridiculous stories while inhibiting serious scholarship.

For many years, the subject of UFOs, or unidentified flying objects, has been widely regarded as disreputable by the general public. This could partially account for Dr. Altshuler's fear of being known as a UFO witness. But what reaction might Altshuler expect from his scientifically trained colleagues and from scientific researchers dedicated to the objective study of natural phenomena? Would these people also be likely to respond to his story with intolerance?

It turns out that the role of science in the story of UFOs is surprisingly complex. Scientists have not simply dismissed the study of UFOs as a fringe subject. In a number of instances, reputable scientists have strongly argued that UFOs involve technology and even physical principles unknown to science. Government-funded scientific studies of UFOs have been made, scientific UFO conferences have been held, and scientific journals have been founded to provide a forum for discussing UFO evidence. But ridicule nonetheless plays a very powerful role in the position of scientists on the UFO issue.

Between 1967 and 1969, the eminent physicist Edward U. Condon headed a scientific study of UFOs under the auspices of the University of Colorado. The study was funded by a government grant of $523,000, and it produced a final report of well over 500 pages. As I will show in Chapter 3, this report—commonly known as the Condon Report—contains strong evidence suggesting that some UFOs may be vehicles exhibiting an unknown technology. However, Condon concluded the report by saying that UFO studies will probably not contribute anything to the advancement of scientific knowledge.

It is instructive to see how this conclusion was communicated to the scientific community. I will begin with an editorial in the prestigious journal *Science*, written by Hudson Hoagland in 1969, just after the publication of the Condon Report. Hoagland was at that time President Emeritus of the Worcester Foundation for Experimental Biology, and he was a member of the Board of Directors of the AAAS, the American Association for the Advancement of Science.

In his editorial, Hoagland compared UFO reports to claims by spiritualistic mediums to produce ectoplasm and movements of objects. He also told an anecdote about how he and the magician Harry Houdini had exposed a bogus medium. Having established this background, Hoagland then made the following remarks about UFOs:

> The basic difficulty inherent in any investigation of phenomena such as those of psychic research or of UFO's is that it is impossible for science to ever prove a universal negative. There will be cases which remain unexplained because of lack of data, lack of repeatability, false reporting, wishful thinking, deluded observers, rumors, lies, and fraud. A residue of unexplained cases is not a justification for

> continuing an investigation after overwhelming evidence has disposed of hypotheses of supernormality, such as beings from outer space or communications from the dead. Unexplained cases are simply unexplained. They can never constitute evidence for any hypothesis. Science deals with probabilities, and the Condon investigation adds massive additional weight to the already overwhelming improbability of visits by UFO's guided by intelligent beings. The Condon report rightly points out that further investigations of UFO's will be wasteful. In time we may expect that UFO visitors from outer space will be forgotten, just as ectoplasm as evidence for communication with the dead is now forgotten. We may anticipate, however, that many present believers will continue to believe for their own psychological reasons, which have nothing to do with science and rules of evidence.[4]

Hoagland was convinced that all reports of unidentified flying objects are products of defective senses, lies, or delusions. His argument was that these negative assessments have been demonstrated in so many cases that we can conclude they apply in all cases. However, it is unfair to ask for proof of this since "science cannot prove a universal negative."

Whether this conclusion is valid or not can only be decided by examining the UFO evidence in detail. But Hoagland's tactic of tying in UFO studies with the antics of bogus spirit mediums is clearly a deliberate strategy of ridicule. His statement that believers will continue to believe for psychological reasons is a way of excluding studies of UFOs from the domain of science: If you study these things you are not a scientist. You are a true believer, and your statements convey irrational beliefs rather than scientific hypotheses.

Edward Condon also lumped in UFO studies with spiritualism and psychical research as pseudoscience, or false science. Immediately after completing the Condon Report, he made the following remarks in an article titled "UFOs I Have Loved and Lost," published in the *Bulletin of the Atomic Scientists*:

> Flying saucers and astrology are not the only pseudo-sciences which have a considerable following among us. There used to be spiritualism, there continues to be extrasensory perception, psychokinesis, and a host of others. . . .

> In ancient times, the future was foretold in many ways that have gone out of favor, such as by examining the entrails of sacrificed animals, or basing omens on the study of the flight of flocks of birds. . . . Before you smile, bear in mind that these views have never really had as much scientific study as have the UFO reports. Perhaps we need a National Magic Agency to make a large and expensive study of all these matters, including the future scientific study of UFOs, if any.
>
> Where the corruption of children's minds is at stake, I do not believe in freedom of the press or freedom of speech. In my view, publishers who publish or teachers who teach any of the pseudo-sciences as established truth should, on being found guilty, be publicly horsewhipped and forever banned from further activity in these usually honorable professions.[5]

Of course, Condon was right in saying that something should not be taught as established truth unless it has been solidly demonstrated. But in the realm of science, opinions will vary as to what is true, and scientific progress is hindered when many different possibilities cannot be freely and openly discussed. Condon was apparently confident enough of his ability to recognize pseudoscience to be convinced that its rigid exclusion would not impede the free pursuit of knowledge.

To understand why UFO investigations are rejected as false science in such strong terms, we must consider how the UFO phenomenon appears to scientists from their own theoretical perspective. To get some insight into this, I will examine some points made by William Markowitz in an article on UFOs published in *Science* in 1967, and reprinted in 1980 in a book entitled *The Quest for Extraterrestrial Life*.

For Markowitz and many other scientists, the starting point for doubt is theoretical. The problem with UFOs is that in many reported cases they don't strike people as mindless natural phenomena. Rather, they seem to be intelligently controlled vehicles not built by human beings. If such vehicles exist, then they must come from somewhere. Science cannot accept anything ethereal, higher-dimensional, or supernatural, and so the vehicles must originate as solid, three-dimensional objects. We don't see facilities for building such things on the earth, and the other planets in the solar system are thought to be uninhabited. This implies that if the UFOs are real vehicles, then they must be visitors from distant stars. That this is so is called the extraterrestrial hypothesis (ETH).

In his article, Markowitz identified himself as an expert on interstellar spaceflight, and he discussed various schemes for accomplishing this. All of these schemes were based on the principle of the rocket, in which matter is expelled from the rear of a craft, and the craft is pushed forward by the resulting reaction. He concluded that interstellar travel is not possible using these methods, and therefore UFOs could not be extraterrestrial craft.

He pointed out that published UFO reports often describe objects 5 to 100 meters in diameter that land and lift off. Arguing that these objects would have to fly on the basis of rocket thrust, he said, "If nuclear energy is used to generate thrust, then searing of the ground from temperatures of 85,000°C should result, and nuclear decay products equivalent in quantity to those produced by the detonation of an atomic bomb should be detected."[6] From this he concluded that the reported objects could not be extraterrestrial spacecraft, unless the laws of physics are wrong. Yet he said, "I do not take issue with reports of sightings and will not try to explain them away. I agree that unidentified objects exist."[7]

He brought up the possibility of reconciling UFO reports with the extraterrestrial hypothesis by assigning "various magic properties to extraterrestrial beings."[8] These include powers of teleportation and antigravity. However, he rejected these out of hand. He also considered "semi-magic hypotheses" that are based on known physical laws but include impractical features such as 100% efficient conversion of matter to energy. These he also rejected.

Markowitz concluded that the extraterrestrial hypothesis is untenable because practical travel between the stars is physically impossible. In contrast, at the AAAS symposium on UFOs in 1969, the astronomer Carl Sagan held out the remote possibility of developing some method of interstellar travel. But he suggested that the chances are vanishingly small that another civilization in this galaxy will launch an expedition that happens to reach the earth.[9]

He argued that out of 10^{10} "interesting places" in this galaxy, at most 10^6 will be solar systems with civilizations that send out interstellar expeditions. That means that for each interesting place to have a good chance of being visited in a given time period, at least 10,000 expeditions per civilization must go out on the average in that period. For example, for the earth to receive one visit per century on the average, an expedition rate of 10,000 expeditions per civilization per century would be necessary. This means 100 expeditions per year in each civilization. Given the great difficulties involved with interstellar travel, Sagan concluded that such expedition rates are not plausible, and therefore UFOs are not likely to be interstellar visitors.

Interestingly enough, Sagan raised the question of why people are so attached to the extraterrestrial hypothesis for UFOs. He asked why people don't propose that UFOs are such things as projections of the collective unconscious, time travelers, visitors from another dimension, or the halos of angels.[10]

As we shall see, people have considered such hypotheses. But for conservative scientists they are all in the same crackpot category as psychical phenomena. The leaders of the scientific community are naturally conservative in outlook, and they are limited to considering hypotheses that seem plausible in the context of accepted physical principles. The idea that people are seeing vehicles built by nonhuman intelligence seems to fly in the face of these principles, and thus the views of Hoagland and Condon on UFOs are naturally attractive to many scientists.

UFO Reports by Scientists and Engineers

Contrary to Hoagland's prediction, UFOs have apparently not been forgotten by scientifically inclined people in the years since 1969. Although the scientific community generally rejects the subject of UFOs as a serious topic of discussion in its formal publications, many scientists seem to take the subject seriously on an individual basis.

For example, in July of 1979, the magazine *Industrial Research/ Development* published an opinion poll on the attitudes of "1200 scientists and engineers in all fields of research and development."[11] In response to the question, "Do you believe that UFOs exist?," 61% responded that they probably or definitely exist, and 28% said they probably or definitely do not exist. Researchers younger than 26 were more than twice as likely to believe in the existence of UFOs than were those older than 65, and there was a continuous shift in belief percentages between these two age limits.

As far as individual sightings are concerned, 8% said they had seen a UFO, and 10% said that perhaps they had seen one. Also, 40% said they believe UFOs originate in "outer space," 2% thought they originate in the U.S.A., and less than 1% thought they are a creation of communist countries. Over 25% thought UFOs are natural phenomena.

Scientists are also actively involved in investigating UFOs, although again, this is done outside of the official institutions of science. In some cases, they carry out such investigations individually, and in other cases they work through UFO research organizations. These organizations are sometimes structured like scientific associations, although they have no standing in the scientific community. An example is MUFON, the Mutual UFO Network. In 1989, this organization had an Advisory Board of Consultants with 96 members. These included 65 people with Ph.D.'s, mostly in the natural sciences, and 16 people with M.D.'s.[12]

One reason for this continuing interest is that credible people, including scientists and engineers, have reported many UFO sightings. To illustrate this, I will begin with a sighting reported by the astronomer Clyde Tombaugh, the discoverer of the planet Pluto. Tombaugh elaborated on his experience in a letter dated September 10, 1957, to a UFO investigator named Richard Hall:

> Dear Mr. Hall:
>
> Regarding the solidity of the phenomenon I saw: My wife thought she saw a faint connecting glow across the structure. The illuminated rectangles I saw did maintain an exact fixed position with respect to each other, which would tend to support the impression of solidity. I doubt that the phenomenon was any terrestrial reflection, because some similarity to it should have appeared many times. I do a great deal of observing (both telescopic and unaided eye) in the backyard and nothing of the kind has ever appeared before or since.[13]

This letter was included in *The UFO Evidence*, an extensive collection of UFO sighting reports that was edited by Hall and published in 1964 by an organization known as the National Investigations Committee on Aerial Phenomena (NICAP). Here is another example from this document of a UFO sighting by an astronomer:

> On May 20, 1950, between 12:15 and 12:20 p.m., Dr. Seymour L. Hess, a meteorologist, astronomer, and an expert on planetary atmospheres, observed a bright, at least partially spherical object in the sky from the grounds of the Lowell Observatory. According to his account of the incident, which he wrote within one hour of the sighting, the object was definitely neither a bird nor an airplane, as it had no wings

> or propellers. Although it appeared to be very bright against the sky, as it passed between Hess and a small cumulus cloud in the Northwest its color appeared dark. Based on the object's elevation and angular diameter as he perceived it through 4-power binoculars, Hess calculated its size to be approximately 3 to 5 feet. Judging from the movement of the clouds, which were drifting at right angles to the motion of the object, he estimated that the object must have been moving at about 100 mph and possibly as fast as 200 mph. However, he neither saw nor heard any sign of an engine. Dr. Hess was head of the Department of Meteorology at Florida State University as of 1964.[14]

It is perhaps significant that neither of these sightings was made during the course of professional astronomical observations. The Condon Report contained a statement by Carl Sagan and five other scientists that "no unidentified objects other than those of an astronomical nature have ever been observed during routine astronomical studies, in spite of the large number of observing hours which have been devoted to the sky." They pointed out that the Mount Palomar Sky Atlas contains 5,000 plates with a large field of view, the Harvard Meteor Project of 1954–58 included 3,300 hours of observation, and the Smithsonian Visual Prairie Network included 2,500 hours. Nonetheless, "Not a single unidentified object has been reported as appearing on any of these plates or been sighted visually in all these observations."[15]

One response to Sagan and his colleagues was given by astrophysicist Thornton Page, who pointed out that "the astronomical telescopes in use have almost no chance of photographing a UFO passing through the telescope field."[16] However, he went on to say that the Prairie Network covered 65 per cent of the sky for bright objects over an area of some 440,000 square miles in the Midwest. It should have been able to pick up UFOs, but didn't.[17]

A possible explanation for this was given by the astronomer Franklin Roach, who had spent over three decades studying the airglow in the night sky. He observed that his photometric records were not routinely examined for UBOs, or unidentified bright objects. In fact, such objects would not be expected to appear in the records because star-like light sources were deliberately "underdrawn" and thereby omitted.

However, at the time of the Condon UFO study, an experiment was made to see what would happen if bright light sources were not omitted:

> During the Colorado Project, Frederick Ayer supervised the detailed study of one night of observations at Haleakala Observatory in Hawaii in which the analysts were instructed *not* to underdraw any deflections at all. All starlike deflections were then compared with the positions of known stars and planets. Somewhat to our surprise, on two of the records near midnight there were unmistakable deflections *not* due to known astronomical objects.[18]

Roach concluded that it is important to distinguish carefully between a lack of reports and a lack of systematic search for anomalous phenomena. He also noted that his records showed that the UBOs were not known objects, but they gave no indication of what they really were. (Although he didn't mention meteors, I presume that he considered this obvious possibility.)

The Condon Report cited a study of over 40 astronomers that was contained in Blue Book Report No. 8 of December 31, 1952. It seems that 5 made UFO sightings— a percentage said to be higher than in the population as a whole. The author of this section of the Condon Report remarked, "Perhaps this is to be expected, since astronomers do, after all, watch the skies. On the other hand, they will not likely be fooled by balloons, aircraft, and similar objects, as may be the general populace."[19] He then commented on some discussions that he had with the astronomers:

> I took the time to talk rather seriously with a few of them, and to acquaint them with the fact that some of the sightings were truly puzzling and not at all easily explainable. Their interest was almost immediately aroused, indicating that their general lethargy is due to lack of information on the subject. And certainly another contributing factor to their desire not to talk about these things is their overwhelming fear of publicity.[20]

So astronomers do sight UFOs, even though it is said that no evidence for UFOs shows up in any astronomical studies. Could fear of publicity be inducing astronomers to avoid reporting UFO observations and to avoid studying or drawing attention to observations that are reported? One begins to wonder when one reads how UFO investigator Jacques Vallee first became interested in the subject of UFOs:

> I became seriously interested in 1961, when I saw French astronomers erase a magnetic tape on which our satellite-tracking team had recorded eleven data points on an unknown flying object which was not an airplane, a balloon, or a known orbiting craft. "People would laugh at us if we reported this!" was the answer I was given at the time. Better forget the whole thing. Let's not bring ridicule to the observatory.[21]

Vallee was working as a professional astronomer at the time, and later he became a computer scientist and, among other engagements, directed a research group working under contract with the Advanced Research Projects Agency in the United States. His experience at the observatory encouraged him to view scientific research from a very radical perspective, and it started him on a career of UFO research that resulted in many influential books on this subject.

In spite of a tendency for data suppression and underreporting, sightings by responsible individuals do add up, and as they are publicized or transmitted by word of mouth, they contribute to an undercurrent of interest in the subject. Here are two additional reports by engineers that appeared in *The UFO Evidence:*

> (1) While on an evening walk in mid-October of 1954, Major A. B. Cox, graduate of Yale University, member of the American Society of Mechanical Engineers, and member of the Society of American Engineers, observed a large, grayish, disc-shaped object in the sky above his farm in Cherry Valley, New York. In a letter to NICAP's Assistant Director of correspondence, Richard Hall, dated December 28, 1955, Cox described the unusual flying patterns of the object, which he estimated to be about 35 feet in diameter and five to six feet thick. It moved like a wheel sliding sideways, but without rotating. At one point, the object suddenly stopped and then continued flying upwards at approximately right angles to its previous course. This was curious to Cox as an engineer, since the turn was shorter and more rapid than he thought possible for any airplane.[22]
>
> (2) A well-attested UFO sighting occurred on April 24, 1949, at about 10:30 a.m. on the White Sands Proving Ground in New Mexico.

> Charles B. Moore, an aerologist and graduate engineer balloonist, along with four enlisted personnel from the Navy White Sands Proving Ground, saw a gleaming white, ellipsoid object while doing work for the Office of Naval Research. Using an ML-47 theodolite including a 25-power telescope, they were tracking weather balloons when they spotted the gleaming object, which subtended an angle of about .02 degrees and was about 2½ times longer than it was wide. (A theodolite is a device for accurately measuring the horizontal and vertical direction of an object sighted through a telescope.) With their naked eyes as well as with the telescope, they viewed the unidentified object for approximately one minute, after which time it disappeared from sight as it suddenly moved up from 25 degrees above the horizon to 29 degrees.
>
> Moore launched another balloon fifteen minutes later to evaluate the wind conditions. This balloon burst after reaching 93,000 feet and traveling only 13 miles in 88 minutes, providing positive proof that the object could not have been a balloon moving at such a rapid, angular speed below 90,000 feet. That day Moore and his group identified every airplane that flew over the launching site by appearance and engine noise. Nothing passed overhead which bore any resemblance to the white, gleaming object they had seen earlier.[23]

The sighting by Cox is noteworthy, coming from an engineer, because the behavior of the object he described was unlike that of any commonly known natural phenomenon or manmade device. However, similar descriptions come up time and time again in UFO reports.

The sighting at White Sands Proving Ground is typical of a whole category of reports emanating from engineers and technical people connected with military research. One such person is Dr. Elmer Green, of the Menninger Clinic in Topeka, Kansas, who informed me personally of his experiences when he was working as a physicist for the Naval Ordinance Test Center at China Lake, California, in the decade from 1947 to 1957. In 1954 or 1955, he was chairman of the Optical Systems Working Group (OSWG), a subsection of the Inter-Range Instrumentation Group (IRIG). This was an organization of professional scientists and engineers, both civilian and military, who were engaged in recording data on weapons tests at several military bases. These included tests of rockets, guided missiles, bombs, and aircraft. OSWG was concerned specifically with metric photography, in which high-speed tracking cameras and photo-theodolites were used to determine the trajectories of rockets and other flying objects. Much of this equipment was custom-made, and it all met the highest professional standards.

In his position, Green heard frequently about incidents in which UFOs flew into camera range during weapons testing and were photographed. He heard about good-quality films that had been made of UFOs, and he personally saw black and white still photos of UFOs that were made by people in his group. He was aware of some 40 to 50 professional people who had some connection with UFO sightings made during weapons testing.

In one case at White Sands, a V2 rocket was about to be fired. Two objects that were two to three feet in diameter came down, circled around the V2 several times, and went back up, vanishing into the sky. The camera crew used up all their film on the UFOs, and so the V2 flight was canceled while they reloaded their cameras.

Green himself made a UFO sighting in the presence of Jack Clemente, who at one time was the photographic officer of the Naval Ordnance Test Center at China Lake. The two men were expecting the arrival of an AJ bomber, which they saw coming in at about 800 feet. As the airplane flew over, they saw an object about 16 feet in diameter flying beneath it at about 400 feet. The object seemed to be a structured, mechanical craft. It had a semicircular forward section marked with what looked like lines of rivets, and a smaller, semicircular back section colored amber like an artist's triangle. In the blink of an eye, the object flipped up to the wing of the plane. It remained there, pacing the plane, for a few seconds, and then it flew away at great speed, vanishing from sight in 2.5–3 seconds. On the basis of his experience with rockets, Green estimated that it accelerated at 10 to 20 g's. The object made no sound and it did not show up on radar (although other UFOs have done so). However, it was photographed, and Jack Clemente wrote a report on the incident.

Clemente later asked to see a copy of his report and the accompanying pictures. But he was told that no trace of such a report could be found in the local base files. He told Green that all films and photographs of UFOs disappeared and were presumably sent to Washington.

I asked Green if he had ever been ordered to keep UFO information secret. He answered that although he had a top secret clearance, he was never told to keep quiet about UFO incidents. He explained that such incidents were simply not discussed by military authorities. There was no need to order secrecy about phenomena that simply did not exist.

Green noted that although UFOs looked mechanical, they seemed to violate the laws of physics. Although they often exceeded the speed of sound, they never produced sonic booms. Their maneuvers reminded him of the movements of a spot of light projected on a wall by a flashlight, and he speculated that they might be structures that were somehow projected into our space-time continuum.

He said that the people in his group experienced many more UFO incidents in the early part of the decade starting in 1947 than in the later part. He noted that in his early days at China Lake he would regularly see the flashes from A-bomb air bursts at the nearby atomic testing range in Nevada, and he remarked that some people speculated on a connection between UFO activity and the atomic testing. He speculated that the reason for the apparent UFO cover-up was that government authorities didn't want to admit their inability to understand UFOs or to prevent them from flying with impunity through our skies.

This is an amazing story, and it again brings us back to questions involving credibility, ridicule, and suppression of information. If the story is true, then at least 40 to 50 professional scientists had definite knowledge of UFOs in the early 1950s. Why is it, then, that UFOs are not openly acknowledged by scientists and leaders of society? The story introduces a new element, governmental secrecy, which I will discuss in Chapter 3. The systematic elimination of "hard" evidence by government authorities, combined with fear of ridicule and loss of career, may explain why none of these scientists ever made strong public presentations of their UFO experiences, either singly or in a group.

On Scientists Who Study UFOs

Gerard Kuiper of the Lunar and Planetary Laboratory at the University of Arizona disagreed with the idea that scientists can be swayed by social pressure. At a meeting of the Arizona Academy of Science on April 29, 1967, he said: "I should correct a statement that has been made that scientists have shied away from UFO reports for fear of ridicule. As a practicing scientist, I want to state categorically that this is nonsense." He pointed out that a scientist "selects his area of investigation not because of pressures but because he sees the possibility of making some significant scientific advance." [24]

Undoubtedly, Kuiper is partially correct. Some scientists may not make anomalous observations involving UFOs, and they may sincerely believe that if they did make such observations, they would openly report them. Others may actually observe UFOs, and then suppress their observations when brought face to face with the fear of losing their careers. This, in turn, reinforces the feeling of the first group that no serious UFO observations are being made.

Some professional scientists have openly engaged in UFO investigations. However, their stories also involve the issues of credibility and data suppression. To illustrate this, I will first discuss the ideas of J. Allen Hynek, an astronomer and long-time consultant to the Air Force on UFOs. Over the years, Hynek's views on UFOs changed greatly, and in the course of this he made a number of seemingly contradictory statements that created doubt and confusion about UFOs for other scientists.

In an article on UFOs, William Markowitz noted a letter to *Science* in which Hynek declared that although scientists are said to never report UFOs, actually "some of the very best, most coherent reports have come from scientifically trained people." [25] Then Markowitz quoted a statement by Hynek in the *Encyclopedia Britannica,* in which he referred to "the failure of continuous and extensive surveillance by trained observers" to produce UFO sightings.[26] Markowitz felt that these apparently contradictory statements called into question the reliability of UFO data.

Markowitz also cited a letter to *Science* on April 7, 1967, by Dr. William T. Powers:

> In 1954, over 200 reports over the whole world concerned landings of objects, many with occupants. Of these, about 51 percent were observed by more than one person. In fact, in all these sightings at least 624 persons were involved, and only 98 of these people were alone. In 18 multiple-witness cases, some witnesses were not aware that anyone else had seen the same thing at the same time and place. In 13 cases, there were more than 10 witnesses. How do we deal

> with reports like these? One fact is clear: we cannot shrug them off. [27]

Powers was making a rather strong claim. Were there really over 100 cases in the United States in 1954 where at least two people saw a UFO land? According to Markowitz, Hynek informed him in 1966 that he had no reliable reports of UFO landings and lift-offs, and no records of cases in which a reliable witness visited an extraterrestrial craft or talked with an occupant.[28] This statement also filled Markowitz with doubts.

However, the statement itself was doubtful. At a symposium on UFOs held under the auspices of the AAAS in 1969, Hynek said the following about close encounters with UFOs:

> I would be neither a good reporter nor a good scientist were I deliberately to reject data. There are now on record some 1,500 reports of close encounters, about half of which involve reported craft occupants. Reports of occupants have been with us for years but there are only a few in the Air Force files; generally Project Bluebook personnel summarily, and without investigation, consigned such reports to the "psychological" or crackpot category. [29]

One might suggest that Hynek regarded these reports as unreliable, even though he didn't say so when he mentioned them before members of the AAAS. But in 1972 Hynek wrote of his meeting with Betty and Barney Hill, two people who claimed to have spoken with aliens on board an extraterrestrial craft. He spoke of their "very apparent sincerity" and said that "there was no question of their normalcy and sanity." [30] This meeting with the Hills occurred in about 1966, close to the time of his reported statement to Markowitz.

The apparent contradictions in Hynek's statements can perhaps be attributed to the gradual evolution of his ideas on UFOs and his caution in making public statements that would damage his credibility. Hynek was a professor of astronomy, and he was chairman of the astronomy department of Northwestern University for many years. He was also a scientific consultant to the U.S. Air Force on UFOs for about 20 years beginning in 1948, and he later served as director of a civilian UFO research organization called CUFOS, or Center for UFO Studies.

Hynek's views on UFOs changed greatly over the years. He began as an avowed skeptic who thought that UFOs were an utterly ridiculous craze or fad that would quickly subside. But by 1979 he was giving serious credence to ideas that would seem outrageous to conservative physical scientists such as Hoagland or Markowitz. In his introduction to Raymond Fowler's book *The Andreasson Affair,* Hynek wrote:

> Here we have "creatures of light" who find walls no obstacle to free passage into rooms and who find no difficulty in exerting uncanny control over the witnesses' minds. If this represents an advanced technology, then it must incorporate the paranormal just as our own incorporates transistors and computers. Somehow, "they" have mastered the puzzle of mind over matter.[31]

One might ask why a professor of astronomy would publish a statement like this. Was he saying what the "creatures of light" might be *if* they exist, while maintaining a healthy skepticism about whether they exist or not? Perhaps, but he also said in his introduction that the book would sorely challenge skeptics who had the courage to take an honest look at it, and he declared that it did not show the slightest evidence of hoax or contrivance.

Hynek's position on UFO humanoids was summed up in his book *The UFO Experience*, published in 1972:

> Our common sense recoils at the very idea of humanoids and leads to much banter and ridicule and jokes about little green men. They tend to throw the whole UFO concept into disrepute. Maybe UFOs could really exist, we say, but humanoids? And if these are truly figments of our imagination, then so must be the ordinary UFOs. But these are backed by so many reputable witnesses that we cannot accept them as simple misperceptions. Are then, *all* of these reporters of UFOs truly sick? . . .
>
> Or do humanoids and UFOs alike bespeak a parallel "reality" that for some reason manifests itself to some of us for very limited periods? But what would this reality be? Is there a philosopher in the house?
>
> There are many such questions and much related information that is difficult to comprehend. The fact is, however, that the occupant encounters cannot be disregarded; they are too numerous.[32]

In 1966, however, the ideas that Hynek was willing to publicly discuss were much less radical. For example, at a hearing of the House Armed Services Committee on UFOs on April 5, 1966, Hynek was asked whether or not UFOs might be piloted by extraterrestrial beings. He answered:

> I have not seen any evidence to confirm this, nor have I known any competent scientist who has, or believes that any kind of extraterrestrial intelligence is involved. However, the possibility should be kept open as a possible hypothesis. . . . But certainly there is no real evidence of intelligent behavior of hardwares.[33]

When asked if he was looking for an explanation of UFOs based on natural phenomena, Hynek answered, "Yes."[34] It appears from these statements that in 1966 Hynek did not think that UFOs were intelligently controlled at all. Yet he said in his 1972 book that on Aug. 1, 1965, a series of remarkable events occurred at U.S. Air Force facilities near Cheyenne, Wyoming. He stated that he was informed of these events at the time through his connection with Project Blue Book. Here are the reports that came in:

> 1:30 a.m. — Captain Snelling, of the U.S. Air Force command post near Cheyenne, Wyoming, called to say that 15 to 20 phone calls had been received at the local radio station about a large circular object emitting several colors but no sound, sighted over the city. Two officers and one airman controller at the base reported that after being sighted directly over base operations, the object had begun to move rapidly to the northeast.
>
> 2:20 a.m. — Colonel Johnson, base commander of Francis E. Warren Air Force Base, near Cheyenne, Wyoming, called Dayton to say that the commanding officer of the Sioux Army Depot saw five objects at 1:45 a.m. . . .
>
> 2:50 a.m. — Nine more UFOs were sighted, and at 3:35 a.m. Colonel Williams, commanding officer of the Sioux Army Depot, at Sidney, Nebraska, reported five UFOs going east.
>
> 4:05 a.m. — Colonel Johnson made another phone call to Dayton to say that at 4:00 a.m., Q flight reported nine UFOs in sight: four to the northwest, three to the northeast, and two over Cheyenne.

4:40 a.m. — Captain Howell, Air Force Command Post, called Dayton and Defense Intelligence Agency to report that a Strategic Air Command Team at Site H-2 at 3:00 a.m. reported a white oval UFO directly overhead. Later Strategic Air Command Post passed the following: Francis E. Warren Air Force Base reports (Site B-4, 3:17 a.m.) — A UFO 90 miles east of Cheyenne at a high rate of speed and descending— oval and white with white lines on its sides and a flashing red light in its center, moving east; reported to have landed 10 miles east of the site.

3:20 a.m. — Seven UFOs reported east of the site.

3:25 a.m. — E Site reported six UFOs stacked vertically.

3:27 a.m. — G-1 reported one ascending and at the same time, E-2 reported two additional UFOs had joined the seven for a total of nine.

3:28 a.m. — G-1 reported a UFO descending further, going east.

3:32 a.m. — The same site has a UFO climbing and leveling off.

3:40 a.m. — G Site reported one UFO at 70° azimuth and one at 120°. Three now came from the east, stacked vertically, passed through the other two, with all five heading west.[35]

Hynek noted with astonishment that when he asked Major Quintanilla, the officer in charge of Blue Book, what was being done to investigate these reports, Quintanilla replied that the sightings were nothing but stars. That seems unlikely, but what were they? The orderly behavior of the objects and their tendency to fly over military installations does suggest intelligent guidance. Oval shapes with flashing red lights in the center are likewise suggestive of intelligent design.

In his book published in 1972, Hynek certainly allowed the reader to interpret this report he received in 1965 as evidence of an unknown intelligence. Yet less than a year after receiving that report, Hynek told Congress that there is "no real evidence of intelligent behavior of hardwares."

I have discussed the development of Hynek's ideas at some length to illustrate both the extreme character of the reported UFO phenomena and the effect that this had on a conservative scientist who was trying to study and understand these phenomena. Hynek's need to protect his credibility apparently led him to make contradictory statements that reduced the credibility of UFO evidence in general. At the same time, his increasing willingness to give serious consideration to the more extreme UFO phenomena is impressive. Hynek showed every sign of being a careful and critical thinker, and so one might wonder what moved him to eventually adopt such a radical position.

Although in the mid 1960s Hynek played down the idea of intelligent control of UFOs, one prominent scientist named James McDonald strongly advocated it. McDonald was a senior physicist at the Institute of Atmospheric Physics and a professor in the meteorology department at the University of Arizona. In a public statement prepared for newspaper editors, he gave the following summary of his views:

> An intensive analysis of hundreds of outstanding UFO reports, and personal interviews with dozens of key witnesses in important cases, have led me to the conclusion that the UFO problem is one of exceedingly great scientific importance. Instead of deserving the description of "nonsense problem," which it has had during twenty years of official mishandling, it warrants the attention of science, press, and public, not just within the United States but throughout the world, as a serious problem of first-order significance. . . .
>
> The hypothesis that the UFOs might be extraterrestrial probes, despite its seemingly low *a priori* probability, is suggested as the least unsatisfactory hypothesis for explaining the now-available UFO evidence.[36]

McDonald's article contains summaries of 18 case studies of UFO sightings, as well as a discussion of the history of the UFO controversy and the role played in it by science, the U.S. Government, and the military. In this regard, he disagreed with the widespread idea that the government is deliberately covering up information on UFOs. Rather, he concluded that there is "*a grand foulup,* accomplished by people of very limited scientific competence, confronted by a messy and rather uncomfortable problem."[37] Hynek also tended to favor this foul-up idea.[38]

One topic that McDonald discussed at some length is the scientific debunking of UFOs. In particular, he mentioned the work of Dr. Donald Menzel, an astronomer who was at one time the director of the Harvard College Observatory and who wrote books dismissing UFOs largely as misperceptions of astronomical or meteorological phenomena.

McDonald discussed how Menzel explained the UFO sighting of astronomer Clyde Tombaugh mentioned above. Menzel's idea was that Tombaugh saw the lighted windows of a house reflected by a ripple in the boundary of an atmospheric haze layer. As this ripple progressed with a wavelike motion, the reflected house would have seemed to move like a flying saucer.[39] McDonald's comments on this are scathing:

> Now this might go down with a layman, but to anyone who is at all familiar with the physics of reflection and particularly with the properties of the atmosphere, . . . the suggestion that there are "haze layers" with sufficiently strong refractive index gradients to yield visible reflections of window lights is simply absurd. But, in Menzel's explanations, light reflections off of atmospheric haze layers are indeed a sight to behold. This, I say, I simply do not understand.[40]

Although McDonald's article was prepared only for newspaper editors and was not published, he did write an article on UFOs in the journal *Astronautics and Aeronautics*.[41] This is a detailed discussion of an episode in July of 1957 in which an Air Force RB-47, manned by six officers, was followed by a luminous, highly maneuverable object for about 1.5 hours as it flew from Mississippi through Louisiana and Texas, and into Oklahoma (see pages 212–13). This case is significant because it involved simultaneous observation of the object by human vision, by radar from the ground and from the airplane, and by electronic counter-measures (ECM) equipment on the airplane.

Unfortunately, the same issue that published this article also contained an obituary notice for McDonald, who died in the desert near Tucson on June 13, 1971, apparently by suicide. The obituary notice included the following statement, which brings us back to the theme of ridicule, science, and UFOs:

> The history of the UFO problem has been full of unusual and tragic events. Men of highest scientific achievements have seen themselves involved in strongly opposing views. Others have become victims of vitriolic attacks or, perhaps worse, of ridicule. McDonald was one of them.[42]

Recent Scientific Studies of UFOs

In more recent years, the tendency of the scientific community to disdain the subject of UFOs has largely continued. However, in the United States the Society for Scientific Exploration was founded in 1982 by 13 professors of science at major universities. The express purpose of this society is to promote the study of anomalous phenomena that scientists tend to neglect, and the society publishes a refereed technical journal entitled *Journal of Scientific Exploration.* This journal has published quite a number of articles on UFOs, and it also publishes articles on paranormal phenomena.

One article published in the *Journal of Scientific Exploration* described in great detail how in 1977 NASA responded to a recommendation from President Carter's science advisor to form a panel of inquiry on UFOs. The author, Dr. Richard Henry, gave some insights into the reasons for NASA's rejection of this recommendation. The main reason was fear of ridicule. As Henry put it, UFOs are a tar baby, and "A scientist who touches the tar baby once, as I have, runs the risk of getting deeper and deeper in goo. I don't have a strong stomach for it." [43] Another important reason was that UFO studies would take already scarce funds away from other important scientific projects.

In France, a fully funded, civilian scientific UFO study group was created by the government in 1977. This is called GEPAN (Groupe d'Etudes des Phīnomznes Aerospatiaux Non-Identifees). GEPAN produced a five-volume, 500-page report, which was summarized as follows by the sociologist Dr. Ronald Westrum in 1978:

> The bulk of the work was devoted to eleven cases of high credibility and high strangeness . . . [which] were studied in great detail; only two proved to have a conventional explanation. In the other nine, it appeared that the distance between the witnesses and the objects was less than 250 meters. Of the five volumes of the report, three were entirely devoted to analysis of these eleven cases, all except one of which was pre-1978. The earliest was 1966. Two of the cases were humanoid sightings.
>
> The analysis and investigation was carried out by a four-person team in each case; the team included a psychologist, who separately carried out a psychological examination relevant to the evaluation of the testimony of the witnesses. The care with which distances, angles, and psychological factors were evaluated makes the bulk of the Con-

> don Report seem very poor by comparison. In many cases, the investigations were textbook models of how such investigations should be carried out.[44]

The ratio of cases with no conventional explanation to the total number of cases will depend on the screening process used to arrive at the initial set of cases. If cases are accepted without discrimination, then this ratio may be very low, and this may be used to argue that the "unexplained residue" of sightings is insignificant. For example, Project Blue Book listed 10,147 sightings in the period from 1947 through 1965, and of these it listed 646 sightings, or about 6%, as unexplained. In his testimony before Congress on April 5, 1966, Air Force Secretary Harold Brown dismissed this small residue by saying, in effect, that given the imperfections in the reports, you can't expect to explain everything:

> The remaining 646 reported sightings are those in which the information available does not provide an adequate basis for analysis, or for which the information suggests a hypothesis but the object or phenomenon explaining it cannot be proven to have been here or taken place at that time.[45]

In France, UFO reports were also handled by the CNES (Centre National d'Etudes Spatiales), the French equivalent of NASA. In 1989, J. J. Velasco reported at a conference of the Society for Scientific Exploration that 38% of the UFO cases studied by CNES remained unidentified as natural phenomena. Thus CNES apparently used stricter screening procedures than the U.S. Air Force, and the cases studied by GEPAN were even more tightly screened.

The conclusions of the GEPAN study stand in sharp contrast to the conclusions of Air Force Secretary Harold Brown:

> In nine of the eleven cases, the conclusion was that the witnesses had witnessed a material phenomenon that could not be explained as a natural phenomenon or a human device. One of the conclusions of the total report is that behind the overall phenomenon there is a "flying machine whose modes of sustenance and propulsion are beyond our knowledge."[46]

Thus instead of seeing a lack of evidence for a natural explanation in the "unknown residue," the GEPAN scientists saw positive evidence for an inexplicable flying machine.

In summary, the aim of this chapter has been to show that the UFO question has engaged the serious attention of quite a few reputable scientists, and it has been discussed in official scientific forums. This suggests that it cannot be simply dismissed as nonsense or pseudo-science. At the same time, many reported UFO phenomena seem to be incompatible with established scientific principles, and others are so bizarre that they violate the norms of common sense in modern society. Even though some reputable scientists have argued that reports of such phenomena should be seriously studied, others have denounced them in very strong terms. This, combined with people's natural tendency to reject bizarre stories, has surrounded the subject of UFOs with an aura of ridicule that makes serious study of the subject difficult.

2

Close Encounters of Various Kinds

In the United States, the story of UFOs is usually said to have begun with the famous sighting by Kenneth Arnold, a businessman from Boise, Idaho. While flying his private plane in Washington State on June 24, 1947, Arnold saw nine flat, shiny objects flying in a line near Mount Rainier, and he compared their peculiar motion to a saucer skipping over water. A newsman, inspired by this description, coined the term "flying saucers," and this became a household word as waves of reports came in of strange objects seen in the skies. Surprisingly, as years passed, these reports did not dwindle away. Rather, they began to be made persistently in countries all around the world, and this continues to the present day.

We have already seen that many of these unidentified flying objects, or UFOs, do not fit very well into accepted scientific theories, and therefore they have proven embarrassing for scientists. Indeed, some were sufficiently strange to outrage practically anyone's world view. In this chapter, I will give an overview of the typical UFO encounters that people have reported.

I should begin by making some observations about my approach to UFO evidence. All of this evidence consists of stories related by witnesses. As I mentioned in the introduction, even "hard evidence" in the form of photographs or landing traces is practically meaningless if not accompanied by personal testimony. For example, suppose someone presents a metal sample of unusual composition and says that it came from a UFO. Given that the testimony is valid, the metal sample may tell us something about what the UFO was made of. But without the testimony it tells us nothing about UFOs. Thus the crucial evidence in UFO cases is always anecdotal—which simply means that it consists of human testimony.

In recounting people's UFO experiences, I will often simply tell their story. However, it should be understood that this is generally the story as told by a witness to an investigator. In other cases, it is the story that an author took from an investigator's report of what a witness told him. In a few cases, it is what a witness directly told me.

The approach of relying on the testimony of others is universally used in science. For example, our knowledge of what Michelson and Morley did in their famous interferometer experiment depends entirely on human testimony and the transmission of that testimony through written reports.

Few people would raise objections to the story of Michelson and Morley. But in the case of the bizarre stories connected with UFOs, one may object that human testimony is not to be trusted and point to the many failings of the human mind and senses. These failings should be considered, but human testimony is still all we have to go on.

I suggest that it is wrong to object to bizarre testimony simply because it is bizarre. To do so is to legislate that only conventionally acceptable statements can be admitted as evidence. This would be all right if nature happened to conform to our notions of what is acceptable, but it is quite possible that nature does not do so. Therefore, my strategy is to give human testimony a chance and recognize that objections to fallible human testimony also depend on fallible human judgment.

In discussing UFO reports, it is important to have a clear understanding of what is meant by a UFO. This term could be used to refer to practically anything that people see in the sky, or on the ground, that seems unusual or unexplainable. However, on the basis of social usage dating back to the days of Kenneth Arnold's sighting, the term generally refers to something that looks like an unknown, intelligently guided vehicle. Here "unknown" means that the observed manifestation doesn't resemble known natural phenomena or known manmade objects.

The phrase "intelligently guided vehicle" means that the manifestation either looks like a manufactured object or behaves in a way that is suggestive of intelligence. For example, if something looks metallic, smoothly curved, and symmetrical, we may say that it looks like a manufactured object. This impression is even stronger if it seems to be equipped with windows, doors, or landing gear. In some cases, only a distant light is seen, but the light's movements may suggest intelligent guidance. Thus, if the light moves about in different directions, witnesses may think that it is not a meteor or a satellite. If the light also seems to move in a way that wouldn't be expected for a balloon or an aircraft, then they might call it a UFO.

From this we can see that a UFO is anything but unidentified, and the term "unidentified flying object" is a misnomer. Calling something a UFO means that it is a particular kind of phenomenon, as defined above. Thus, we will sometimes find someone saying, "It was not an aircraft or a star. It was a genuine UFO." This does not mean that the observed phenomenon was genuinely unidentified. Rather it means that the person wants to identify the phenomenon as an unknown manifestation that appears to be an intelligently guided vehicle.

Some UFO sightings involve distant lights seen at night, solid-looking objects seen during the day, or objects detected by radar. J. Allen Hynek has classified these as nocturnal lights, daylight discs, and radar cases.[1] The latter include radar-visual sightings in which a visual sighting was found to correlate strongly with radar observations. In addition, there are the so-called close encounters, which Hynek broke down as follows:

> *CE1:* Objects seen on the ground or at a close distance to the observer.
> *CE2:* The same, with definite effects on the environment, observers, or instruments, such as burned, baked, or impacted areas of ground, temporary paralysis of the witness, or interference with electrical apparatus.
> *CE3:* Sighting of alien entities, either by themselves or in association with a UFO.

The close encounters of the third kind (or CE3s) involve extremely strange material. It is customary in cartoons to associate flying saucers with "little green men." It is less widely known that humanlike beings of a variety of shapes and sizes have been regularly reported in connection with UFOs since the late 1940s. These beings are often short in stature but are rarely green. Since they tend to be roughly human in form, they are called humanoids.

In some cases, these beings are simply seen, and in others they are said to communicate, often by telepathy. In most CE3 cases, the beings do not physically capture the witnesses, although they are sometimes reported to stun them or temporarily paralyze them. But in recent years, great publicity has been given to a subset of the CE3 cases in which UFO beings are reported to aggressively abduct humans and take them on board their vehicles. These are called UFO abductions or close encounters of the fourth kind (CE4s). I will discuss CE3 cases without abduction in this chapter and leave UFO abductions to Chapter 4.

There is also another kind of UFO encounter, not included in the four close-encounter categories, in which the witness enters into a friendly relationship with UFO humanoids, engages in extended conversations with them, and may even be taken on rides in their vehicles. These so-called contactee cases are infamous among UFO investigators, and they are often branded as hoaxes. Many probably are hoaxes, but I will argue in Chapter 5 that there seems to be a continuum of cases stretching from CE4s to full-fledged contactee cases. It is very hard to draw a neat line separating these two types of cases, and so a review of the UFO evidence must consider both types.

The statistics on numbers of UFO sightings are extremely variable. I mentioned in Chapter 1 that between 1947 and the end of 1965 the U.S. Air Force accumulated 10,147 UFO reports. One might think that this gives a good idea of how many UFO sightings actually take place. However, by 1981 the Center for UFO Studies in Illinois had compiled a computer-coded file, called UFOCAT, of about 60,000 UFO reports from 113 countries.[2] This file was started by Dr. David Saunders after he joined the staff of the Condon UFO study, and thus it covers the interval from 1967 to 1981. Out of the 60,000 UFOCAT cases, about 2,000 involved CE3s, and 200 involved CE4s.[3]

Clearly, the number of UFO reports in any given collection will depend on selection criteria and the number of people who are engaged in collecting reports. It is therefore very hard to estimate the total number of UFO sightings in any given time period or part of the world.

Vallee, writing in 1990, estimated that the number of close-encounter cases known at that time was between 3,000 and 10,000. He went on to argue that an average of 1 close encounter in 10 will actually be reported. Taking 5,000 as an estimate of the number of known cases, this means that 50,000 close encounters may have actually occurred. Since the known cases are concentrated in Europe, North and South America, and Australia, Vallee argued that twice as many cases might turn up if one could fully survey the entire world. This yields an estimate of 100,000 close encounters.[4]

Vallee noted that close encounters tend to be nocturnal, with a high peak of activity at 9 p.m. and a lesser peak at about 3 a.m. However, people tend to be in bed between 9 p.m. and 3 a.m., and thus there are fewer potential witnesses in this period. By using statistics for the number of people active outdoors at different times of day, one can compute a curve for UFO encounters per available observer. This rises steadily throughout the night and peaks at about 3 a.m. Vallee suggested that this curve might give a true picture of UFO activity as a function of time of day, and it implies an overall activity level 14 times higher than the actual reported level. He also noted that CE4's show a pronounced peak between 10 p.m. and midnight.[5]

UFO False Reports

One notorious feature of the UFO controversy is that people will often mistake various known objects or phenomena for UFOs. A good discussion of this is found in Raymond Fowler's *Casebook of a UFO Investigator*.[6] He pointed out some causes of false UFO reports:

> *Manmade flying objects:* Aircraft lights, advertising planes with signs made of many lights, the Goodyear blimp, military refueling exercises, amateur aircraft (hang gliders), kites, fireworks, children's homemade hot air balloons, weather balloons, research balloons of various kinds, rocket launches and reentries, sodium and barium clouds released from rockets for atmospheric tests, satellites (and their reentry), and drops of flares from military planes.
>
> *Natural phenomena:* Mirages, ball lightning, birds, meteors (fireballs, bolides), stars (for example, Sirius, Capella, and Arcturus in the Northern Hemisphere), planets (Venus, Mars, Jupiter, and Saturn), and the moon (often when full and near the horizon).

To give an idea of how often these mistaken identifications are made, Fowler pointed out that in the first 6 months of 1978, the CUFOS hotline received 452 UFO reports attributable to ordinary objects. These were aircraft in 210 cases, stars or planets in 127 cases, and meteors in 54 cases.[7]

Stars may appear to move about due to autokinesis, a process caused by the movements of the eye. Celestial bodies may also appear to follow a moving car because, on a straight road, their position relative to the car will stay the same. Other causes of false UFO reports are hallucinations due to drugs, alcohol, or mental derangement. There are also hoaxes, including hot air balloons, frisbees, and models photographed by children, as well as elaborate frauds created by adults. Fowler pointed out that hoaxes make up a very small percentage of UFO reports, and these are mostly pranks by schoolboys.

Sightings of relatively distant objects could be caused by mild misperceptions or simple hoaxes. But spurious close encounters seem to require something stronger. If a person reports a detailed, close-range view of a humanoid, then it would seem that either (1) he actually saw an unusual being, (2) he saw an illusion of such a being projected by an unknown cause, (3) he saw a manmade hoax, (4) he experienced a hallucination, or (5) he is a liar. To evaluate options (4) and (5), it is important to be able to assess the character, mental health, and personal motivations of the witnesses in close-encounter cases.

In some cases, witnesses are mentally imbalanced, and in others they turn out to be "con men," out to gain money and influence by exploiting gullible people. However, there are large numbers of cases in which the witnesses are sane, responsible people who gain neither profit nor fame from their experiences, and who often try to conceal them in order to avoid ridicule. These cases provide some of the most persuasive evidence for the reality of close-encounter experiences. But one can still propose that failings of perception and memory in mentally sound people might generate reports of bizarre experiences. I will discuss this possibility in more detail in the next section.

Reports by mentally sound people might also be due to very elaborate hoaxes, which may be backed by considerable funding and manpower. It is possible that human beings might, for some nefarious reason, abduct someone and try to disguise this as a UFO abduction. The abductors could wear UFO alien costumes, alter the victim's consciousness with the aid of drugs and hypnosis, and even take the victim into a Hollywood-style UFO set.

There are, in fact, cases in which something like this may have happened. For example, in one abduction case reported by the British UFO investigator Jenny Randles, a woman named "Margary" (a pseudonym) was apparently abducted, drugged, and programmed with posthypnotic suggestions. When she later began to remember details of the episode, she recalled one of her abductors saying in an amused tone, "They will think it's flying saucers."[8]

In this case, the abductors were entirely human-looking, and the site to which Margary was taken seemed to be an ordinary house. It is not at all clear what was going on here. For example, why go to all the trouble to stage a phony UFO abduction and then spoil it by making such remarks to the victim?

In recent years accusations have been made linking UFO abductions with military and intelligence establishments in the U.S. and Europe. Researcher Martin Cannon has hypothesized that the CIA is carrying out extensive mind control experiments on U.S. citizens and covering its tracks by disguising these experiments as abductions by aliens.[9] Cannon argues that CIA-developed techniques of hypnosis aided by drugs and radio control could well account for some of the phenomena of mind control reported by abductees. He would attribute the Margary case to a sloppy job done by novices in hypnotic manipulation.[10]

Although Cannon's theory is nearly as extreme as the hypothesis of alien abduction itself, it does explain why many abductees report harassment by what appear to be human secret agents. It might explain why abductees such as Leah Haley[11] and researchers such as Karla Turner[12] report apparent abductions involving military personnel. It also fits in nicely with Jacques Vallee's charge that the abduction of Franck Fontaine in France was a test of mind control techniques carried out under the orders of highly placed officials in the French government.[13]

Of course, it is possible that a certain percentage of UFO abductions are being carried out by human beings and others are due to some other agency. Thus the psychologist Richard Boylan argues that some UFO abductions are "conducted as psychological warfare (PSYWAR) operations by military/intelligence/special operations figures," but he attributes other abduction accounts to intervention by alien beings.[14]

As we will see later on, many UFO close-encounter cases have features that would be quite difficult to simulate by human beings. If UFO close encounters in general are being staged by human conspirators, then this would require a tremendous covert investment in manpower and Hollywood special effects. At present, it seems unlikely that such an enormous effort is being carried out, although there does seem to be evidence of a covert human component in some UFO abduction accounts. I will say more about the possible role of government agencies in UFO encounters in Chapter 3 (pages 110–15).

On Misperceptions and Failings of Memory

The psychiatrist Ian Stevenson has made some observations about misperception and failings of memory that are applicable to the evaluation of UFO reports. Stevenson has spent many years studying what he calls spontaneous cases in the field of parapsychology. These are cases in which a person reports some ostensibly paranormal experience outside the confines of a controlled, laboratory situation. These include telepathic and precognitive impressions, out-of-body experiences, memories of past lives, poltergeist cases, and apparitions. Stevenson has specialized in the study of past-life memories in young children, and he has carefully studied the use of interviews with witnesses as the main method of researching these cases.

I will briefly summarize some remarks that Stevenson made about evaluating the evidence for spontaneous cases. Although he did not mention UFOs in his discussion, his observations are quite relevant to the evaluation of UFO reports.

One of his first points was that the adjectives "authentic" and "evidential" are applied to spontaneous cases. A case is authentic if the witnesses and the reporting are highly reliable, so that one can justifiably believe that the events in question happened as reported. It is evidential if it is authentic and there is justification for thinking that the case has paranormal features.[15]

J. Allen Hynek expressed similar ideas. He spoke of a credibility index and a strangeness index.[16] The credibility index measures the reliability of UFO witnesses, as indicated by their reputations, medical histories, occupations, sharpness of eyesight, and other factors. He also said that single witness cases should be given "no more than quarter-scale credibility." The strangeness index measures how far the reported events seem to defy explanation in normal physical terms. Hynek felt that there are UFO cases of high credibility and strangeness, and Stevenson similarly felt that there are spontaneous cases that are authentic and evidential.

Stevenson pointed out that one defect in many spontaneous cases is that the case was not described in writing until considerable time had elapsed. This is also true of many (but by no means all) UFO cases. It leads to the problem that human memories may erode with time and that accounts may be filled in with reconstructions or supplemental material. However, Stevenson pointed out that retention of detail in memory depends on the emotional intensity of the experience, on repetition, and on motivation to remember. Many paranormal experiences involve high emotional intensity and motivation to remember.[17] The same is reported by many UFO close-encounter witnesses.

Stevenson went on to point out four cases in which it could be demonstrated that witnesses retained good memory of paranormal experiences over several years. In one example, a man and his wife wrote detailed accounts in 1909 of an apparently precognitive dream that he had in 1902. Eight years later, the woman wrote another account without consulting any memoranda or discussing the case with her husband. This account differed in only one minor detail from her husband's earlier account.[18] Stevenson pointed out that in all four cases there was not only little loss of detail but also little elaboration of new detail.

It is often charged that percipients in spontaneous cases tend to embellish their memories as time passes, and this makes it impossible to find out later what they originally experienced. Although Stevenson acknowledged that embellishment does happen, he said, "In my own experience embellishment of the *main* features of an account occurs very rarely." [19] He said that he had checked this many times by coming back unexpectedly after one or several years and requestioning a witness about his experiences. I am not aware of any feature of UFO witnesses that would make them more prone to embellishment than witnesses of paranormal events not involving UFOs.

Stevenson noted that embellishment is more apt to occur in accounts given by secondhand reporters of a case than it is by primary witnesses. However, even these reporters do not always embellish the case. He commented, "Quite as often, if not more so, they drop important details and thus diminish its evidentiality." [20]

These tendencies could have a serious effect on UFO reports presented in secondary literature. The authors of UFO books may be more likely to distort testimony than many original witnesses. The only way to guard against this is to be aware of the reputations of UFO authors and identify the biases of particular presentations by surveying a wide variety of books and reports. My own impression after making a broad survey of the literature is that certain popular UFO authors do tend to introduce their own biases into UFO accounts. Often they do this by omitting features of UFO accounts that do not fit into their favored hypotheses.

Another problem with reports of spontaneous cases is malobservation. There have been many studies by lawyers and forensic psychologists in which an event is staged before witnesses, who are later asked to tell what happened. It is observed that the witnesses will frequently make many errors in their accounts of what they saw. For example, in a staged confrontation with guns, they may fail to correctly identify which party pulled out his gun first.

Stevenson commented that, "Such experiments certainly have some relevance to our field, but again I resist their use to reject all human testimony in spontaneous cases."[21] One reason he gave for this is that witnesses may be confused about details that are crucial in a court of law, such as who drew his gun first. But they are not confused about the basic fact that the main event occurred—in this case an argument in which guns were drawn.

The False Memory Syndrome

Can a person falsely remember a complete event—such as a bank robbery—which never actually occurred? When we are dealing with a clear, conscious memory by a sane adult of an event which occurred a no more than a few years ago during adult life, it seems unlikely that this will occur. However, memory does have its gray areas. If a person does not have a clear memory of a particular event, then persuasive social pressure may induce the person to "remember" that event, at least vaguely, even though it never happened. This may take place when the person is dominated by an authority figure, such as a psychotherapist, who strongly believes that the person has repressed memories of certain experiences and is able to recover them. It is even more likely to happen when a highly suggestible person is hypnotized for the purpose of recovering lost memories.

In recent years these failings of human memory have become a topic of heated controversy. Many people undergoing certain forms of psychotherapy have supposedly recovered repressed childhood memories of sexual abuse by parents or close relatives. Families have been disrupted when these recovered memories led to bitter accusations and expensive lawsuits directed at family members.

This has resulted in a strong backlash in which accused family members have charged that the recovered memories of abuse are really fantasies generated in the accusers' minds by the psychotherapeutic process. This generation of pseudomemories has been named the false memory syndrome (FMS), and it has become the focus of a great deal of psychological research.

Proponents of the false memory syndrome argue that human memory is a highly malleable, reconstructive process. Some maintain that repressed memories may not even exist and that the apparent recovery of lost memories is an illusion. Thus sociologist Richard Ofshe maintains that "The notion of repression has never been more than an unsubstantiated speculation tied to other Freudian concepts and speculative mechanisms."[22]

Others say that repressed memories can be recovered, but false memories may also be generated by the recovery process. Extreme statements abound in this controversial area, but the following statement from the American Psychological Association gives a moderate summary of the basic FMS hypothesis: "It is possible for memories of abuse that have been forgotten for a long time to be remembered. . . . It is also possible to construct convincing pseudo memories for events that never occurred. . . . There are gaps in our knowledge of the processes that lead to accurate or inaccurate recollection of childhood sexual abuse."[23]

A review of the evidence suggests that false memories can, indeed, arise in peoples' minds under the influence of suggestion. This observation can always be used to cast eyewitness testimony into doubt, particularly if the testimony may have been influenced by social pressures. However, the general rejection of human testimony has serious consequences. Child victim expert Lucy Berliner observed: "I don't think all eyewitness accounts should be discredited as a result [of the false memory syndrome]. Many of the cases in our criminal justice system depend on eyewitness accounts. If an environment is created in which we say not to listen to those accounts, then what do we do?"[24]

I would suggest that a reasonable position is that human memory is imperfect, but not totally imperfect. The existence of false memories does pose complications for the interpretation of eyewitness testimony, but it does not imply that all eyewitness testimony should be disregarded.

Unfortunately the idea of the false memory syndrome can be used to totally rule out certain categories of testimony as invalid. This is shown by some of the examples of false memories cited by FMS researchers. For example, Nicholas Spanos and his colleagues cited three categories of false memories in an article published in *The International Journal of Hypnosis.*[25] These are (1) hypnotically induced memories of past lives, (2) memories of UFO close encounters, and (3) memories of ritual abuse carried out by members of satanic cults. The article took it for granted that the memories in the first two categories must be false since, after all, past lives and UFOs do not exist. The memories of satanic abuse were dismissed as unreal because investigations by law enforcement agencies have failed to show the existence of the alleged satanic cults.

Ian Stevenson has pointed out that hypnotically induced memories of past lives are often spurious.[26] But he also maintained that "rarely—very rarely—something of evidential value emerges during attempts to evoke previous lives during hypnosis," and he cited two studies of his own in which this happened.[27]

It would seem that memories of past lives recovered through hypnosis are not necessarily false. In a few cases, such memories may actually be evidential. If so, then these are presumably part of a larger set of cases that are genuine but not strong enough to be considered evidential. In still other cases, the memories may contain genuine elements as well as elements produced by imagination.

The situation of reports of satanic ritual abuse may be similar. FBI investigator Kenneth Lanning, who has spent years investigating the sexual victimization of children, has pointed out that many accusations of satanic sexual abuse are being made that law enforcement agents have been unable to corroborate. However, Lanning does not dismiss all of these satanic abuse reports as false. He stated that, "Some of what the victims allege may be true and accurate, some may be misperceived or distorted, some may be screened or symbolic, and some may be "contaminated' or false. The problem and challenge, especially for law enforcement, is to determine which is which."[28]

I would suggest that similar observations might be made about reports of UFO close encounters. The failings of human memory make it likely that many of these reports may contain spurious material. This is particularly true of uncorroborated reports in which hypnosis was used by a zealous investigator to recover lost memories from a highly suggestible witness.

However, since human memory is not completely imperfect, and sometimes works remarkably well, it also seems reasonable to suppose that many reports of UFO close encounters contain realistic material and some may be quite accurate. This is particularly true of multiple witness cases and cases involving responsible adults with clear, conscious memories. The use of hypnosis to recover memories of UFO encounters is controversial, and I will discuss it in Chapter 4.

I should also note that, according to Stevenson's analysis, suggestion and social pressure often tend to suppress rather than encourage the reporting of unusual phenomena. Some researchers argue that human errors in paranormal cases "are nearly all in the direction of reinforcing previously held favorable beliefs about paranormal events." [29] Stevenson said he has encountered this kind of amplification and it is particularly common among people seeking to cash in on their experiences. But he pointed out that many people report paranormal events with great reluctance due to fear of ridicule. And "many subjects also insist that prior to their experiences they had no settled convictions or knowledge about the experiences which parapsychology studies." [30] He suggested that these people are not likely to amplify normal events into paranormal ones and may do just the opposite. Very similar observations have been made by investigators of UFO close encounters.

A Well-Corroborated Close Encounter

Now I will look at a few close-encounter cases in detail. The first case is typical of UFO close encounters in content, but it has an unusual number of apparently independent eyewitnesses who offer corroborating testimony. It was originally investigated by the New York City UFO researcher Budd Hopkins, and I will summarize his account.[31]

In January of 1975, George O'Barski, an astute, 72-year-old New Yorker, was traveling home to North Bergen, New Jersey, after closing his Manhattan liquor shop and doing some bookkeeping and replenishing of shelves. The time was around 1 or 2 a.m. In North Hudson Park, across the Hudson from Manhattan, his car radio began to pick up static, and a brilliantly lit object passed the car 100 feet or so to the left. It emitted a quiet humming sound and stopped in a playing field ahead of the car. As O'Barski incredulously drew closer, he saw a roundish, 30-foot-long ship that was now hovering 10 feet above the ground. The ship had evenly spaced windows about one foot wide by four feet high. As it sank to a height of about four feet, a door opened between two of the windows, and nine to eleven small, helmeted figures in one-piece garments emerged and descended using a ladderlike apparatus. They were about 3.5 to 4 feet tall and looked like children in snowsuits.

As O'Barski drove slowly by, watching in terror, the beings apparently ignored him and proceeded to use spoonlike implements to scoop dirt into bags they carried. After doing this, they quickly reentered the craft, and it ascended, moving north. O'Barski estimated that the whole episode took about four minutes. He recalled it consciously, without recourse to hypnotism.

The next morning, O'Barski returned to the site, saw the holes made by the digging entities, and felt one with his hand to convince himself that they were real. As he put it, "A man my age telling a story like this—why, they'd put you away. If you'd come in here a year ago and told me the same story, I wouldn't have believed you either." [32]

As it turns out, however, Hopkins was able to find other people who may have seen the same flying object. His second witness was a man named Bill Pawlowski, who was a doorman at the Stonehenge Apartments, a high-rise apartment building near the landing site in Hudson Park. Pawlowski testified that he was on duty in the apartment building at 2 or 3 in the morning one day in January of 1975. He looked up into the adjacent park and saw a row of 10 to 15 brilliant, evenly spaced lights, which appeared to be about 10 feet off the ground, with a dark mass surrounding the lights. He walked to the window for a better view and then turned back to call a tenant in the building on the phone. At that moment, he was startled by a high-pitched vibration and cracking sound. A glance at the window showed that it was broken, and later inspection showed that it had been hit from the outside.

As Pawlowski glanced up, the lights had disappeared. He duly reported the incident to the North Bergen police, who came to inspect the window. But he discreetly avoided mentioning the strange lights in the park. Later, however, he did relate the incident to police lieutenant Al Del Gaudio, who lived in the building. Del Gaudio told Hopkins that he remembered hearing Pawlowski's story about the "big thing with lights on it" and had dismissed it as unbelievable.[33]

Another witness located by Hopkins was Frank Gonzalez, a doorman who worked at Stonehenge on Pawlowski's nights off. He had sighted a similar object at the same location between 2 and 3 a.m. on January 6, six days before O'Barski's sighting. He described his experience as follows: "I saw something round, very bright, you know . . . with some windows. I hear some noise . . . it's not like a helicopter, nothing like that. Like a plane, no, no. Something different. . . . Then, you know, I see that light go straight up and I said, 'Oh God!'"[34]

Then there is the experience of the Wamsley family. After Hopkins's colleague Gerry Stoehrer gave a talk on UFOs to a North Bergen PTA group, he was approached by 12-year-old Robert Wamsley and his mother, Alice. They said that while the family watched the Bob Newhart show one Saturday night in January, Robert looked out the window and saw a round, domed, brilliantly lit craft outside the window. It had rectangular windows that gave off a yellowish glow, and it floated forty to fifty feet off the ground. Four members of the family, including barefooted Mrs. Wamsley in her bathrobe, then ran out into the street and followed the slowly moving object for about two minutes.

The Wamsley family lived about 14 blocks from the Stonehenge apartments, and as the UFO was lost from view it was moving in that direction. Hopkins mentions that this apparently occurred on the same day as O'Barski's sighting; one piece of corroborating evidence for the date was that both O'Barski and the Wamsleys noted that the weather was unusually mild for January.[35] It would seem, then, that O'Barski, Pawlowski, and the Wamsleys may have seen the same craft on the same date. Gonzalez may have seen the same or a similar craft in another visitation six days earlier.

This story is typical of many UFO close-encounter reports. There is the strange craft, which looks like a piece of flying architecture with no evident means of propulsion. The craft makes a humming sound, and it carries bright lights. Little human figures dressed in suits emerge from it, engage in some apparently meaningless action, and then depart.

The UFO definitely seems to be intelligently controlled. It doesn't operate according to known physical principles in any immediately obvious way. But, at the same time, the story contains no direct evidence that the UFO is extraterrestrial. This conclusion could only be inferred indirectly by saying that if the "little men" do not live on the earth, then they must come from another planet. But this is surely not the only possibility.

Could the story be a hoax, a hallucination, or a misperception of natural phenomena? Certainly, natural phenomena seem to be ruled out. The hypothesis of hallucination or hoax runs into difficulty because several people testified to seeing the strange object.

One might argue that the witnesses influenced one another, perhaps on a subconscious level, so that they all came up with mutually supportive stories. This might be true of Pawlowski and Gonzalez, who both worked at the Stonehenge apartments. But the three groups consisting of (1) O'Barski, (2) Pawlowski and Gonzalez, and (3) the Wamsleys were supposed to be mutually unknown to one another. If they did influence one another, they must have known one another and that suggests a deliberate conspiracy. Or one might propose that when Hopkins visited Stonehenge, he influenced Pawlowski and Gonzalez to imagine their stories. Then Stoehrer's talk influenced the Wamsleys to concoct their story.

Of course, only O'Barski saw the little suited figures, and one would have to ask if he had a history of mental derangement. However, Hopkins characterized him as intelligent, "street-wise," reflective—and a strict teetotaler. Also he was not a UFO "believer" before his experience.

The digging up of soil samples by the little figures is puzzling. This activity has been reported in large numbers of UFO cases, and it has been popularized in movies like *ET.* O'Barski might have heard about it, but why would he invite ridicule by claiming to have seen such a thing himself?

Now, it turns out that the Stonehenge encounter story has additional features that I haven't yet mentioned. We can get another perspective on this story by turning to the testimony of the psychiatrist and UFO researcher Berthold Schwarz:

> I also saw four of the Stonehenge protagonists in cursory psychiatric and paranormal surveys. . . . EU, the day doorman, and a leading experient, has had lifelong high-quality psi: e.g., possible precognition—he claimed foreknowledge of the UFO activity—apparitions and telekinesis. His son and wife also had unusual presumed psi experiences. EU and the apartment electrician shared a close daytime sighting. They noted how the top floor of their apartment was unique, and might have resembled the stereotyped concept of a conventional UFO by its circular shape, dome, and flashing lights on the sides. EU wondered: "Is there an attraction to this building?" [36]

Apparitions and telekinesis? Four protagonists? It begins to look as though the Stonehenge apartment building was a hotbed of psychic and UFO activity. The fact that several persons at Stonehenge had paranormal and UFO interests might cause one to speculate that the whole sighting story was concocted by people with overheated imaginations. But we should consider that Hopkins's first lead was the story of O'Barski, who—barring conspiracy—had no connection with Stonehenge. How, then, could the story have originated at Stonehenge? And was the Wamsley family driven to lies or hallucinations by stories originating in that apartment building?

The information added by Schwarz illustrates two important points about the UFO phenomenon. The first is that no matter how much one knows about a given UFO encounter case, there is likely to be other important information about which one is unaware. In many instances, this information simply hasn't come to light during the investigation of the case. In other instances, the investigator may be able to believe and report certain aspects of the case, but he finds other aspects so incredible that he decides not to mention them. Or he may not mention certain aspects because he fears people may dismiss the whole case on hearing of them.

The second point is that UFO encounter cases tend to be connected with paranormal phenomena. Sometimes the witnesses, or people associated with them, have a past history of paranormal experiences. In others, a person will begin to have paranormal experiences after the UFO encounter, involving telepathy, poltergeist phenomena, or psychic healing. This is an empirical observation that I will gradually document with a number of examples. Later on I will consider what it might mean.

A Report to Congress

Next I turn to a UFO close-encounter case that was reported to Congress on April 5, 1966, during the hearings on Unidentified Flying Objects by the Committee on Armed Services. The report on the case was submitted to Congress by its investigator, Raymond Fowler, who was identified as a project administrator and engineer in the Minuteman missile program. This case involves a close-range sighting of what seemed to be a strange flying machine, and it also involved several eyewitnesses. The total report in the Congressional record occupies about 33 pages.

The story unfolded near Exeter, New Hampshire, during the early-morning hours of September 3, 1965. The first sign that something strange was happening came at 1:30 a.m., when police officer Eugene Bertrand investigated a parked car and found a distraught woman (some reports say two women) who claimed that her car had been followed for some 12 miles by a flying object encircled with a brilliant red glow. She stated that the object dived at her moving automobile several times.

Bertrand rejected this story but was soon summoned back to his police station to investigate a similar story by 18-year-old Norman Muscarello. The teenager had burst into the station at 1:45–2:00 a.m. in a state of near shock. He stated that while he was hitchhiking along Route 150, a glowing object with pulsating red lights suddenly came floating across a nearby field in his direction. He said that the object was as big as a house and that it was completely silent as it moved toward him. After he dove for cover, the object backed away and disappeared over the trees. After banging on the door of a nearby house with no response, he flagged down a car, which took him to the police station.

Bertrand and Muscarello returned to the scene, and at about 2:25–2:40 a.m. both saw the object rise silently from behind a row of trees. As Bertrand later described it, the object was as big as a house. It seemed compressed, as if it were round or egg-shaped, with no protrusions like wings, rudder, or stabilizer. The object had a row of four or five blinding red lights that blinked cyclically, casting a blood-red glow over the field and a nearby farmhouse. Bertrand said that the lights were brighter than any he had ever seen, and he had the impression that he and Muscarello might have been burned if they did not run from the object as it approached them.

The lights seemed to be part of a large, dark, solid object.[37] As nearby horses kicked in their stalls and dogs howled, the object floated about two hundred feet off the ground, yawing from side to side with a fluttering motion like a falling leaf. The total time of the sighting was about ten minutes.

This testimony was confirmed by officer David Hunt, who arrived on the scene in time to observe the object for five or six minutes as it departed in the direction of Hampton. The police also received a phone call from an excited man in Hampton, who reported seeing a "flying saucer" but whose line went dead before he could be identified.

At one point, this sighting was identified in a newspaper as an advertising plane owned by the Sky-Lite Aerial Advertising Agency of Boston. However, some checking showed that this plane was not flying on the night in question, and it carried a sign made of 500 white lights—with no red lights.[38]

The Air Force initially proposed that Muscarello, Bertrand, and

Hunt had seen high-flying airplanes in a refueling exercise called Big Blast "Coco." However, the timing of the sighting ruled this out, and Air Force officials concluded:

> The early sightings by two unnamed women and Mr. Muscarello are attributed to aircraft from operation Big Blast "Coco." The subsequent observation by Officers Bertrand and Hunt occurring after 2 a.m. are regarded as unidentified.[39]

The high intensity of the lights reported by the witnesses seems to be crucial for the interpretation of this sighting. It would seem that an advertising plane or military planes in a refueling exercise would not produce such an overpowering impression of blinding light to observers at ground level. And if the witnesses were seeing an unknown natural phenomenon, why would the lights be arranged in a row and flash in sequence?

Un Disco Volante

Most of the well-publicized cases recorded in the UFO literature in the forties, fifties, and early sixties involved sightings of strange flying objects from a distance. However, close encounters were also being reported during this time. We have already cited Dr. William Powers's statement in *Science* that there were over 200 reports of UFO landings in 1954, many with occupants. The actual level of reporting may have been even higher. Thus Edward J. Ruppelt, the head of U.S. Air Force UFO investigations in 1952, wrote in *The Report on Unidentified Flying Objects* that he felt plagued by reports of landings, and his team conscientiously eliminated them.[40]

It is not clear why UFO reports became so much more prominent after 1947 than they were before. There are earlier reports, but these are relatively few in number. For example, Raymond Fowler, while explaining the origin of his interest in UFOs, pointed out that his mother had a UFO encounter in Bar Harbor, Maine, when she was a child in 1917. Fowler relates that one evening his mother and her sister were returning home with friends from a church club meeting. As they took a shortcut across a field, a huge silent object suddenly appeared overhead, and "hues of reds, blues, greens, and yellows reflected off their frightened faces."[41] In this case, the terrified children ran for home, and the incident was over.

This encounter is striking because it shares features with the much more elaborate visitations at Fatima, Portugal, which also occurred in 1917. The events at Fatima also involved multicolored lights from the sky which reflected from peoples faces, and they are described in Chapter 8 (pages 293–96).

Going back to 1947, a classical encounter with diminutive humanoids was said to have taken place in Italy on August 14th of that year. This report is interesting because it antedates all other well-known reports of "little men" from UFOs. At the same time, it was never widely publicized, and so it is hard to see how it could have influenced the many similar cases that took place after it.

The witness in this case was Rapuzzi Luigi Johannis, a well-known Italian painter and science fiction writer. His encounter occurred near Villa Santina, north of Venice and near the borders of Austria and Yugoslavia. Johannis said he first related his story in confidence to two people in America when he visited there in 1950. He tried in 1952 to publish an account in *L'Europeo* but was turned down since he lacked proofs. He finally published it in an Italian magazine, *Clypeus,* No. 2–5, in May 1964 under the title "*Ho visto un disco volante.*" [42] I mention these dates to show that Johannis *could* have made up the story, based on UFO literature available by 1964. Of course, this doesn't mean that he did make it up.

On the day of the encounter, Johannis, who was interested in geology and anthropology, was hiking up a mountain stream, looking for fossils. He saw, wedged into a transverse cleft in the mountainside, a red, metallic, disc-shaped object with a low central cupola and no apertures. This object had a telescoping antenna and was some 10 meters wide. He noted in his account that at that time he knew nothing of flying saucers.

While looking to see if anyone else was around, he saw two "boys" at a distance of some 50 meters. On approaching them, he realized that they were not human, and he felt paralyzed and devoid of strength. They were not over 90 centimeters in height (about 3 feet) and wore translucent, dark blue coveralls, with red collars and belts. Their heads were bigger than those of a normal man. Johannis said that their facial features, described in anthropomorphic terms, included enormous, protruding round eyes, a straight, geometrically cut nose, and a slitlike mouth shaped like a circumflex accent. The "skin" was of an earthy green color.

After gaping in astonishment for a couple of minutes, he waved his geological pick and shouted something. In response, one entity touched his belt, sending forth a "ray" that left Johannis prostrated on the ground, devoid of all strength to move. He managed to roll over slowly in time to see one entity make off with his pick. The entities then returned to their craft, which shortly flew off, dislodging a cascade of stones from the mountainside. He testified that the disc, as it hovered in the air, suddenly grew smaller and vanished. This was accompanied by a blast of wind which rolled him over on the ground.

After some three hours, Johannis felt strong enough to painfully make his way home. He recalled that he resolved to say nothing about the incident, as he didn't want to be considered a crazy visionary, or worse. On traveling to New York two months later, he heard for the first time about the flying saucers seen by Kenneth Arnold, and at that time he decided to reveal the story in confidence. He also said that two local people testified to seeing a red ball rising in the sky and vanishing at about the time of the incident.

In this account, there are several features that occur repeatedly in UFO close encounters. The paralyzing ray is standard, and so is the abrupt disappearance of the disc at the time of its departure. In his testimony before Congress on April 5, 1966, J. Allen Hynek made the observation that if UFOs are tangible objects, then they should be seen flying from point to point over considerable distances. Hynek found it puzzling that this is not observed.[43] Instead, UFOs often seem to appear abruptly, maneuver about in a localized area, and then abruptly disappear. After we review more examples of this, I will consider some ideas as to what may be happening.

The little men seen by Johannis were also typical in a number of ways. The uniforms, small stature, big heads and eyes, and slitlike mouths show up repeatedly in close-encounter cases. The green skin, however, is somewhat unusual—despite jokes about little green men.

Martians, Fertilizer, and Psychiatry

The validity of the Johannis account depends entirely on the integrity of Johannis himself. He was clearly an intelligent and talented individual, but one might suspect that he was too talented. Did he just make up the story? It is not clear why he would do this or what he had to gain by it, but we cannot rule out the possibility.

In contrast to the Johannis case, Dr. Berthold Schwarz has recounted an equally bizarre story in which a careful evaluation of the character of the witness makes it seem highly unlikely that he was concocting a tall tale. Since the credibility of this story rests on the reputation of Schwarz, I note that he is a psychiatrist who has written books on both child psychiatry and psychical research. He has also written a book on psychiatric aspects of the UFO phenomenon, and that book is the source of the story I am about to present.

At about 10 a.m. on April 24, 1964, a 26-year-old farmer named Gary Wilcox was spreading manure in a field on his dairy farm in Newark Valley, New York. He saw a white, shiny object above the field, just on the inside edge of the woods, and he drove up on his tractor to investigate it. At first he thought it was a fuselage or fuel tank of an airplane, and he went up and touched it. There then appeared from underneath the object two four-foot-tall men holding a metal tray filled with alfalfa, roots, soil, and leaves. They wore whitish metallic suits that left no part of their bodies exposed.

As Wilcox stood there anxiously, wondering what kind of trick was being played on him, one of the men said, "Do not be alarmed," in an eerie voice that seemed to emanate from the general vicinity of his body. They then asked Wilcox questions about farming and fertilizer and claimed they came from Mars, which is made of rocky substances not fit for growing crops. They made comments about air pollution in congested areas, and predicted the deaths of astronauts Glenn and Grissom from exposure in space. The men then ducked under the craft and disappeared. The craft produced a noise like a car motor idling, glided away for about 150 feet, and disappeared into the air.

In response to the men's request for some fertilizer, Wilcox later brought a bag of it to the site. The next morning he noticed that it was gone.[44]

Now, what kind of person would tell a story like that? Berthold Schwarz carried out a psychiatric examination of Wilcox and found that he "had no past history for neonatal disturbances, serious illness in the formative years, neurotic character traits, dissociative or amnesic experiences, fugues, sociopathic behavior, school problems, head injury, encephalopathy, surgery, or any kind of aberrant behavior." [45] He was in good health and had been a good student in school. He had no previous interest in UFOs or exotic subjects, and his reading was limited to newspapers and popular magazines. He sometimes attended services at a local Baptist church.

Schwarz tested Wilcox, using the Cornell Medical Index Health Questionnaire, the Rotter Incomplete Sentences Test, and the computer-automated Minnesota Multiphasic Personality Inventory (MMPI). The results were consistent with physical and emotional health. On the MMPI "a configural search for positive traits and strengths showed correlations for describing the subject as compliant, methodical, orderly, socially reserved, and sincere."[46]

Schwarz concluded, "It would be most unusual . . . for Gary Wilcox to concoct such a fantastic story without some clues for this from his psychiatric examination or from interviews with his friends, acquaintances, and family."[47] At the same time, he noted that there is no reason to suppose that the beings in the story came from Mars just because they said so.

Cases Involving Children

In a survey of UFO cases, it is important to note that UFO encounters are also reported by children. Here I will give three examples of such reports. One might object that children are prone to lies and fantasy, and therefore their testimony carries little weight. However, as shown by the traditional story of the boy who cried wolf, adults are able to distinguish between honesty and dishonesty in children. One lie may be hard to detect, but a child is unlikely to stop at one, and adults who inquire about the child will learn about his pattern of dishonest behavior.

The stories that I will recount in this section are not as well attested as the Wilcox case, but I think they deserve to be considered. As with all UFO stories, they do not constitute proof. But a satisfying approximation to proof can come only by understanding the overall pattern in large bodies of data, and then judging individual cases by how well they fit into the pattern. If we exclude large sections of data from consideration, we may miss important clues to the pattern.

The first of the three children's reports is given in an article entitled "The Landing at Villares del Saz," by Antonio Ribera, a long-time UFO researcher in Spain.[48] The principal witness in this case was an illiterate 14-year-old cowherd boy named Máximo Muñoz Hernáiz. He had an encounter while tending cows in the early part of July 1953 near the village of Villares del Saz, Cuenca, Central Spain. Here are excerpts from an interview of the boy, conducted by the editor of the newspaper *Ofensiva:*

What you saw doesn't exist. So how do you explain it?
I *did* see it. I *did* see the little chaps.
At what time did you see the machine?
At one o'clock.
What were you doing at that moment?
I was sitting down, watching the cattle to see that they didn't get on to the crops.
Did you hear any sound beforehand?
Yes, but slight. So I didn't turn round.
You had your back turned in that direction?
Yes, sir.
What did you hear?
(Máximo Hernáiz said that he had heard a faint, muted, intermittent whistling. When he turned in that direction, the machine had already landed.)
What did you do when you saw it?
Nothing. I thought it was a big balloon— one of those that they let off at fairs. Then I realized it wasn't. It glowed very brightly.
Did it glow the whole time?
Less when it was stationary than when it moved off.
What was its colour?
[The boy indicated that the object was grey in color, about 1 meter 30 centimeters high (51 inches), and shaped like a small water jug.]
Did it remain there on the ground long?
A very short time. As I thought it was a balloon, I went over to grab hold of it. Before I had time to reach it, a door opened and little chaps started coming out of it.
What were the little chaps like?
They were tiny. Like this (about 65 centimetres [26 inches]).
Were their faces like ours?
Their faces were yellow, and their eyes were narrow.
(The painter Luis Roibal, who was with the newspaper editor, made a number of sketches of little men according to the lad's description.)
Yes, like that, but more *chaparrete.*
(The features of the faces are completely oriental.)
How many little men came down out of the balloon?
Three.
Where did they come out?

> Through a little door that the *thing* had on top.
> *How did they get down?*
> They did a little jump.
> *Then what did they do?*
> They came over to where I was.
> *Did they speak?*
> Yes, Sir, but I couldn't understand them.
> *How did they stand?*
> One on one side of me, one on the other, and the one who spoke to me was in front of me.
> *Did they do anything to you?*
> When I didn't understand what he said to me, the one standing in front of me smacked my face.
> *And then what?*
> Nothing. They walked off.
> *How did they get up into the machine?*
> They grabbed hold of a *thing* that was on the balloon, and jumped, and in they went.

The boy said that the men were dressed in smart blue suits like musicians at a fair, and they wore flat hats with visors. They also wore metal sheets on their arms, but he couldn't describe these clearly. After the men reentered the object, it glowed very brightly and flew off rapidly, leaving no exhaust trail and making the same whistling sound as before.

Ribera said that the boy's father went to the site with the officer in charge of the local police station, and they found footprints plus four holes forming a square 36 centimeters (14 inches) on each side. Other witnesses, including the police constable of the Honrubia Police Post near Villares, reported seeing a flying grayish-white sphere coming roughly from the site in Villares de Saz at the time of the encounter.

Strange though this story is, it is similar to many others told all over the world. If it really does originate with an illiterate cowherd boy from central Spain, it is hard to see why he would have thought of commonly reported details such as the oriental-looking faces, the whistling sound, the glowing of the globe, and its flight without a visible exhaust trail. Presumably, he would have needed coaching by someone knowledgeable, and this doesn't seem plausible for an illiterate 14-year-old boy from a family of farmers. If the story is false, then it seems likely that the entire description of the boy and his situation must be false.

Another example of testimony from a child comes from 12-year-old John Swain, the son of a farmer living near Coldwater, Kansas. He had a UFO encounter in September of 1954, and wrote a letter about it to one Reverend Baller on Oct. 3, 1954:

> You ask me about the saucer I saw. I was disking the field when I saw it. We had tractor trouble. It was late when we got it finished. It was cooled off some, so I worked till 8 p.m. Then I unhitched from the disk and came in. I met it about 400 feet and didn't see it. I came on a . . . [terrace?]. He was crouched behind it. He jumped up and looked at me, and kind of floated. He jumped into the saucer and it lighted up and took off. It went out of sight. I told Mom and Dad about it. We talked it over. Then Mom called the sheriff. He came down that night and questioned me. He said he would come again in the morning and look and see if there were any tracks around. There was. He sent the reports to Washington, D.C. Signed, John Swain.[49]

The tracks in question were said to be wedge-shaped and unlike those made by ordinary shoes. The floating of UFO entities is commonly reported, and once again we see a reference to a flying object that lights up or glows when it takes off. If this is a fantasy, it shows adherence to standard UFO themes rather than free imagination.

For the third example of a UFO encounter reported by children, I turn to South Africa. On Oct. 2, 1978, at about 11:15 a.m., four teenage schoolboys were waiting for the mother of the eldest to pick them up at an isolated spot in the Groendal Nature Preserve in South Africa. The boys became aware of a silver object protruding above the bush several hundred meters away, on the other side of a valley. At this moment, another of the boys noticed two men dressed in silver coveralls about 275 meters to the west of the object. Shortly after this, the two men were joined by a third, and the boys noticed that their mode of walking was peculiar. "They moved only from the knees downward and used their legs like a fin," one of the boys said.[50]

After the matter was reported some 10 days later, the boys were separately interviewed and gave similar stories. They also separately made comparable drawings of the men.

Several investigators, including a major in the police, spent about 90 minutes cutting through dense brush with machetes to reach the site.

They found "a depressed area of 6 by 18 meters where the bush had been flattened to ground level and on the outside perimeter of the oval depression, there were 9 marks, each containing 3 or 4 tiny imprints." [51]

Physical Traces and Effects

In the cases just described, we have seen several instances in which landed UFOs were reported to leave physical traces of their presence on the earth and on vegetation. Since such traces can be evaluated in a laboratory, they provide one of the main lines of scientific evidence for the physical reality of UFOs.

One case involving measurable ground traces was investigated by the UFO study group called GEPAN, established by the French space agency CNES in 1977 (see page 38). This case was described as follows by the head of GEPAN, Jean-Jacques Velasco:

> At about 5 p.m. on February 8, 1981, Monsieur Collini was working quietly in his garden at Trans en Provence. Suddenly his attention was attracted by a low whistling sound that appeared to come from the far end of his property. Turning around, he saw in the sky above the trees something approaching a terrace at the bottom of the garden. The ovoid object suddenly landed. The witness moved forward and observed the strange phenomenon behind a small building.
>
> Less than a minute later, the phenomenon suddenly rose and moved away in a direction similar to that of its arrival, still continuing to emit a low whistle. M. Collini immediately went to the apparent scene of the landing and observed circular marks and a clear crown-shaped imprint on the ground. The Gendarmerie arrived the next day to report, and, following our instructions, took samples from the ground and surrounding vegetation. On D+39 [39 days after the sighting], a GEPAN team was sent to investigate. The first results of the analysis proved interesting, with soil and vegetation samples showing significant effects, in particular biochemical disturbances to plant life.[52]

The investigation showed that the witness had no psychological problems and that his testimony was internally and externally consistent. It also found signs of ground heating at the landing site to between 300° and 600°C, and the probable deposition of trace quantities of phosphate and zinc.

A biochemical analysis was carried out by the biochemistry lab of INRA (Institute National de la Recherche Agronomique) under one Prof. Bounias. This study dealt with the chlorophyll and carotenoid pigment content of a species of wild alfalfa growing in the area of the landing site. At 39 days after the sighting, a 30–50% reduction in chlorophyll pigments A and B was observed, with young leaf shoots showing the highest losses along with signs of premature senescence. The strength of the effect correlated strongly with distance from the center of the landing site. However, there were no signs of residual radioactivity.[53]

This investigation seems to show that some UFO phenomena are amenable to serious scientific investigation. In this case, the integrity of the witness and the empirical measurements at the landing site combine to indicate that something unusual but physically real did actually happen. Perhaps the simplest hypothesis to explain the observed data is that an unknown type of flying machine did land in M. Collini's garden, pause for a minute or so, and then fly away.

UFO ground trace cases have been studied extensively by Ted R. Phillips, who is a longtime UFO investigator and has participated in several scientific UFO studies in the United States. In 1981, he reported on a 14-year study of 2,108 physical trace landing cases from 64 countries.[54] Of these, he personally investigated over 300. He summed up his study by saying:

> 1. The cases show significant statistical patterns.
> 2. The UFOs observed by multiple witnesses appear to have been solid, constructed vehicles under intelligent control.
> 3. They produced physical traces that, in many cases, have no natural or conventional explanation.
> 4. There has been very little scientific investigation of these reports.

Phillips presented a number of statistics on UFO landing reports with physical traces. For example, in the first four decades of this century, his records showed about 6 physical trace cases per decade. This shot up to 43 in the 1940s, and in the 1970s there were 1,001 physical trace cases.

Out of his total of 2,108 cases, he said that about 275 involved two witnesses, and about 430 involved three or more. Humanoids were reported in 460 cases, and in 310 of these the humanoids were small compared to normal humans. In 87 cases, they were of normal size, and in 63 they were considered large.

Phillips gave the following data on the external features of the observed UFOs in his cases. Each feature is accompanied by the number of cases in which it was reported:

sound	214	ports or windows	207
external lights	207	vertical ascent	184
light beam	183	landing gear	159
dome	144	vapor	128
UFO rotated	125	antenna	117
heat	117		

The landing traces were described as circular, oval, or irregular in shape. Vegetation in the traces was burned, depressed, or dehydrated, and there were often symmetrically arranged marks that were suggestive of landing gear imprints.

Statistics can give us some idea about what is typical, but they cannot explain it. However, statistical analysis can reveal some interesting patterns in UFO data. For example, some statistics on humanoid reports were compiled by the UFO investigators Coral and Jim Lorenzen in a book published in 1976.[55] In a collection of 164 reports dating from 1947 to 1975, they classified humanoids as small (under 40 inches) and large (over 40 inches). They also divided the reported UFOs into different categories, including large and small discs, and large and small egg shapes.

In reports mentioning large discs, 24 featured large humanoids and five featured small humanoids. In reports mentioning small discs, there were nine featuring large humanoids and 28 featuring small humanoids. Thus the size of the beings seems to roughly parallel the size of the discs. The same pattern showed up in the cases with egg-shaped UFOs. I don't know the explanation of this pattern, but it would be interesting to see if it shows up in Phillips's physical trace cases as a correlation between humanoid size and landing track spacing.

Electromagnetic Effects on Cars

In addition to producing physical effects on the ground, UFOs are also well known for producing transient electromagnetic effects on motor vehicles. This was described as follows by Roy Craig in the famous Condon Report:

> Of all physical effects claimed to be due to the presence of UFOs, the alleged malfunction of automobile motors is perhaps the most puzzling. The claim is frequently made, sometimes in reports which are impressive because they involve multiple independent witnesses. Witnesses seem certain that the function of their cars was affected by the unidentified object, which sometimes reportedly was not seen until after the malfunction was noted. No satisfactory explanation for such effects, if indeed they occurred, is apparent.[56]

Car interference cases are reported in many different countries around the world. Here is an example from Australia:

> On October 20, 1986, near Edmonton, Queensland, a 41-year-old local woman driving home from Cairns along the Kamma Pine Creek road began to experience extreme difficulty in controlling her car. It was pulling to the right side of the road. About 400 meters further along the road the dashlights and headlights almost faded out. The woman then heard a buzzing sound and the vehicle lost power from the engine. The witness looked up to see a bright oval, blue-green light ahead. She had her "foot flat on the floor," but the motor seemed only to be idling and the car continued to travel forward at a fairly slow speed. These phenomena continued for approximately 4 kilometres.[57]

The lady said that the whole experience took some 8 to 10 minutes, and that the UFO seemed to be traveling roughly parallel to the road. While passing over a one-lane bridge, the UFO "suddenly took off," and she regained full control of her car. She testified that the car was all right before and after the episode.

One explanation that is sometimes offered for such incidents is that they are caused by a natural phenomenon involving a vortex of electrified plasma. This idea has recently been elaborated by the English meteorologist Terence Meaden in an effort to explain the celebrated English crop circles, and UFO researcher Jenny Randles has suggested that it might account for some reported UFO phenomena.[58] The "bright oval, blue-green light" that paced the Australian lady's car might be interpreted as a glowing plasma that produced a buzzing sound and interfered with her car's electrical system. But for this to happen, very high energies would be required within the electrified mass, and it is very hard to see what natural process might produce such energies and contain them within a limited volume of space for longer than a fraction of a second. From an orthodox scientific point of view, it is almost as hard to account for a plasma vortex that can interfere with a car for 8 minutes as it is to account for a flying machine of nonhuman origin.

It turns out that UFO car-encounter cases are not all the same. They fall naturally into several distinct groups, and this includes a group in which some kind of electrified plasma may be involved. This breakdown into groups can be accomplished by performing a statistical analysis on a large collection of cases.

Such a study was carried out by Dr. Donald Johnson, a psychologist and statistician who is director of a New Jersey management consulting firm.[59] Johnson performed a cluster analysis of 200 car-encounter cases ranging in date from 1949 to 1978. This analysis used the following three variables: duration of the event, estimated distance from the automobile to the UFO, and estimated size of the UFO.

For each case, these three variables can be thought of as the x, y, and z coordinates of a point in space. (Actually Johnson used the "standard deviates of base-2 log transformations" of the variables.) We can imagine that the 200 points corresponding to the 200 cases may fall into a number of clusters in space. The cases in a cluster all share something in common, and they differ from the cases in another cluster. Cluster analysis does not tell why the cases should fall into distinct groups, but it can help researchers identify such groups for further study.

Johnson found that his 200 cases fell into the following seven clusters:

> *Cluster 1:* [19] Small objects (2 meters) that appear for one or two minutes at very close distances. Some are red, white, or yellow-orange balls of light that may be related to ball lightning. But eight involve reported landings.

Cluster 2: [48] Larger than average objects (12–20 meters in diameter), maintaining a distance of 200–300 meters. Encounters last 15–20 minutes on the average. Over half are domed discs, and a third are described as metallic. Other features of these cases are: hovering (58%), light beams (21%), falling leaf motion (25%), and repair activity, in which entities emerge from a landed UFO and seem to work on it.

Cluster 3: [33] Objects slightly smaller than average size (six meters), at medium distance (90 meters), with encounter times usually under one minute. Over half are metallic, and they typically depart rapidly from a landing (42%) or a position close to the ground.

Cluster 4: [11] Objects of average size, but approaching closely to an average distance of about 15 meters. Encounter times average to about one hour. Many of these cases involve pursuit (82%), landing (45%), and abduction (27%). Nearly 66% involve physiological effects such as paralysis, electrical shock, tingling, or heat. There are often noises (e.g., humming) from the UFO (45%), a light beam (45%), and the UFO often has multiple colors (45%). The UFO often shot up or away very quickly when departing (64%), and 75% of the witnesses experienced fear or panic during or after the event.

Cluster 5: [62] Objects 9–10 meters in diameter, with average distance of 25 meters and average encounter time of 5 minutes. Over 60% are classic domed discs, with 40% described as metallic, and 58% described as whitish in color, or as having a white light. Other features are hovering (68%), landing (40%), and departing at incredible speed (39%). Half of the witnesses expressed fear.

Cluster 6: [12] Large objects (typically 60 meters in length), sometimes cigar shaped, that are seen only for a couple of minutes at moderately large distances (110 meters). Half are glowing in red or orange colors (and none are green). Other characteristics are: blocking the roadway and quickly departing when detected (30%), hovering (50%), landing (50%), shooting away rapidly (50%), and silent flight (75%). They generally leave no physical traces.

Cluster 7: [5] Small objects (1 meter or less in diameter), at average distance of 150 meters, with an average encounter time of 45 minutes. There are no electromagnetic effects, and none of the objects are described as metallic. Johnson suspected that these may be natural phenomena.

Once a number of clusters have been identified based on the three variables, one can ask if the clusters differ significantly in other ways. For example, do reports of domed discs tend to fall mainly in certain clusters and not others? If this happens, then it indicates that the clusters are meaningful. Different clusters involve different kinds of phenomena.

According to Johnson's analysis, there are indeed a number of features that tend to be strongly present in some clusters and not others. It would seem that there are different kinds of UFO car-encounter cases, with some involving natural phenomena and others involving different kinds of structured flying machines.

Johnson concluded, "My advice is to watch out for the noisy, domed-disks with the bluish-white light beams, because the odds are that if you encounter one of those you are likely at the very least to suffer some physiological effects. This appears to be particularly true if the object is hovering over the roadway in front of your car, and seems to take an interest in your course of travel!" [60]

Photographic Evidence

The topic of photographic evidence for UFOs is vast, and I will be able to touch on it only briefly. Over the years, many photographs have allegedly been made of UFOs, and there are also movies and videotapes. In some cases, these photographic records show only points of light, and thus they may represent airplane lights or natural phenomena. However, there are many cases where the photograph or movie clearly shows a structured, metallic craft. In these cases, there is always the question of whether or not the images have been hoaxed. Unfortunately, it is practically impossible to prove in any given case that a hoax has not been perpetrated.

The status of photographic evidence for UFOs was summed up by William K. Hartmann in the Condon Report. After surveying various photographic cases, he admitted that "A very small fraction of potentially identifiable and interesting photographic cases remain unidentified." Regarding these cases, he said,

> 1. None of them conclusively establishes the existence of "flying saucers," or any extraordinary aircraft, or hitherto unknown phenomenon. For any of these cases, no matter how strange or intriguing,

it is always possible to "explain" the observations, either by hypothesizing some extraordinary circumstance or by alleging a hoax. That is to say, none of the residual photographic cases investigated here is compelling enough to be conclusive on its own.

2. Some of the cases are sufficiently explicit that the choice is limited to the existence of an extraordinary aircraft or to a hoax.
3. The residual group of unidentifieds is not inconsistent with the hypothesis that unknown and extraordinary aircraft have penetrated the airspace of the United States, but none yields sufficient evidence to establish this hypothesis.[61]

Cases that must involve either an unknown flying machine or a hoax are perhaps not as rare as Hartmann suggests. A large collection of such cases can be found in the book *UFO Photographs,* by Stevens and Roberts.[62] Here is one example of a case from that book.

On July 29, 1952, George Stock, a lawnmower repairman in Passaic, New Jersey, was in his yard working on a mower. At about 4:30 p.m., he saw an unusual object flying in the sky and yelled to his father to bring their box camera. With the camera, Stock shot seven black and white photos of a solid-looking, metallic disc with a semitransparent dome on the top. This object seemed to be about 20 to 25 feet in diameter, and it was traveling slowly about 200 feet above the ground. Stock had the photographs developed by a local photo developer named John H. Riley and had them published in the *Morning Call*, Vol. CXLI, No. 28 of Paterson, New Jersey, on Aug. 3, 1952.

Stock was contacted by George Wertz of OSI (the Air Force Office of Special Investigation), who strongly insisted on obtaining the negatives of Stock's photographs. These were retained by the Air Force for six months, and then, after many complaints by Stock, five of the negatives were returned.

The case was investigated by August Roberts, who found that a woman living three and a half blocks from Stock's house also saw a disc-shaped flying object at about the time Stock took his pictures. There were other witnesses, but these suddenly became silent. Stock himself seemed to be undergoing considerable harassment, and at one point he turned over the remaining five negatives to Roberts, saying, "You tell them you took them, or that they are fakes. I don't need all this."[63]

What are we to make of this case? If we examine the pictures, we see that they clearly must be either fakes or images of a genuine craft of the typical domed-disc variety. It seems that both Roberts and OSI officer Wertz heard a rumor that some neighbor saw Stock "throwing up a model." But they were not able to trace it down, and Wertz felt on the basis of his study of the photos that "the object in question was quite high and that therefore a very large model must have been used, if it was a model." [64] According to Roberts, Wertz was noncommittal about the Stock photos but tended to lean in favor of the idea of their being genuine.

One feature of both UFO reports and photographs is that although it is widely acknowledged that UFO shapes largely fall into a few basic categories (such as disc, sphere, and ellipsoid), their detailed forms show great variety. As a result, there are relatively few cases where identical-looking UFOs have been independently photographed in different locations. The Stock case provides one example, since UFOs looking very similar to Stock's were photographed in mid-1952 in Chauvet, France, and Anchorage, Alaska.[65]

Another example of two apparently independent pictures of nearly identical UFO shapes was reported by Dr. Bruce Maccabee, a Navy physicist who is the chairman of the Fund for UFO Research in Maryland. During the evening of July 7, 1989, Mr. Hamazaki of Kanazawa, Japan, videotaped an object that passed nearly over his house. Maccabee described the shape of the object as a bright square with a dark hemisphere extending below it, giving the impression of the planet "Saturn with a square ring." [66] It seems that a virtually identical object was photographed by Michael Lindstrom in Hawaii on January 2, 1975. According to Maccabee, the only notable difference between the Hamazaki and Lindstrom UFOs was that the Hamazaki UFO had a bright ring with a dark hemisphere, and the opposite was true of the Lindstrom UFO.

In recent years, with the proliferation of home video cameras, a number of interesting videotapes have been made of UFOs. Two such tapes were made by an anonymous witness in Pensacola Beach, Florida, on March 24 and 31, 1993.[67] The second tape is particularly interesting, since it shows a UFO hovering for some time and then abruptly disappearing. When the tape is played frame by frame, the UFO was seen to move quickly off the screen in a series of elongated images after its apparent disappearance. Analysis by researchers Bruce Maccabee and Jeffrey Sainio indicated that the UFO was moving off the screen with an angular acceleration of about 911 degrees per second squared.

The videotape did not make it possible to determine the distance of the UFO from the camera. But if the object was 1000 feet away, it would have been 8–9 feet wide and its acceleration would have been about 500 times the acceleration of gravity. (At closer or further distances from the camera, these figures should be changed proportionately.) This acceleration was maintained uniformly for 4 video frames or about 4/30 of a second. Maccabee calculated that at a distance of 1000 feet, the UFO would have reached a speed of 2,650 feet/sec. when it went off the edge of the screen. This is about 2.3 times the speed of sound.

As always, it can be argued that the videotape was hoaxed. But if it was genuine then it provides quantitative evidence for the well-known ability of UFOs to make abrupt changes in velocity and to suddenly appear and disappear. In this case, the disappearance of the UFO was due to a high rate of acceleration which caused it to become invisible to the eye.

My last illustration of photographic evidence is taken from another important category: photographs and movies that were reportedly taken by military personnel but are not available to the public. In Chapter 1 (pages 29–30), I presented testimony from Dr. Elmer Green regarding UFO photos and films taken on military bases by scientists and engineers who were members of OSWG, the Optical Systems Working Group.

Here is a similar case. It is based on the testimony of Dr. Robert Jacobs, a former first lieutenant in the Air Force, and now Assistant Professor of Radio-Film-TV at the University of Wisconsin. Jacobs claimed that on September 15, 1964, he was in charge of filming an Atlas F missile test at Vandenberg Air Force Base in California. He said that a couple of days after making the film he was summoned by his superior officer, Major Florenz J. Mansmann. The major asked Jacobs to view the film and drew his attention to what it showed at a certain point:

> Suddenly we saw a UFO swim into the picture. It was very distinct and clear, a round object. It flew right up to our missile and emitted a vivid flash of light. Then it altered course, and hovered briefly over our missile . . . and then there came a second vivid flash of light. Then the UFO flew around the missile twice and set off two more flashes from different angles, and then it vanished. A few seconds later, our

> missile was malfunctioning and tumbling out of control into the Pacific Ocean, hundreds of miles short of its scheduled target.[68]

He was told by Major Mansmann, "You are to say nothing about this footage. As far as you and I are concerned, it never happened! Right . . .?" Jacobs says that he waited 17 years to tell the story.[69]

Dr. Green said that members of his group were told by military authorities that there were no records that the UFO incidents they witnessed had ever happened. But they weren't ordered not to talk about them. Green also pointed out that none of the high-quality UFO photos and movies taken by OSWG were made available for study to the scientists preparing the Condon Report. The Condon Report's conclusions on photographic evidence, mentioned above, were based entirely on photographs shot by civilians on the spur of the moment with amateur camera equipment.

3

The Role of the Government

The public clamor over "flying saucers" that began in 1947 was paralleled by growing concern in U.S. military circles. Within two years after the end of World War II, a controversy had developed within the newly formed U.S. Air Force on whether or not UFOs constituted a threat to national security. According to Edward Condon, opinions were sharply polarized:

> Within the Air Force there were those who emphatically believed that the subject was absurd and that the Air Force should devote no attention to it whatsoever. Other Air Force officials regarded UFOs with the utmost seriousness and believed that it was quite likely that American airspace was being invaded by secret weapons of foreign powers or possibly by visitors from outer space.[1]

In this chapter, I will briefly summarize the history of the involvement of the U.S. Government and its military forces with the UFO issue. I will begin with the official story, as presented by Edward Condon in the *Scientific Study of Unidentified Flying Objects*.

One of the earliest official efforts to deal with flying saucer reports was initiated on September 23, 1947, by Lt. General Nathan Twining, the Chief of Staff of the U.S. Army and the Commanding General of the Army Air Force. Twining wrote a letter recommending the formation of a study group to investigate the problem of the "Flying Discs." In this letter, he ventured the opinion that:

1. The phenomenon reported is something real and not visionary or fictitious.
2. There are objects probably approximating the shape of a disc, of such appreciable size as to appear to be as large as man-made aircraft.

> 3. There is a possibility that some of the incidents may be caused by natural phenomena, such as meteors.
> 4. The reported operating characteristics such as extreme rates of climb, maneuverability (particularly in roll), and action which must be considered evasive when sighted or contacted by friendly air craft and radar, lend belief to the possibility that some of the objects are controlled either manually, automatically or remotely.[2]

Twining went on to say that piloted aircraft of the reported type might be constructed on the basis of current U.S. technology, but an effort along these lines would be extremely time-consuming and expensive. He allowed the possibility that the unknown objects might be products of a secret American project not known to his command, and he considered that they might also be produced by some foreign nation.

The study group recommended by General Twining was designated as Project Sign, and it continued to operate until February of 1949. The project's work was carried out by the Air Technical Intelligence Center (ATIC) at Wright-Patterson Air Force Base near Dayton, Ohio.

The project's final report seemed to indicate an ambivalent attitude toward continued UFO investigations:

> Future activity on this project should be carried on at the minimum level necessary to record, summarize and evaluate the data received on future reports and to complete the specialized investigations now in progress. When and if a sufficient number of incidents are solved to indicate that these sightings do not represent a threat to the security of the nation, the assignment of special project status to the activity could be terminated.[3]

This mood of doubt also came across in two appendices to the report, written by Prof. George Valley of M.I.T. and Dr. James Lipp of the Rand Corporation. These scientists argued that the existence of the flying discs was unlikely from a theoretical point of view, and suggested that psychological explanations should be seriously considered.

Both commented that the behavior of the flying objects seemed senseless, and Valley suggested that they might be some kind of animal, even though he humorously admitted that "there are few reliable reports on extra-terrestrial animals."[4] Both men also pointed out that UFOs might possibly be piloted by extraterrestrials who were alarmed by our testing of atomic bombs. Valley then echoed another common

UFO theme by saying, "In view of the past history of mankind, they should be alarmed. We should, therefore, expect at this time above all to behold such visitations."[5]

After February 11, 1949, the work at ATIC on UFOs was called Project Grudge. Apparently, this version of the UFO study created some grudges among participating personnel. For example, the astronomer J. Allen Hynek, who did a great deal of case analysis for the project, later said:

> The change to Project Grudge signaled the adoption of the strict brush-off attitude to the UFO problem. Now the public relations statements on specific UFO cases bore little resemblance to the facts of the case. If a case contained some of the elements possibly attributable to aircraft, a balloon, etc., it automatically became that object in the press release.[6]

Likewise, another participant, Captain Edward J. Ruppelt, said, "This drastic change in official attitude is as difficult to explain as it was difficult for many people who knew what was going on inside Project Sign to believe."[7]

Project Grudge produced one report in August of 1949 with the following conclusions:

> There is no evidence that objects reported upon are the result of an advanced scientific foreign development; and, therefore they constitute no direct threat to the national security. In view of this, it is recommended that the investigation and study of reports of unidentified flying objects be reduced in scope. Headquarters AMC will continue to investigate reports in which realistic technical applications are clearly indicated.
>
> NOTE: It is apparent that further study along present lines would only confirm the findings presented herein.[8]

The report further concluded that all UFO reports are due to (1) misinterpretation of conventional objects, (2) mild mass hysteria and war nerves, (3) fabrications, and (4) psychopathological persons. Furthermore, it stated that the Psychological Warfare Division should be informed of the results of the study, since it indicated that systematic planting of UFO hoaxes and false UFO stories could cause mass hysteria.[9] A press release announcing the closure of Project Grudge was published on December 27, 1949.

Regarding point (4), Condon himself stated that "only a very small proportion of sighters can be categorized as exhibiting psychopathology." [10] As I pointed out in connection with the Gary Wilcox case (pages 61–63), many thoroughly "far out" UFO encounter stories are told by completely sane and levelheaded people. I will discuss this point in greater detail in Chapter 4 (pages 153–157).

One would think this report might mark the end of UFO study within the Air Force. But according to Condon, on September 10, 1951, a mistake was made at the Army Signal Corps radar center at Fort Monmouth, N.J. An object was clocked on radar at a speed much faster than any existing jet plane. The object later turned out to be a conventional jet, but before this was discovered, General C. B. Cabell, the director of Air Force Intelligence, responded to the incident by reactivating Project Grudge in a new and expanded form.[11]

For a project that was seemingly resurrected by a fluke, this version of Grudge showed remarkable longevity. It was initially headed by Captain Edward J. Ruppelt, and it was renamed Project Blue Book in March of 1952. Under this name, it continued until the publication of the Condon Report in 1969, at which point the Air Force finally terminated its official involvement with UFO investigations.

The CIA and the Robertson Panel

At a certain level of the government, it seems that reports of UFOs, rather than the UFOs themselves, were regarded as a threat to national security. Thus on September 24, 1952, the Assistant Director for Scientific Intelligence, H. Marshall Chadwell, wrote a memo to CIA Director Walter Smith. The memo indicated that apart from a huge volume of letters, phone calls, and press releases, ATIC had received about 1,500 official UFO reports since 1947 and 250 official reports in July of 1952 alone.

The Air Force regarded about 20 percent of these reports as unexplained. However, Chadwell was concerned with a more pressing issue than explaining reports. His main points included the following:

> 1. The public concern with the phenomena, which is reflected both in the United States press and in the pressure of inquiry upon the Air Force, indicates that a fair proportion of our population is

mentally conditioned to the acceptance of the incredible. In this fact lies the potential for the touching-off of mass hysteria and panic.

2. The U.S.S.R. is credited with the present capability of delivering an air attack against the United States, yet at any given moment now, there may be current a dozen *official* unidentified sightings plus many unofficial ones. At any moment of attack, we are now in a position where we cannot, on an instant basis, distinguish hardware from phantom, and as the tension mounts we will run the increasing risk of false alerts and the even greater danger of falsely identifying the real as phantom.
3. A study should be instituted to determine what, if any, utilization could be made of these phenomena by United States psychological warfare planners and what, if any, defenses should be planned in anticipation of Soviet attempts to utilize them.[12]

What could be done? Some method had to be devised to get people to stop reporting UFOs, and it is perhaps for this reason that the CIA convened a special panel of eminent scientists, who met to discuss the UFO issue during January 14–17, 1953.

The panel was named after its chairman, Dr. H. P. Robertson, director of the Weapons Systems Evaluation Group in the office of the Secretary of Defense. It included Dr. Luis Alvarez, a physicist who worked on the atomic bomb project and later received the Nobel Prize in physics; Dr. Samuel Goudsmit, a physicist at Brookhaven National Laboratories; Dr. Thornton Page, former professor of astronomy at the University of Chicago and deputy director of the Johns Hopkins Operations Research Office; and Dr. Lloyd Berkner, a physicist and a director of Brookhaven National Laboratories.

After deliberating for four days (for a total of 12 hours), the panel delivered a secret report, which was finally declassified in 1966. The report presented the following conclusions:

2. As a result of its considerations, the Panel *concludes*:
   a. That the evidence presented on Unidentified Flying Objects shows no indication that these phenomena constitute a direct physical threat to national security.

      We firmly believe that there is no residuum of cases which indicates phenomena which are attributable to foreign artifacts capable of hostile acts, and that there is no evidence that the phenomena indicate a need for the revision of current scientific concepts.

3. The Panel further *concludes*:
   a. That the continued emphasis on the reporting of these phenomena does, in these parlous times, result in a threat to the orderly functioning of the protective organs of the body politic.[13]

This threat was thought to involve the clogging of communication channels by UFO reports, the ignoring of real signs of hostile action, and the "cultivation of a morbid national psychology in which skillful hostile propaganda could induce hysterical behavior and harmful distrust of duly constituted authority." As a consequence, the Panel recommended that "the national security agencies take immediate steps to strip the Unidentified Flying Objects of the special status they have been given and the aura of mystery they have unfortunately acquired."[14] The method prescribed by the Panel for eradicating this aura of mystery was "debunking," a term defined by Condon as "to take the bunk out of a subject." Here is the Panel's debunking strategy:

> The "debunking" aim would result in reduction in public interest in "flying saucers" which today evokes a strong psychological reaction. This education could be accomplished by mass media such as television, motion pictures, and popular articles. Basis of such education would be actual case histories which had been puzzling at first but later explained. As in the case of conjuring tricks, there is much less stimulation if the "secret" is known. Such a program should tend to reduce the current gullibility of the public and consequently their susceptibility to clever hostile propaganda.[15]

Robertson and his colleagues seemed confident that since the extraordinary flying discs are plainly impossible, reports of such things must reflect irrational thought processes. They instinctively associated UFO reports with conjuring tricks, much as Hudson Hoagland did years later when he associated UFOs with bogus spirit mediums in the pages of *Science* (Chapter 1). The underlying conviction here is that science knows the truth, and people make statements contrary to that truth only because they are gullible, easily manipulated fools. They are not necessarily crazy, but in their normal, sane state they are prone to believing pseudoscientific nonsense.

There is no reason to think that the panelists were cynical manipulators. It is quite possible that they were fully sincere in their conclusions and were simply trying to carry out their patriotic duty to protect the United States by adjusting the volatile consciousness of the masses.

What Was Happening Meanwhile

While these activities were going on within the Air Force and the government, UFO sightings and encounters continued to be reported by military personnel. In 1964, the National Investigating Committee on Aerial Phenomena (NICAP) published a compilation of information on UFOs entitled the *The UFO Evidence.* This document included a table of 92 UFO sightings by U.S. Air Force personnel dating from 1944 to 1961, with a heavy concentration in 1952 and 1953.[16]

The 92 cases in this table include 24 in which an Air Force plane chased a UFO or was chased or repeatedly buzzed by one. In an additional 20 cases, a UFO seemed to deliberately follow an Air Force plane (but not chase it) or fly at low altitude over a military base. These statistics are hard to reconcile with the official Air Force conclusion that UFOs have never been seen to pose a military threat. If this is true, we must suppose that Air Force pilots have repeatedly thought they were being chased by weather balloons, meteors, or the planet Venus, and they have repeatedly gone scrambling after such objects with afterburners blazing.

Richard Hall, the editor of the *The UFO Evidence,* noted that after the passage of Air Force Regulation 200-2 on August 6, 1953, the number of sighting reports emanating from the Air Force greatly decreased. This regulation established Air Force procedures for handling evidence on unidentified flying objects. An important feature of the regulation was its rule for disclosure of UFO reports to the public:

> 9. Exceptions. In response to local inquiries resulting from any UFO reported in the vicinity of an Air Force base, information regarding a sighting may be released to the press or the general public by the commander of the Air Force base concerned only if it has been *positively identified as a familiar or known object.* . . . If the sighting is unexplainable or difficult to identify, because of insufficient information or inconsistencies, the only statement to be released is the fact that the sighting is being investigated and information regarding it will be released at a later date.[17]

This could be interpreted as a sound procedure for preventing the release to the public of misleading, half-baked stories. Certainly if unknown flying machines do not actually exist, then one should not release UFO reports until conventional explanations can be found. But if they do exist, then this regulation has the effect of suppressing important evidence that could help people properly understand them.

Examples of Military UFO Cases

Many accounts are available of UFO activity that was regarded as threatening by military pilots. For example, on February 10, 1950, one Lt. Smith, a U.S. Navy patrol plane commander, was conducting a routine security patrol near Kodiak, Alaska. He saw an object off his starboard bow that had a radar range of five miles. Within ten seconds the object was directly overhead, which indicates a speed of approximately 1,800 miles per hour. To Smith and his crew, the object appeared as two orange lights slowly rotating around a common center. Here is a description of Smith's interactions with this object:

> Lt. SMITH climbed to intercept and attempted to circle to keep the object in sight. He was unable to do this, as the object was too highly maneuverable. Subsequently the object appeared to be opening the range, and SMITH attempted to close the range. The object was observed to open out somewhat, then to turn to the left and come up on SMITH's quarter. SMITH considered this to be a highly threatening gesture, and turned out all lights in the aircraft. Four minutes later the object disappeared from view in a southeasterly direction.[18]

This is one of a series of UFO encounters described in a U.S. Navy report that was obtained from FBI files using the Freedom of Information Act. A comment appended to the end of the document says that the objects sighted could not have been balloons, since no weather balloons were known to have been released within a reasonable time before the sightings. One commentator suggested that they were "phenomena (possibly meteorites), the exact nature of which could not be determined by this office." [19] Another commentator said that they *might* be jet aircraft.

On March 8, 1950, Capt. W. H. Kerr, a Trans World Airways pilot, and two other TWA pilots reported that they had seen a UFO near

Dayton, Ohio. At that time there were over 20 other reports from the area, which was near Wright-Patterson AFB. Control tower operators and personnel of the Air Technical Intelligence Center on the base also sighted the UFO in the same position, and four interceptors were sent up. Two F-51 pilots saw the UFO and described it as being round in shape, huge, and metallic. When clouds moved in, the pilots had to turn away. The Master Sergeant who tracked the object on radar said, "The target was a good solid return . . . caused by a good solid target." Witnesses said that the UFO departed by flying vertically into the sky at great speed.[20]

Another case involving a UFO chase took place in Japan. On October 15, 1948, a UFO traveling at about 200 mph between 5,000 and 6,000 feet was detected on radar by an F-61 "Black Widow" night fighter. Each time the F-61 tried to close in on the object, it would accelerate to approximately 1,200 mph., outdistancing the aircraft interceptor before slowing down. In one of their six chase attempts, the crew got close enough to the object to see its silhouette. They described the object as being about 20–30 feet long and shaped "like a rifle bullet."[21] This case was reported to the original Project Sign.

Regarding such encounters, Dr. J. E. Lipp, one of the scientific consultants to Project Sign, said:

> The lack of purpose apparent in the various episodes is also puzzling. Only one motive can be assigned: that the space men are "feeling out" our defenses without wanting to be belligerent. If so, they must have been satisfied long ago that we can't catch them. It seems fruitless for them to keep repeating the same experiment.[22]

We can gather from this that Dr. Lipp must have examined quite a number of UFO chase reports. His remarks are intended to cast doubt on the reality of the reported events. However, his argument that the "space men" could have only one possible motive is not correct. One can think of many other possible motives. For example, the motive might be to repeatedly drive home the message that beings exist with technology superior to our own.

Cases Involving Radar

It is significant that many military UFO encounters involve the observation of UFOs using radar. The Air Force apparently took these cases seriously, at least in the 1950s, since Air Force Regulation 200-2 contained instructions for handling radarscope photos of UFOs:

> (5) Radar. Forward two copies of each still-camera photographic print. Title radarscope photographic prints in accordance with AFR 95-7. Classify radarscope photographs in accordance with section XII, AFR 205-1, 1 April 1959.[23]

The Condon Report contains a section on radar cases written by Gordon Thayer of the U.S. Environmental Science Services Administration. There we find a typical ambivalent statement, which tries to explain the unexplainable and then admits the inadmissible:

> (5) There are apparently some very unusual propagation effects, rarely encountered or reported, that occur under atmospheric conditions so rare that they may constitute unknown phenomena; if so, they deserve study. This seems to be the only conclusion one can reasonably reach from examination of some of the strangest cases. . . .
> (6) There is a small, but significant, residue of cases from the radar-visual files (i.e., 1482-N, Case 2) that have no plausible explanation as propagation phenomena and/or misinterpreted man-made objects.[24]

The subject of radar is extremely technical, and I won't be able to discuss it in detail here. Radar operates by reflecting high-frequency radio waves off objects, and this process can be affected by many different atmospheric conditions that cause the waves to refract or reflect in unusual ways. These are known as anomalous propagation effects. However, Thayer's analysis does indicate that in a significant number of cases, radar sightings of UFOs cannot be explained by such effects.

The unknown phenomena that he mentions are worth noting. These include atmospheric temperature gradients in the order of 10°C to 15°C in one centimeter.[25] Such unheard of gradients are needed to explain some UFOs in terms of mirages and anomalous radar propagation.

An example of a UFO sighting limited strictly to radar occurred off the shores of Korea in the fall of 1951. Lt. Cmdr. M. C. Davies had an encounter with a UFO while deployed with an Anti-Submarine Squadron aboard a CVE class carrier. The incident occurred while he was flying at night at 5,000 feet.

He picked up a target, which had been circling the fleet, on his radar scope. Upon leaving the fleet, it took up a position behind his wingman, flying about 3 miles astern, and held about the same position relative to Davies's plane as the wingman. The ship also reported the target on their radars. After approximately 5 minutes, the target left at a speed of over 1,000 mph and was observed on the radar scope by Davies out to 200 miles, the maximum range of his radar. After his flight, Davies learned that the target had been held for about 7 hours on the ship's radars.[26] Here it seems strange that an anomalous propagation effect would first appear to circle the fleet for hours, then trail a plane for 5 minutes, and then fly off at a high speed.

In another case involving combined visual and radar observation, the Ground Observer Force spotted a UFO hovering in the eastern skies near Rapid City, South Dakota, on August 12, 1953. Ground radar began tracking the object, along with the F-84 that was vectored upon it. The F-84 chased the UFO for 120 miles. As the pilot abandoned the chase, heading back toward the base, the UFO followed him. When another F-84 was scrambled on the object, chasing it for 160 miles, it obtained a radar lock-on (automatically guiding the plane toward the UFO). However, the pilot became frightened and asked to break the intercept when a red light began blinking on his radar-ranging gunsight, indicating that a solid object was ahead of him. The climax of the sighting came when both the UFO and the F-84 were clearly seen on the GCI radar screen, and the pilot saw a white, unidentified light speeding in front of him. In this case, gun camera photographs supplemented the pilot's testimony as to what he saw.[27]

In yet another case, a U.S. Navy plane had taken off from an aircraft carrier off Korea in September of 1950 and was headed for an attack on an enemy truck convoy about a hundred miles from the Yalu River. The radar operator on the plane made the following report:

> I was watching the ground below for the convoy, reported . . . and was startled to see two large circular shadows coming along the ground from the Northwest at a high rate of speed. . . . When I saw the shadows I looked up and saw the objects which were causing them. They were huge. I knew that as soon as I looked at my radar screen. They were also going at a good clip—about 1,000 or 1,200 miles per hour. My radar display indicated one and a half miles between the objects and our planes when the objects suddenly seemed to halt, back up and

> begin a jittering, or fibrillating motion. My first reaction, of course, was to shoot. I readied my guns, which automatically readied the gun cameras. When I readied the guns, however, the radar went haywire. The screen bloomed and became very bright. . . . I realized my radar had been jammed and was useless. I then called the carrier, using the code name. I said the code name twice, and my receiver was out—blocked by a strange buzzing noise. I tried two other frequencies, but I couldn't get through. Each time I switched frequencies the band was clear for a moment, then the buzzing began.[28]

The witness described the objects as having a silvered mirror appearance and a surrounding red glow. They were shaped like coolie hats, and had oblong, glowing ports. They had a shiny red ring encircling the top portion, and when they maneuvered over the plane, a coal black, circular area was said to be visible.

One very curious feature of this report is that the UFO supposedly jammed the plane's radar just when the witness readied his guns. Now, how could the UFO pilots know when the guns were readied? This feature might seem to detract from the credibility of the report. However, it turns out that many UFO reports seem to involve apparent direct responses of the UFO to the thoughts of the observer. For another military example, see the Iran UFO case below (pages 100–1).

I conclude this section with a combined radar-visual encounter that took place near Lakenheath, England, on August 13–14, 1956. My summary of this case is taken from the Condon Report.

A radar target was initially observed traveling at 4,000 mph by air traffic control radar at the USAF-RAF stations near Lakenheath, and it was also reportedly seen as a blurry light by control tower personnel and a C-47 airplane flying over the base. Subsequently, a radar target was observed that would remain stationary for some time and then move at a constant speed of about 600 mph to another point, where it would again remain stationary. Its speed was described as constant from the moment it would start to the moment it would stop.

At this point a RAF interceptor was vectored toward the UFO:

> Shortly after we told the interceptor aircraft he was one-half mile from the UFO and it was twelve-o'clock from his position, he said, "Roger, . . . I've got my guns locked on him." Then he paused and said, "Where did he go? Do you still have him?" We replied, "Rog-

> er, it appeared he got behind you and he's still there. . . ." The pilot of the interceptor told us he would try to shake the UFO and would try it again. He tried everything— he climbed, dived, circled, etc., but the UFO acted like it was glued right behind him, always the same distance, very close, but we always had two distinct targets.[29]

The conclusion of the Condon Report on this case was that "although conventional or natural explanations certainly cannot be ruled out, the probability of such seems low in this case and the probability that at least one genuine UFO was involved appears to be fairly high." [30] The Condon Report also cites the conclusion of the Project Blue Book report on this case:

> The maneuvers of the object were extraordinary; however, the fact that radar and ground visual observations were made on its rapid acceleration and abrupt stops certainly lend credence to the report. It is not believed these sightings were of any meteorological or astronomical origin.[31]

As a final point, in the Congressional hearing on UFOs held in April, 1966, Major Hector Quintanilla, the director of Project Blue Book, was asked if the project had any reports of objects seen on radar that could not be conventionally explained. Quintanilla replied, "We have no radar cases which are unexplained." [32] But Dr. J. Allen Hynek wrote that there are unidentified radar cases in Blue Book files.[33]

The Condon Report

After the formation of Project Blue Book and the deliberations of the Robertson Panel in the early 1950s, UFO sightings and encounters continued to occur. For over a decade, government and military authorities took no new public action on the UFO issue. Then, in 1965, Maj. Gen. E. B. LeBailly, the head of the Office of Information of the Secretary of the Air Force, proposed that a panel of physical and social scientists should be organized to review Project Blue Book. His reasoning was that out of 9,265 UFO reports processed by Blue Book, 663 could not be explained. But many of these "have come from intelligent and well qualified individuals whose integrity cannot be doubted. In addition the reports received officially by the Air Force include only a fraction of the spectacular reports which are publicized by many private UFO organizations."[34]

This formal request resulted in the formation of the Ad Hoc Committee to Review Project Blue Book, consisting of physicist Brian O'Brian, psychologists Launor F. Carter and Jesse Orlansky, electrical engineers Richard Porter and Willis H. Ware, and astronomer and space scientist Carl Sagan. In their conclusions, the committee members emphasized that there is no evidence that UFOs pose a threat to national security, and there are no cases that are clearly outside the framework of presently known science and technology. Also, most unidentified UFO sightings are "simply those in which the information available does not provide an adequate basis for analysis."[35]

But they also pointed out that many sightings were listed as identified without adequate justification. They therefore recommended that Blue Book should be strengthened by negotiating contracts for scientific UFO research with a number of universities. This research would require perhaps 1,000 man days per year for about 100 selected sightings. It would be coordinated by one university or nonprofit organization, which would keep in close touch with Project Blue Book. The research would be published in improved Blue Book reports, and the Committee recommended (for unstated reasons) that "anything which might suggest that information is being withheld . . . be deleted" from these reports.[36] They maintained that such scientific reports would help strengthen the public position of the Air Force on UFOs.

Shortly after the Ad Hoc Committee issued its report, a highly publicized series of UFO sightings took place near Dexter, Michigan, and they were explained by Dr. J. Allen Hynek with his famous swamp gas theory. Congressman Gerald Ford objected to the notoriety that Michigan was getting as the "swamp gas state" and pressed for a congressional investigation. This culminated in a one-day hearing of the House Armed Services Committee on the UFO issue on April 5, 1966.

In this hearing, Secretary Harold Brown of the Air Force recommended that a scientific study of UFOs be set up along the lines of the Ad Hoc Committee report, and he was supported in this by J. Allen Hynek. Hynek expressed the urgency of the problem by saying:

> During this entire period of nearly 20 years I have attempted to remain as openminded on this subject as circumstances permitted, this

> despite the fact that the whole subject seemed utterly ridiculous, and many of us firmly believed that, like some fad or craze, it would subside in a matter of months. Yet in the last 5 years, more reports were submitted to the Air Force than in the first 5 years.
>
> Despite the seeming inanity of the subject, I felt that I would be derelict in my scientific responsibility to the Air Force if I did not point out that the whole UFO phenomenon might have aspects to it worthy of scientific attention.[37]

The engineer Raymond Fowler presented testimony on the Exeter sighting at the hearings (pages 57–59). He stated, "After years of study, I am certain that there is more than ample high-quality observational evidence from highly trained and reliable witnesses to indicate that there are machinelike solid objects under intelligent control operating in our atmosphere." He also suggested that the Air Force must be withholding important information supporting his conclusion about these objects: "I am reasonably sure that if qualified civilian scientists and investigators are able to come to this conclusion, that the USAF, supported by the tremendous facilities at its disposal, [must] have come to the same conclusion long ago."[38]

After the congressional hearing, the Air Force Office of Scientific Research (AFOSR) was given the responsibility of implementing the recommendations of the Ad Hoc Committee. They decided that a UFO study should be undertaken by one university, rather than several. In the summer of 1966, AFOSR asked the University of Colorado to undertake the study, and they asked the eminent physicist Dr. Edward U. Condon to head it up.

For Condon, this was a difficult assignment. He was accustomed to the enlightened, rational world of physics, where subatomic particles dance elegantly in obedience to rigorous equations. But in the UFO field he was bombarded by bizarre, unscientific nonsense. The wilder aspects of the UFO phenomenon seemed to both repel and fascinate him, and to color his attitude toward UFOs in general.

For example, Condon devoted a full page of his final report to a discussion of the Cisco Grove Robot. On Labor Day weekend in 1964, three men went bow hunting near Cisco Grove, California. One man, called "Mr. S" to protect his identity, got stranded in the wilderness at dusk and built signal fires to attract rangers to show him the way out. Then he noticed a moving light that seemed unusual, and, frightened, he climbed a tree.

He noticed a "dome-shaped affair" about 400–500 yards away, and then two strange figures came toward the tree and appeared to look up at him. These were about 5 feet 5 inches tall, clothed in silvery-gray material, and they had no visible neck or facial features. They were shortly joined by another more ominous figure that seemed to lurch *through* the bushes, rather than go around them:

> The third "entity" was grey, dark grey, or black. It, too, had no discernible neck, but two reddish-orange "eyes" glowed and flickered where the "head" would be. It had a "mouth" which, when it opened, seemed to "drop" open, making a rectangular hole in the "face." [39]

This apparition, by issuing "smoke" from its "mouth," tried to "gas" the witness, who had belted himself to the upper branches of the tree. This smoke would make the man temporarily unconscious, after which he would awaken, sick and retching, only to meet another blast of "smoke." After a final gas attack, he awakened tired, cold, and sick, to find that the entities had gone.

With some dismay, Condon noted that this information had been gathered by a professional man, Dr. James A. Harder, an associate professor of civil engineering at the University of California at Berkeley.[40] However, stories of this kind were too much for Condon, and he was therefore inclined to reject the stories and the UFO phenomenon in general.

The meteorologist James McDonald criticized Condon for this, saying, "I fail to understand how Dr. Condon's repeated allusion to crackpot cases he has examined can be justified in the face of his seemingly scant interest in energetically digging into the serious aspects of the problem." [41]

However, Condon's approach brings out a serious problem affecting the scientific study of UFOs. The story of UFOs begins with accounts of unknown flying machines that seem able to outperform military aircraft. Such reports might seem outlandish to a scientist, but what does he find if he seeks further information about these strange machines? He finds that they are piloted by weird humanlike beings. This is even worse, but if he inquires into the nature of these beings, he finds that they are endowed with mysterious powers reminiscent of the superstitions rejected long ago by science. The further one pursues the investigation, the deeper one enters into scientifically forbidden territory.

So one way to look at Condon's position is that he recognized that the UFO phenomenon threatened his scientific belief system and he chose instinctively to do what was logically required to keep that system intact. In any event, in 1969 Condon submitted the *Final Report of the Scientific Study of Unidentified Flying Objects*. His conclusions were as follows:

> 1. Our general conclusion is that nothing has come from the study of UFOs in the past 21 years that has added to scientific knowledge. Careful consideration of the record as it is available to us leads us to conclude that further extensive study of UFOs probably cannot be justified in the expectation that science will be advanced thereby.[42]
> 2. The question remains as to what, if anything, the federal government should do about the UFO reports it receives from the general public. We are inclined to think that nothing should be done with them in the expectation that they are going to contribute to the advance of science.[43]
> 3. We know of no reason to question the finding of the Air Force that the whole class of UFO reports so far considered does not pose a defense problem.[44]
> 4. Therefore we strongly recommend that teachers refrain from giving students credit for school work based on their reading of the presently available UFO books and magazine articles. Teachers who find their students strongly motivated in this direction should attempt to channel their interests in the direction of serious study of astronomy and meteorology, and in the direction of critical analysis of arguments of fantastic propositions that are being supported by appeals to fallacious reasoning or false data.[45]

I have cited the Condon Report on a number of occasions, and on examining these citations it should be clear that there is a substantial difference between the main body of the report, which is mainly authored by members of Condon's staff, and Condon's own conclusions. This has been observed by a number of people. For example, a UFO subcommittee set up by the American Institute of Aeronautics and Astronautics (AIAA) said the following about the Condon Report:

> There are differences in the opinions and conclusions drawn by the authors of the various chapters, and there are the differences between

> these and Condon's summary. Not all the conclusions contained in the report itself are fully reflected in Condon's summary. [46]

Likewise, Dr. Claude Poher, a French UFO researcher and one of the directors of the French Space Committee, told J. Allen Hynek that he got interested in UFOs because of the Condon Report. Hynek replied that most people had the opposite response to it. Poher responded, "Well, if you really read the report from cover to cover, and don't just stop with Condon's summary, you will realize that there is a problem there." [47]

Although some scientists have seen a real phenomenon in the UFO data, the prevailing scientific opinion has always been that if this data cannot be explained in orthodox scientific terms, then it is simply unexplained. In some cases, this is because adequate evidence is lacking. In many others, it is because the evidence disagrees with the accepted scientific view of what is possible and is therefore set aside.

An example of this is the Condon Report's treatment of a close-encounter case in Beverly, Massachusetts, on April 22, 1966. Here is a summary of that case, as presented in the Condon Report itself.

On the night of the 22nd, Nancy Modugno, age 11, saw a bright blinking light through her window shortly after 9 p.m. On looking out, she saw a flying football-shaped object the size of an automobile that made a whizzing, ricocheting sound and carried flashing colored lights. This object was headed toward a large field behind nearby Beverly High School. The girl alerted her mother, Claire, who was visiting her friends Barbara Smith and Brenda Maria in an adjoining apartment. The three women could see a flashing light near the school, and they went to the edge of the high school field about 300 yards from the school building to check things out. There they saw three brilliantly lighted flying objects that were circling, halting, and again circling over the school building and other nearby buildings.

Thinking that these might be planes or helicopters, the three crossed the field to get a better look. At this point the following events unfolded:

> Still thinking they might be planes or helicopters, one of the women beckoned the nearest light with an arm motion, whereupon it came directly toward her. She said that as it approached nearly overhead, she could see that it was a metal disc, about the size of a large automobile, with glowing lights around its top. She described the object as

> flat-bottomed and solid, with a round outline and a surface appearance like dull aluminum. The other two women ran. Looking back, they saw their friend directly beneath the object, which was only 20–30 ft. above her head. She had her hands clamped over her head in a self-protective manner and later reported that she thought the object was going to crush her. The object tilted on edge and returned to a position about 50 ft. over the high school as the women ran home to call more neighbors.[48]

Later on, two policemen arrived and observed the UFOs. One officer said in an interview that what he saw was "neither an airplane nor helicopter, but he did not know what it was. The object seemed to the officer to be shaped like a half dollar, with three lights of different colors in indentations at the "tail end,' something like back-up lights."[49]

On the next page of the Condon Report, the investigators Roy Craig and Norman Levine gave the following explanation for these events: First, "Review of all reports indicated that all observers other than the young girl and the group of three women had seen something that looked like a star."[50] This contradicts their statement about the policeman's observations on the previous page of the report, but they did not mention that. They said that the changing colors of the "objects" could have been due to ordinary twinkling of starlight. Their apparent motions could have been due to autokinesis, in which movements of the eye create the illusion of movement in a stationary light source. The star could have been the planet Jupiter, since this planet was visible in the sky at that time and was situated in the right direction to be seen over the school by the witnesses.

What should one do with the testimony of the three women? Craig stated:

> While the current cases investigated did not yield impressive residual evidence, even in the narrative content, to support an hypothesis that an alien vehicle was physically present, narratives of past events, such as the 1966 incident at Beverly, Mass., (Case 6), would fit no other explanation if the testimony of witnesses is taken at full face value.[51]

But one can always choose to disregard such testimony if one desires. In this case, Craig did this by labeling the evidence "anecdotal" and saying that it was too late to subject the witnesses to significant psychological testing. In a letter to Raymond Fowler, one of the original investigators of the Beverly case, Craig also commented that "I will not speculate, here or elsewhere, as to what the women saw." [52]

A review board organized by the prestigious National Academy of Sciences evidently felt satisfied with this approach to the UFO evidence. In its Annual Report for the Fiscal Year 1968–69, the NAS gave the following endorsement of the Condon Report:

> We are unanimous in the opinion that this has been a very creditable effort to apply objectively the relevant techniques of science to the solution of the UFO problem. The Report recognizes that there remain UFO sightings that are not easily explained. The Report does suggest, however, so many reasonable and possible directions in which an explanation may eventually be found, that there seems to be no reason to attribute them to an extraterrestrial source without evidence that is much more convincing.[53]

More Ongoing Events

Shortly after the publication of the Condon Report, the Air Force officially divorced itself from the study of UFOs. But military UFO encounters continued to take place. In later years, civilian UFO research groups made efforts to use the Freedom of Information Act (FOIA) to demonstrate that the military and various intelligence agencies were covertly continuing to document UFO cases. These efforts resulted in the release of large amounts of UFO-related material from government files, and this has been discussed extensively in books such as *Above Top Secret,* by Timothy Good,[54] and *The UFO Cover-up,* by Lawrence Fawcett and Barry Greenwood.[55]

A great deal of material is also available on UFO sightings and encounters in China and the former U.S.S.R. Some of this can be found in *Above Top Secret,* along with UFO material from Canada, Australia, and various Western European countries. UFO material from the former Soviet Union can be found in *A Study Guide to UFOs, Psychic and Paranormal Phenomena in the U.S.S.R.* by Antonio Huneeus.[56] This book includes Soviet military encounters with UFOs, as well as civilian close encounters with UFOs and associated humanoid entities. Jacques Vallee's *UFO Chronicles of the Soviet Union*[57] and Bryan Gresh's "Soviet UFO Secrets,"[58] also contain information indicating that Soviet scientists and military personnel have been involved in extensive studies of the UFO phenomenon.

An extensive collection of sighting accounts from Communist China can be found in *UFOs over Modern China*, by Wendelle Stevens and Paul Dong.[59] Among other things, this book includes stories of a serious border dispute between China and the Soviet Union in 1970 that was allegedly precipitated by massive UFO sightings over the northern Mongolian frontier. Supposedly, the Russians interpreted the UFOs as weapons deployed by the Chinese, and the Chinese thought they were Russian weapons.[60] If this story is true, it is a practical realization of some of the fears of the early 1950s, when U.S. military planners worried that wars might be started by misinterpreted UFO sightings.

Turning to the UFO information obtained through FOIA, here is an example provided by Raymond Fowler. Fowler stated that during October and November 1975, several major Air Force bases experienced UFO visitations, and it proved possible to use the Freedom of Information Act to obtain edited government documents describing these incursions. The following incidents are exerpted from the 24th NORAD Region senior director's log (Malmstrom AFB, Montana):

> *07 November / 1035Z (5:35 A.M.):* Received a call from the 341st Strategic Air Command Post (SAC CP) saying that the following missile locations reported seeing a large red to orange to yellow object: M1, L-3, LIMA and L-6. The general object location would be 10 miles south of Moore, Montana, and 20 miles east of Buffalo, Montana. Commander and Deputy for Operations (DO) informed.
>
> *07 November / 1203Z (7:03 A.M.):* SAC advised that the Launch Control Facility at Harlowton, Montana, observed an object which emitted a light which illuminated the site driveway. . . .
>
> *08 November / 0635Z (1:35 A.M.):* A security camper team at K-4 reported UFO with white lights, one red light 50 yards behind white light. Personnel at K-1 seeing same object.
>
> *08 November / 0645Z (1:45 A.M.):* Height personnel [i.e., radar] picked up objects 10–13,000 feet. . . . Objects as many as seven.
>
> *08 November / 0753Z (2:53 A.M.):* Unknown . . . Stationary/ seven knots/ 12,000 . . . Two F-106 . . . notified.
>
> *08 November / 0820Z (3:20 A.M.):* Lost radar contact, fighters broken off.

> *08 November / 0905Z (4:05 A.M.):* L-sites had fighters and objects (in view); fighters did not get down to objects.
>
> *08 November / 0915Z (4:15 A.M.):* From SAC Command Post: From four different points: Observed objects and fighters; when fighters arrived in the area, the lights went out; when fighters departed, the lights came back on. . . .[61]

It is important to note that intercontinental nuclear missiles are deployed at these Strategic Air Command (SAC) bases. Through his position as a project administrator for the Minuteman Project, Fowler claims to have received information from acquaintances assigned to Minuteman bases indicating that at Malmstrom AFB, during the week of March 20, 1967, a full flight of ten nuclear missiles became inoperative. Radar confirmed the coincidental presence of a UFO, and jet fighters attempted to intercept it. An incident of the same kind occurred in the early spring of 1966. Again ten missiles were simultaneously inoperative due to a fault in their guidance and control systems, and UFOs were reported by above-ground personnel at the time of the failures.[62]

Reports of this kind go on and on. In the *New York Times* of June 17, 1974, science writer Barry J. Casebolt stated:

> Last August [1973], the Air Force launched a Minuteman ICBM from Vandenberg AFB . . . targeted for a point near . . . Kwajalein Missile Range. . . . The nosecone had separated from the third stage of the missile and it was coming in at about 22,000 feet-per-second. . . . At about 400,000 feet, radar picked up an inverted saucer-shaped object to the right and above the descending nosecone. . . . The object was described as being 10 feet high and about 40 feet long.[63]

According to Casebolt, Army missile experts, who asked not to be identified, assured him that the UFO had been tracked independently by two radar systems, that it was not a product of natural phenomena (such as temperature inversions), and that it was not pieces of the missile's stages.

Another incident occurred in Iran in 1976 during the reign of the Shah and involved an encounter between a UFO and jet fighters of the Imperial Iranian Air Force. This is an example of a report in which a pilot chasing a UFO claimed that his weapons control systems were jammed at the precise moment he tried to use them against the UFO. I have already given an example of this involving a pilot flying over Korea in the Korean War (see pages 89–90).

The incident was described in a Defense Intelligence Agency report that is reproduced in *Above Top Secret.*[64] Here is a partial transcription of that report, beginning at the point when an F-4 jet fighter was scrambled from Shahrokhi AFB near Tehran to investigate the reported UFO:

> B. At 0130 hrs on the 19th [of September, 1976], the F-4 took off and proceeded to a point about 40 NM [nautical miles] north of Tehran. Due to its brilliance, the object was easily visible from 70 miles away. As the F-4 approached a range of 25 NM he lost all instrumentation and communications (UHF and Intercom). He broke off the intercept and headed back to Shahrokhi. When the F-4 turned away from the object and apparently was no longer a threat to it, the aircraft regained all instrumentation and communications. At 0140 hrs, a second F-4 was launched. The backseater [radar operator] acquired a radar lock on at 27 NM, 12 o'clock high position, with the VC (rate of closure) at 150 NMPH [nautical miles per hour]. As the range decreased to 25 NM, the object moved away at a speed that was visible on the radar scope and stayed at 25 NM.
>
> C. The size of the radar return was comparable to that of a 707 tanker. The visual size of the object was difficult to discern because of its intense brilliance. The light that it gave off was that of flashing strobe lights arranged in a rectangular pattern and alternating blue, green, red and orange in color. The sequence of the lights was so fast that all the colors could be seen at once. The object and the pursuing F-4 continued on a course to the south of Tehran when another brightly lighted object, estimated to be one half to one third the apparent size of the moon, came out of the original object. This second object headed straight toward the F-4 at a very fast rate of speed. *The pilot attempted to fire an AIM-9 Missile at the object but at that instant his weapons control panel went off and he lost all communications (UHF and Interphone).* At this point the pilot initiated a turn and negative G dive to get away. As he turned, the object fell in trail at what appeared to be about 3–4 NM. As he continued in his turn away from the primary object, the second object went to the inside of his turn and then returned to the primary object for a perfect rejoin.[65]

A final example involves military UFO encounters in Belgium in 1990. The following account is from the July 5, 1990, *Paris Match* and was translated by R. J. Durant in the *International UFO Reporter* (15:23, July/August 1990):

> On the night of March 30th, one of the callers reporting a UFO was a Captain of the national police at Pinson, and [Belgian Air Force] Headquarters decided to make a serious effort to verify the reports. In addition to the visual sightings, two radar installations also saw the UFO. One radar is at Glons, southeast of Brussels, which is part of the NATO defense group, and one at Semmerzake, west of the Capitol, which controls the military and civilian traffic of the entire Belgian territory. . . . Headquarters determined to do some very precise studies during the next 55 minutes to eliminate the possibility of prosaic explanations for the radar images. Excellent atmospheric conditions prevailed, and there was no possibility of false echoes due to temperature inversions.
>
> . . . At 0005 hours the order was given to the F-16s to take off and find the intruder. The lead pilot concentrated on his radar screen, which at night is his best organ of vision. . . .
>
> Suddenly the two fighters spotted the intruder on their radar screens, appearing like a little bee dancing on the scope. Using their joy sticks like a video game, the pilots ordered the onboard computers to pursue the target. As soon as lock-on was achieved, the target appeared on the screen as a diamond shape, telling the pilots that from that moment on, the F-16s would remain tracking the object automatically.
>
> [Before the radar had locked on for six seconds] the object had speeded up from an initial velocity of 280 kph to 1,800 kph, while descending from 3,000 meters to 1,700 meters . . . in one second! This fantastic acceleration corresponds to 40 Gs. It would cause immediate death to a human on board. The limit of what a pilot can take is about 8 Gs. The trajectory of the object was extremely disconcerting. It arrived at 1,700 meters altitude, then it dove rapidly toward the ground at an altitude under 200 meters, and in so doing escaped from the radars of the fighters and the ground units at Glons and Semmerzake. This maneuver took place over the suburbs of Brussels, which are so full of man-made lights that the pilots lost sight of the object beneath them. . . .
>
> Everything indicates that this object was intelligently directed to escape from the pursuing planes. During the next hours the scenario repeated twice. . . .

> This fantastic game of hide and seek was observed from the ground by a great number of witnesses, among them 20 national policemen who saw both the object and the F-16s. The encounter lasted 75 minutes, but nobody heard the supersonic boom which should have been present when the object flew through the sound barrier. . . . Given the low altitude and the speed of the object, many windows should have been broken.[66]

These events were part of a UFO wave in Belgium that involved hundreds of well-attested sightings. At the International Symposium on UFO Research held in Denver, Colorado, in May, 1992, a report on this wave was presented by Patrick Ferryn, a documentary videographer who heads the Belgian UFO organization called SOBEPS.[67] He discussed many close-range UFO sightings by gendarmes (the Belgian police), and he showed a videotape of the F-16-to-UFO radar lock-on mentioned above. He said the tape was made available by the Belgian Air Force, which is openly cooperating with civilian Belgian UFO researchers.

It is noteworthy that no sonic boom was reported when the UFO dove out of range of the F-16s. In the Condon Report, it was pointed out that there are many cases where a UFO was reported to move at supersonic speeds without producing a sonic boom. In a chapter devoted to this topic, William Blumen observed:

> Some meteorological factors occasionally could reduce sonic boom intensities or, even more rarely, prevent sonic booms from reaching the ground at all. However, the reported total absence of sonic booms from UFOs in supersonic flight and undergoing rapid accelerations or intricate maneuvers, particularly near the earth's surface, cannot be explained on the basis of current knowledge. On the contrary, intense sonic booms are expected under such conditions.[68]

Blumen also noted that efforts were being made at the Northrop Corporation to avoid sonic booms by modifying the airstream around the plane by means of an electromagnetic field. Conceivably, UFOs might use some kind of field effect to gently divert air around the body of the craft.

Deep, Dark Conspiracies

It is often charged that the U.S. Government is engaged in a massive and unjustifiable cover-up of UFO information. This accusation has often been made regarding the UFO material processed by Project Blue Book, but J. Allen Hynek and James McDonald have both suggested that mishandling of this material is due to a grand foul-up rather than a cover-up (pages 35–37). The charge of cover-up has been made in connection with UFO records released from government agencies through use of the Freedom of Information Act. But it can be argued that this information, intriguing though it may be, reflects run-of-the-mill bureaucratic secrecy rather than a grand conspiracy. Likewise, it can be argued that military secrecy in connection with UFOs might simply reflect standard military procedures, and it might also be justified by quite ordinary defense considerations. For example, a movie of a UFO zapping a missile may reveal secrets regarding the missile.

But there are deeper undercurrents in the UFO controversy. Since the early 1950s, there have been claims that UFOs have crashed and that the U.S. Government has retrieved the downed vehicles along with the bodies of their alien pilots, both dead and alive. Going further, there are stories of research projects to learn the operating principles of captured UFOs and stories about covert government organizations that direct this research and keep it strictly secret. Going still further, there are even stories of clandestine deals between alien forces and the U.S. Government.

These stories are often quite wild, and some tend toward extreme paranoia. Many are probably false, and they constitute pitfalls of which we should be aware. But some may be true, and it is a curious fact that some crashed-disc accounts do seem to be backed up by respectable evidence. My own position is that none of this material is essential to the thesis that I will develop in the second part of this book. But I think that it should be mentioned since it plays such a prominent role in the current literature on UFOs.

The Roswell Crash

To my knowledge, the most substantial crashed-disc story is the well-known case in which anomalous wreckage was reportedly recovered by U.S. military personnel on a ranch near Roswell, New Mexico, in early

July of 1947. I will first summarize this story and then discuss some of the evidence connected with it.

The story began when ranch manager William "Mac" Brazel found metallic debris strewn over a wide area near Corona, about 75 miles northwest of Roswell. This was on the day after townspeople reportedly saw a bright, disc-shaped object flying northwest over Roswell. Military authorities were eventually alerted, and some of the wreckage was recovered by Major Jesse Marcel, a staff intelligence officer of the 509th Bomb Group Intelligence Office at Roswell Field. This wreckage included small beams that were very light, like balsa wood, but extremely tough and nonflammable. Some of these were said to bear strange hieroglyphic writing consisting of geometric symbols. There were also sheets of thin, light metal that looked like tinfoil but couldn't be dented with a sledgehammer.

A press statement reporting recovery of a crashed flying disc was issued by Roswell base commander Colonel William Blanchard, and the debris was loaded onto a B-29 to fly it to Wright Field in Ohio for examination. Later, however, a second press statement was released on the orders of General Roger Ramey, the commander of the 8th Air Force. This stated that the wreckage was from a crashed weather balloon with an attached radar reflector, and photographs of USAF officers looking at balloon fragments were published along with the story. This remains the official story to this day.

The Roswell case was first written up in *The Roswell Incident* by Charles Berlitz and William Moore in 1980.[69] Additional discussions of the Roswell evidence can be found in articles by Stanton Friedman and William Moore in 1981[70] and William Moore in 1985.[71] A book discussing extensive research on the case was published by Kevin Randle and Donald Schmitt in 1991 under the title *UFO Crash at Roswell*.[72]

The most striking feature of the Roswell case is that a number of eyewitnesses and secondhand witnesses have allowed their testimony about the case to be videotaped and publicly distributed. In a popular videotape, "UFO's Are Real," there is a section of monologue by Jesse Marcel lasting about two minutes. Marcel testified that he observed very unusual wreckage at the Roswell crash site, and he said that he was ordered to conceal this by his commanding officer. He spoke about the thin metal that couldn't be burned or dented with a 16-pound sledgehammer and the beams marked with hieroglyphics. Referring to his position as an intelligence officer, he also said, "One thing I was certain of, being familiar with all our activities, is that it was not a weather balloon, nor an aircraft, nor a missile."

The advantage of videotaped testimony is that it provides direct evidence that the people involved are voluntarily making public statements. The only plausible alternative is that hired actors are being videotaped, and such a fraud could easily be exposed.

Another videotape containing Roswell testimony is "Recollections of Roswell," sponsored by The Fund for UFO Research of Washington, D.C.[73] The table on page 107 lists some of the persons who testified on this tape, accompanied by brief summaries of what they said. Additional witnesses gave testimony supporting the general story told by the persons listed in the table.

"Mac" Brazel's son reportedly told Randle and Schmitt that he found a few small scraps of crash debris himself and mentioned them in the pool hall in the nearby town of Corona. He said that as soon he did this, "lo and behold, here comes the military," and he was asked to surrender the material. He also said that the crash left a track that took a year or two to "grass back over and heal up."[74] Randle and Schmitt also said that, according to Major Marcel, the wreckage covered an area about three-quarters of a mile long and 200–300 feet wide. This area was littered with metallic fragments, and there was also a gouged area of ground about 500 feet long and 10 feet wide.[75]

Randle and Schmitt introduced some testimony from an Air Force brigadier general named Arthur E. Exon. In 1947, Exon was a lieutenant colonel and was assigned to Wright Field. He testified that he was present at Wright Field when the wreckage from Roswell came in, and he said that he also flew over the Roswell crash site. He said that he heard about the analysis of the crash debris: "The metal and material was unknown to anyone I talked to. Whatever they found, I never heard what the results were. A couple of guys thought it might be Russian, but the overall consensus was that the pieces were from outer space."[76]

All in all, this testimony conveys the impression that something unusual crashed to the earth near Roswell, New Mexico, in 1947. It does seem odd that debris from a crashed balloon, a missile, or an airplane would seem so strange to people, including Jesse Marcel, a military intelligence officer. But what was it? Some have suggested that it might have been an advanced, manmade experimental device. Regarding the unusual metal, Jacques Vallee stated:

Some of the Videotaped Roswell Testimony

WITNESS	TESTIMONY
Major Jesse Marcel, the former Army Intelligence Officer who investigated the crash site.	His comments about the debris he saw were similar to his comments in "UFOs are Real."
Dr. Jesse Marcel (M.D.), the son of Major Jesse Marcel.	He was 11 at the time and saw some of the debris when his father brought it home to show his mother. This included an "I-beam" with writing of a violet hue on it, made of curved, textured geometric shapes unlike any symbols he had ever seen. He said there was "no way" they could have been Russian or Japanese.
Lewis "Bill" Rickett, retired, Army Counter Intelligence Corps.	He accompanied Counter Intelligence Officer Sheridan Cavitt on a visit to the crash site. He testified that the pieces of debris were extremely hard and as light as a feather.William Brazel, "Mac" Brazel's son.
William Brazel, "Mac" Brazel's son.	He was 12 years old at the time and was shown some of the debris by his father. He said, "It was something like balsa wood but it wouldn't burn and I couldn't cut it with my knife."
Loretta Proctor, "Mac" Brazel's neighbor.	Brazel showed her some of the debris. This included something that looked like a piece of tape with printing on it: "It wasn't writing as we knew it, and it wasn't Japanese writing."
Robert Shirkley, former Assistant Base Operations Officer.	He testified that Colonel Blanchard asked him if a B-29 airplane that had been summoned was ready, and he said it was. Five or six people carrying parts what he understood to be a flying saucer quickly boarded the B-29 to fly to Fort Worth, and he saw them as they walked by. He briefly saw the piece of I-beam with the unusual writing on it.

> The material recovered in the crash itself, while it remains fascinating, was not necessarily beyond human technology in the late Forties. Aluminized Saran, also known as Silvered Saran, came from the technology already available for laboratory-scale work in 1948. It was paper-thin, was not dented by a hammer blow, and was restored to a smooth finish after crushing.[77]

However, Jesse Marcel swore that the thin material could not be burned and could not be dented by a 16-pound sledge hammer. This doesn't sound like Aluminized Saran, which should be a composite of aluminum and a kind of plastic. (The term Saran refers to any of a number of thermoplastic resins used to make fabrics, acid-resistant pipes, and transparent wrapping material.) It is just possible that the Roswell wreckage did come from some aerial device of nonhuman origin.

The Vexing Question of the Alien Bodies

According to Randle and Schmitt, General Exon also said, "There was another location where . . . apparently the main body of the spacecraft was . . . where they did say there were bodies. . . . They were all found, apparently, outside the craft itself but were in fairly good condition."[78] This remarkable statement is corroborated by a remark made on the "Recollections of Roswell" videotape by Sappho Henderson, the widow of an Air Force pilot named "Pappy" Henderson. She said that her husband told her, "I'm the pilot who flew the wreckage of the UFO to Dayton, Ohio." He mentioned dead alien bodies, and described these as being small, as having large heads for their size, and as wearing suits of a strange kind of material.

This brings us to the controversial issue of recovered alien bodies. For years, there have been rumors about crashed UFOs accompanied by dead bodies of aliens. Exon's testimony is the first example I have seen in which a responsible public figure is said to have openly confirmed such rumors.

One story of alien bodies that may be related to the Roswell affair involves an employee of the U.S. Soil Conservation Service named Grady L. "Barney" Barnett. On July 3, 1947, about 120 miles from the Brazel ranch on the plains of San Agustin, Barnett is said to have stumbled upon another craft that had crashed.[79] This date is close to that of the Roswell crash, which is said to have taken place on the evening of July 2.[80]

Barnett died in 1966 without publicly testifying about this case, but he did speak about it to his friend Vern Maltais. Here is a summary of what Maltais said on the "Recollections of Roswell" videotape:

> Mr. Barnett said that while returning from a field trip he came upon a craft which had burst open, and he saw that there were four beings on the ground. As he and a four- or five-member archeology group from the University of Pennsylvania began to investigate further, the military moved in and told them to go away and to be quiet about what they had seen for the sake of national security. Barnett said that he had no doubt that the beings were from outer space. The beings were not exactly like human beings. They were 3½ feet to 4 feet tall. They were slim in stature and hairless, their heads having no eyebrows, no eyelashes, and no hair. Their heads were pear-shaped and the top of the head was larger. Their uncovered hands had four fingers.

This version of the story places the wreckage seen by Barnett quite far from the Roswell crash. However, according to Randle and Schmitt, Barnett was not necessarily 120 miles from the Roswell site. The debris littering the field near Roswell could have come from a disintegrating craft that came to rest at a second site two and a half miles away, and was later seen by Barnett.[81] They cited testimony from an anonymous intelligence operative who had been assigned to Roswell Army Airfield in 1947 and who confirmed the existence of this nearby second site.[82]

Randle and Schmitt also presented drawings by Glenn Dennis, who was a mortician in Roswell in 1947. Dennis said they were based on drawings done in his presence by a nurse who participated in the autopsies of the alien bodies at Roswell one day after their recovery.[83] The nurse allegedly told Dennis that the bodies were short and of delicate build, with unusually large heads and four-fingered hands. The eyes were large, and the nose, ears, and eyes were slightly concave. She also said that the bodies were partially decomposed and had been gnawed by predators. They gave off a powerful stench.[84]

Barnett's story of the archeologists was apparently confirmed in October of 1989 by Mary Ann Gardner, a nurse who had worked in a cancer ward in St. Petersburg Hospital in Florida. According to her, a woman patient who was dying of cancer said that when she was in school in the late 1940s she was involved in an archeological site survey in New Mexico. The dying woman then went on to tell of discovering the wreck of an alien ship and the dead bodies of its crew.[85]

In recent years there have been many new developments in the Roswell case. The story of five-year-old Gerald Anderson and his family seeing a crashed disk and alien bodies in the New Mexico desert has been repudiated. Anderson's honesty was cast in doubt when it was found that he had faked a number of documents connected with his case.[86] It is interesting to contrast this case, in which continued investigation showed the witness to be a liar, with other cases in which the witness turns out to have a well deserved reputation for honesty.

Randle and Schmitt have come out with a new book on the Roswell affair that introduces many new witnesses and supports the theory that there were two crash sites in close proximity to one another near Roswell.[87] This book identifies the archeologists mentioned by Barnett and maintains that they did see a crashed craft with alien bodies. But it places the craft a couple of miles away from the debris field near Roswell and not in the plains of San Agustin. The eyewitness testimony is impressive, but the authors do not make clear how one aerial disaster could result in a compact downed vehicle in one place and a broad area of metallic fragments a couple of miles away. In one sense, this book makes the Roswell case more mystifying than ever.

On September 8, 1994, the U.S. Air Force announced the completion of a search for Air Force records that might help explain the Roswell incident. The conclusion of the search effort was that "the material recovered near Roswell was consistent with a balloon device of the type used in a then-classified project. No records indicated or even hinted at the recovery of 'alien' bodies or extraterrestrial materials."[88] In view of the extensive eyewitness testimony that disagrees with the balloon explanation, it would appear that the official Roswell cover-up may be continuing.

Disinformation and MJ-12

Some have claimed that government intelligence agents will spread false UFO stories for purposes of disinformation—a technique whereby people are diverted from fruitful but undesirable lines of research and sent off on the track of false leads. For example, Howard Blum, award-winning journalist and author of *Out There,* swears that the following story is true: Paul Bennewitz, president of Thunder Scientific Corporation in Albuquerque, New Mexico, had been making films and recording radio transmissions of UFOs that seemed to be flying near the Sandia National Labs complex, a classified Department of Energy facility on Kirtland Air Force Base. Meanwhile, a well-known UFO researcher named William Moore had formed a liaison with certain agents from AFOSI (Air Force Office of Special Investigations) as part of his effort to gain access to inside UFO information.

According to Blum, the AFOSI agents had been systematically feeding Bennewitz bogus UFO stories in an effort to confuse, discourage, and discredit him. Through a variety of sophisticated covers, they passed on to Bennewitz fake government documents "detailing the secret treaty between the U.S. government and evil aliens, the existence of underground alien bases, the exchanges of technology, the wave of brain implants, and even the tale about the spaceship that had crashed into Archuleta Peak."[89]

Moore was supposedly recruited by the AFOSI agents as part of their disinformation campaign against Bennewitz. This story is confirmed by Jacques Vallee, who pointed out that Moore publicly disclosed it at a MUFON (Mutual UFO Organization) conference in Las Vegas in 1989.[90]

Blum said that after making friends with Bennewitz, Moore watched carefully as a steady diet of paranoid fantasy drove him into madness and nervous collapse. Then Moore got his reward. One day in December, 1984, Moore's friend, Jaime Shandera, received a mysterious roll of unexposed film in the mail. When developed, it revealed the famous MJ-12 document.

This document is ostensibly a briefing paper prepared by Admiral Roscoe Hillenkoetter for President-elect Dwight D. Eisenhower on November 18, 1952. It informs the president of the existence of a secret group of 12 scientists and top government officials involved with the recovery of a crashed vehicle and four alien bodies near Roswell Army Air Base in July, 1947. The document recommended that this group should continue to operate on a top secret basis during the Eisenhower administration.[91]

There has been a great deal of controversy about the authenticity of this document. One of the main proponents of authenticity is Stanton Friedman, who argues that extensive research into government records has failed to reveal any information contradicting the document's statements about persons, times, and places.[92] He also argues that numerous details of style in the document can be confirmed by checking other government documents of the same time period.

None of this conclusively establishes that the MJ-12 document is genuine. But it does seem to show that if the document is a forgery, then it was created by a consummate expert—of the kind one might find in an intelligence agency.

Professor Roger Wescott, an expert in linguistics at Drew University in New Jersey, compared the MJ-12 document with known examples of Admiral Hillenkoetter's writing and concluded that there is no compelling reason to think the document was written by anyone other than Hillenkoetter himself.[93] However, when I asked Wescott if he would explain his reasons for this conclusion, he replied, "I doubt that *either* the authenticity *or* the fraudulence of the document can be conclusively demonstrated." [94]

The alternatives regarding MJ-12 seem to be roughly as follows: (1) The MJ-12 document is genuine, and there is a high-level government cover-up involving crashed UFOs and alien bodies. (2) The MJ-12 document is a fake ordered by high government authorities. This implies that there is a high-level government policy for spreading disinformation about UFOs. (3) The MJ-12 document is a fake produced by a small group of intelligence agents for their own reasons. This group may have been responsible for feeding disinformation to Bennewitz, and it may include the AFOSI agents involved with Moore. (4) The MJ-12 document is a fake perpetrated by people not connected with the government or the armed forces.

Whether the MJ-12 document is genuine or not, there is some evidence suggesting the existence of a high-level government UFO cover-up. An example is the following letter, said to be from Dr. Robert Sarbacher to UFO investigator William Steinman. Sarbacher was Dean of the Georgia Tech Graduate School from 1945 to 1948, and he was a consultant to the U.S. Government's Research and Development Board. In 1983, he was president and chairman of the board of the Washington Institute of Technology, Oceanographic and Physical Sciences, in Palm Beach, Florida. On November 29th of that year, he wrote the following in response to persistent letters by Steinman inquiring into the matter of crashed and recovered UFOs:

> 1. Relating to my own experiences regarding recovered flying saucers, I had no association with any of the people involved in the recovery and have no knowledge regarding the dates of the recoveries. If I had I would send it to you.
>
> 2. Regarding verification that persons you list were involved, I can only say this:
>
> John von Neumann was definitely involved. Dr. Vannevar Bush was definitely involved, and I think Dr. Robert Oppenheimer also.
>
> My association with the Research and Development Board under Doctor Compton during the Eisenhower administration was rather limited so that although I had been invited to participate in several discussions associated with the reported recoveries, I could not personally attend the meetings. I am sure that they would have asked Dr. von Braun, and the others that you listed were probably asked and may or may not have attended. This is all I know for sure. . . .
>
> About the only thing I remember at this time is that certain materials reported to have come from flying saucer crashes were extremely light and very tough. I am sure our laboratories analyzed them very carefully.
>
> There were reports that instruments or people operating these machines were also of very light weight, sufficient to withstand the tremendous deceleration and acceleration associated with their machinery. I remember in talking with some of the people at the office that I got the impression these "aliens" were constructed like certain insects we have observed on earth, wherein because of the low mass the inertial forces involved in the operation of these instruments would be quite low.
>
> I still do not know why the high order of classification has been given and why the denial of the existence of these devices.[95]

In support of the authenticity of this letter, Steinman cited a UFO newsletter, *Just Cause*, published by Lawrence Fawcett and Barry Greenwood. In issue 5 of September, 1985, Greenwood wrote that he had reached Sarbacher by phone. Greenwood said, "First, and most importantly, Sarbacher confirmed to me that the information in the Steinman letter was based on his recollection and was not a hoax."[96] It is no longer possible to obtain direct confirmation from Sarbacher since he died on July 26, 1986.[97]

Sarbacher's remarks indicate that he was not directly involved with crashed UFO recoveries. Stanton Friedman told me that he had personally discussed these matters with Sarbacher, and he thought that

Sarbacher was simply relating scuttlebutt that was circulating among government scientific consultants. Still, it is curious that such rumors would be circulating in those circles.

Victor Marchetti provided another example of testimony supporting the scenario of a government UFO cover-up. Marchetti was once the executive assistant to the Deputy Director of the CIA, and he coauthored an expose of the CIA entitled *The CIA and the Cult of Intelligence*. In an article entitled "How the CIA Views the UFO Phenomenon," in the May, 1979, issue of *Second Look,* Marchetti said that although he heard rumors "at high levels" about crashed UFOs and alien bodies during his time in the CIA, he had not seen any conclusive evidence for the reality of UFOs. Nonetheless, he felt moved to speculate as follows:

> There are UFOs or there have been contacts— if only signals—from outer space, but the evidence reveals the aliens are interested only in observing us. . . . But public knowledge of these facts could become a threat. If the existence of UFOs were to be officially confirmed, a chain reaction could be initiated that would result in the collapse of the Earth's present power structure. Thus, a secret international understanding—a conspiracy—has been agreed to by the world powers to keep the public ignorant of and confused about contacts or visitations from beyond the Earth.[98]

Whatever the status of Marchetti's speculations, it is clear that the public is now being subjected to a great deal of confusion and ignorance regarding UFOs. Lurid propaganda of the kind supposedly given to Bennewitz is now being spread throughout the United States, and it has a negative effect on both the credibility of UFO research and the credibility of government authorities. Persons such as William Cooper and John Lear (son of the inventor of the Lear jet) are lecturing extensively on the topic of alien deals with the government and other UFO-related conspiracy theories. Flyers are being widely circulated that describe alien underground bases and warn of an alien takeover. Researcher Linda Howe has written that one of the AFOSI agents connected with the Bennewitz story showed her secret documentation detailing contacts between the U.S. Government and aliens.[99]

On Oct. 14, 1988, a television documentary entitled "UFO Cover-Up? Live" was broadcast across the United States. This show presented testimony by the Falcon, an alleged intelligence agent involved in the Moore/Bennewitz affair, who spoke with a computer-disguised voice about relations between aliens and the government.[100] In a television program entitled "UFOs: The Best Evidence," narrated by George Knapp, a physicist named Robert Lazar made extraordinary claims about being employed to reverse-engineer alien technology in U.S. Government hands at a secret base in Nevada.[101] The stories go on and on.

It is not the aim of this book to try to answer the many questions involving the role of secret government agencies in the UFO issue. My purpose is simply to point out that organized disinformation coming from an unknown source may be involved in some UFO reports.

To solidly verify rumors of covert government activities, it would be necessary to undertake a counterespionage effort requiring resources of the kind generally available only to national governments. It is interesting that Edward Condon recognized this problem, and he reacted to it in the following pragmatic way:

> We adopted the term "conspiracy hypothesis" for the view that some agency of the Government either within the Air Force, the Central Intelligence Agency, or elsewhere knows all about UFOs and is keeping their knowledge secret. Without denying the possibility that this could be true, we decided very early in the study that we were not likely to succeed in carrying out a form of counter-espionage against our own Government, in the hope of settling this question. We therefore decided not to pay special attention to it, but instead to keep alert to any indications that might lead to any evidence that not all of the essential facts known to the Government were being given to us.[102]

Condon did not believe that a secret government UFO project existed, although he admitted that he couldn't prove this. However, he recognized that the government has cloaked the subject of UFOs in secrecy, and he deplored this, saying, "Official secretiveness also fostered systematic sensationalized exploitation of the idea that a government conspiracy existed to conceal the truth."[103]

4

UFO Abductions

Thus far I have considered UFO close-encounter reports in which (1) unusual flying objects were seen at close range, (2) these objects were said to leave tangible traces of various kinds, and (3) humanoid beings were seen in connection with the objects. The humanoids are a stumbling block for many people, including some who seem to take intelligently maneuvering, unknown objects in stride. Thus J. Allen Hynek said, "It would be helpful . . . if we could demonstrate that Close Encounters of the Third Kind differ systematically from the other five UFO categories. Then we could, with some comfort, dismiss them."[1] He went on to say, however, that he knew of no criteria by which these cases could be separated from the general body of UFO reports.

The cases discussed in this chapter may seem even more repugnant to our sensibilities than the humanoid reports we have seen thus far. These are cases in which strange-looking beings were said to intervene forcefully in the lives of human subjects. People have reported being captured, taken on board UFOs, and subjected to humiliating physical examinations. These cases are called close encounters of the fourth kind or UFO abductions.

In examining this data, my suggestion is to suspend both belief and disbelief, and simply try to get an overview of the available evidence. All of the cases that I will mention are based on human testimony. As such, they do not provide proof. In this field, we are forced to make use of inductive reasoning in which understanding of a general phenomenon comes from a study of patterns appearing in large numbers of examples. I suggest that the UFO abduction reports will begin to make sense when they are seen in the light of broader categories of recent and ancient reported phenomena that I will introduce in the course of this book.

I should make a comment here on methodology. I will often point out certain features that appear frequently in close-encounter cases. Many of these features are of interest because they also show up in Vedic accounts of encounters with humanlike beings. Others are of interest because they seem to aid in the interpretation of UFO cases, and still others simply strike me as puzzling. In discussing these features, I will often refer to certain well-publicized cases in the UFO literature that exhibit them. These are not the only cases in which these features appear, and other illustrations could also be used. My aim is not to pick out these particular cases as uniquely significant.

The Buff Ledge Case

I will begin the discussion of UFO abductions by giving a classical example. This case was studied by Walter N. Webb, a longtime UFO investigator and director of the planetarium at the Boston Museum of Science. The following summary of the case is based on a report that he presented in 1988.[2]

The encounter was reported to have taken place on August 7, 1968, on Lake Champlain, north of Burlington, Vermont. The two primary witnesses, "Michael" and "Janet," were working at Buff Ledge Camp, a private summer camp for girls located on the lake shore. Michael, who was 16, was employed to ferry water-skiers from the dock and to maintain the waterfront equipment. Janet, a 19-year-old Smith student from New Hampshire, was a water-skiing instructor. Webb used pseudonyms to protect the identities of all the witnesses in the case.

One of the key features of this case is that Michael and Janet went their separate ways immediately after their UFO experience and did not communicate with each other until the case was investigated by Webb ten years later. When this investigation began, the first person to testify was Michael. Here is a summary of his version of what happened that evening, as remembered without the use of hypnosis:

Michael and Janet were relaxing on the end of the dock after an afternoon of sunbathing. A bright, starlike light suddenly swooped down in an arc and became visible as a cigar-shaped object. Then it emitted three small white "lights" from its body and flew quickly away. The three lights executed zigzags, falling-leaf descents, spirals, and other remarkable maneuvers, and as they came closer, Michael could see that they were domed discs. After about five minutes, two of the discs flew away, while the third began approaching them across the lake. He could see that the disc had a band of colored, plasmalike light rotating around its edge and that it produced complex tones synchronized with the pulsation of this light. It seemed to be 40–50 feet across and "as big as a small house."

The UFO then shot straight into the sky, dropped, entered the lake, and then emerged. As it came within 60 feet of the dock, Michael could see two entities seated under its transparent dome. The beings appeared to be very short with large heads, big oval eyes, double nasal openings, and small mouths. They were wearing grayish or silvery skin-tight uniforms. At this point, Janet seemed to be in a trancelike state. Telepathically, one of the beings assured Michael that he would not be hurt. They were from another planet and had made trips to the Earth in the past. They had returned after the first nuclear explosions.

As the UFO moved directly over their heads, Michael tried unsuccessfully to touch its bottom to confirm its solidity. A bright beam of light then turned on, and Michael next remembered holding onto Janet as he fell down with her onto the dock. He recalled losing consciousness while under the beam but also felt as though he was floating up. He remembered alien voices, machinelike sounds, and "soft lights in a dark place."

When they became conscious again, the UFO was hovering over them, and it was now completely dark out. He could hear the sounds of other campers returning from a swim meet. "Susan" and "Barbara," two camp girls, came running toward the bluff overlooking the lake, and at that point the UFO departed. The two girls then conducted Janet to her cabin. After the incident, both witnesses experienced a time/memory lapse and extreme fatigue, and quickly went to sleep. However, before falling asleep, Michael went to the male staff quarters, where he met 20-year-old "Patrick," who, he said, had apparently observed part of the close encounter. Patrick encouraged Michael to call the nearby Plattsburgh Air Force Base. He recalled the Air Force spokesperson saying that the base had received several reports of UFOs that evening but that military aircraft were not responsible for them.

Michael woke up once that evening (about one hour after falling asleep), at which time he made his only attempt to contact Janet about their experience. However, she was sound asleep, and her cabin was off limits to males. Later, Michael quickly became disillusioned by the disbelief shown to his story by family and friends, and he maintained that he did not try to discuss it further with Janet while they were attending the camp. He and Janet went their separate ways after the camp closed a few weeks later.

Ten years later, in October of 1978, Michael became inspired to understand more about what actually happened that night, and he was advised by UFO researchers at CUFOS to contact Walter Webb. Webb agreed to investigate the case, and he initially located Janet, who was now married and living in the Southeast. She remembered being on the dock with Michael and seeing the moving lights in the sky, followed by the close approach of a "big light." She said that she and her companion got down on the dock when it seemed that the object would hit them, but after that her mind was blank. She remembered wanting very badly to speak with Michael about something the day after the experience, but she was at a loss to think of what she wanted to talk about, and the conversation never did take place.

Webb ascertained that Michael and Janet had not met since leaving the camp ten years before. He arranged to have them separately hypnotized by two professional clinical hypnotists, Harold Edelstein and Claire Hayward, in an effort to recover their lost memories of the incident. Between September 1979 and April 1980, Michael underwent five hypnotic interviews totaling five hours, and between February and December of 1980, Janet underwent three interviews totaling six hours.

The use of two hypnotists is significant. Webb's strategy was to obtain two independent accounts of the witnesses' experiences. This was rendered possible by the fortunate circumstance that Michael and Janet had apparently not communicated since the UFO incident, and thus they had not had any opportunity to influence one another's memory of what took place. By having them separately hypnotized by different hypnotists, the chances that the story of one witness could directly or indirectly influence the story of the other were minimized.

Here is a summary of what emerged in Michael's memory during the hypnosis sessions: While in the beam of light, he heard a whining noise and felt as though "filled with light." He seemed to be floating upwards. He saw streams of colored lights and seemed to be flying through space. Next, he remembered standing next to one of the alien beings on an upper deck inside the UFO. As he looked out through the transparent dome, he saw the earth, the stars, the moon, and a huge cigarshaped vessel. Below him, Janet was lying on a table and being examined by two other aliens. On one of the walls, a console filled with various screens seemed to display data from the examination. The beings had large, oval heads, greenish skintight clothing, long, thin limbs, and webbed, three-fingered hands. Their eyes were large and oval with large black pupils, their mouths seemed to consist of a small slit, and their noses seemed to consist simply of two holes. Their skin was greenish-blue.

One being acted as Michael's telepathic interpreter during the abduction. He was surprised at how alert Michael had been throughout the experience and warned him that this would make things difficult for him afterwards. Michael remembered feeling close to this being.

On the lower deck, he watched as examiners scraped skin from Janet's body, drew blood from her arm with a syringe, and "sucked" fluids from her body through two openings, using a machine that retracted into the ceiling. When Michael questioned his guide about the latter procedure, he was told that they were "spawning consciousness." When it was his turn to be examined, the beings moved him toward a table near Janet's, at which time he went unconscious. Just before this, though, he saw that their ship was approaching the big cigar-shaped vessel outside.

Upon awakening, Michael had the impression that their ship was now inside the big ship. He and his guide floated through the bottom of the craft into a tube of light and were pulled through a hangarlike room in the larger ship by this light beam. They passed *through* the far wall of this room, went up an elevatorlike device, and eventually ended up in a room with many other similar-looking entities. Here a helmet-like device was placed on Michael's head, and the entities applauded and emitted audible sounds to each other as they looked up at a bubble-shaped screen. Then he was taken to another room, where he saw a strange scene with a purple sky, grass, trees, fountains, and ordinary but dazed-looking humans. Janet was crying in fright near him. Then, he fell asleep.

His next memory was of falling through space toward a globe faceted with TV-like screens. The picture on the screens was one of Michael and Janet on the floor of the dock with the UFO hovering above. After passing through one screen, he became conscious on the dock. At this point, Michael received a telepathic message from his guide, saying that they cared for him, that there would be much he wouldn't understand about the encounter, and that Janet wouldn't remember anything. Another voice assured him that Janet was okay, and he finally heard a voice say, "Goodbye Michael."

Before undergoing hypnosis, Janet could remember very little of the experience consciously, but under hypnosis she recalled events that corroborated Michael's account. Webb emphasized that during the period of her hypnotic sessions she had no knowledge of Michael's abduction story.

The following is a summary of her experience that night, as recalled under hypnosis. She remembered the original light dropping down from the sky. From it, she thought, other lights may have emerged. After displaying various aerial movements in the sky, one of the lights passed in front of them and then disappeared. The object was oval-shaped, encircled by lights, and made a high-pitched sound. It was "larger than a car or house," and it looked "like a spaceship." She also saw in the object alien figures that looked out at them. These had unusual heads and wore one-piece uniforms.

She recalled that the object stopped and hovered over them at the exact spot mentioned by Michael and that a bright beam of light came from underneath it. Next, she remembered lying on a table under a transparent dome, surrounded by "people." She, like Michael, could not recall how the transfer into the UFO was made. She recalled that a being was in charge of her, assuring her in her mind that she was going to be all right, and at this time she felt very relaxed and calm.

She recalled that she was examined by various beings and was told not to look or move while they performed their tests. On taking a peek at them, she was horrified by what she saw, and she recalled that her guide was scolded by the other beings for allowing this to happen. With great reluctance, she described the appearance of the beings. Her description was similar to Michael's, but she thought that their skin was a whitish, unhealthy-looking color and that they wore smocklike garments. Her recollection of the procedures used in her examination differed somewhat from Michael's, but she agreed with him in describing the instrumented monitoring panel. She sensed Michael's presence on another table at some point and remembered seeing him twice on other occasions within the UFO.

Then, she remembered waking up on the dock next to Michael, who appeared scared and fascinated. She recalled not understanding why he was so enthralled over "a few lights." But while in trance she remembered seeing a dark disc hovering above them and vaguely recalled observing it go away. She recalled climbing the steps up the bluff with Michael and seeing Susan and Barbara at the top. At this point she felt very tired and lightheaded, and immediately went to bed in the staff cabin.

Webb said that Janet verified 70 percent of Michael's descriptions of what occurred outside on the dock and 68 percent of his claimed on-board descriptions. Questioning of other people present at Buff Ledge in the summer of 1968 led Webb to locate Barbara and Susan. Independently, both women vaguely remembered seeing a dark, silent, circular object with lights around its edge swiftly leaving the waterfront. However, neither of them could recall seeing Michael or Janet in the area at the time.

Webb also contacted Michael's friend Patrick, who did not confirm Michael's claim that he had watched part of the close encounter. Patrick could only recall that Michael claimed to see a UFO that summer. Patrick said that after this he and others on the beach observed strange lights maneuvering in the sky at long range, and they later saw several Air Force jets flying across the lake. In response to this, Michael denied that any jets were involved and contended that his friend's memory was confused. Webb also noted that Patrick's testimony was questionable, since he had been undergoing psychiatric treatment for many years.

Webb tried to confirm that Plattsburgh Air Force Base had received calls about UFOs that evening. However, all telephone logs had been destroyed after a year, and all UFO sighting reports (kept at SAC Headquarters) were destroyed after 6 months.

Webb also got in touch with "Elaine," who was 25 years old and directing a play at the camp early in August of 1968. She recalled that someone came to the playhouse shouting something about lights in the sky. All the kids went running out toward a clearing in the bluff, and she recalled seeing a silvery glow moving over the trees as she followed at a slower pace. This may have occurred at the time of Michael's and Janet's encounter, but the date could not be accurately pinned down.

Various psychological tests were administered to both Michael and Janet, including the MMPI (Minnesota Multiphasic Personality Inventory) and the PSE (Psychological Stress Evaluator). Both subjects proved to be normal, although Michael's tests indicated some intellectual rebellion toward traditional ideas and parental/societal rules. On the basis of these tests and background character checks, Webb firmly believes that both witnesses were credible, honest people who did not create a hoax and did not share some type of hallucination. Michael graduated in 1978 with a B.A. in religion and went on to pursue a modeling and acting career in New York. Janet graduated in 1971 with honors in psychology, worked as an administrator at an Ivy League school, and then married a physician and became a mother of two children.

History and Frequency of Abduction Cases

Over the last 30 years, so-called UFO abduction experiences such as the one at Buff Ledge have been repeatedly reported. Although abduction cases have been uncovered that date back to the 1940s, UFO abductions of the modern pattern have become widely recognized only in recent years.

The earliest of these experiences to be well-publicized was the case of Betty and Barney Hill, who reported having a close encounter with a UFO on September 19, 1961, while driving home from a vacation trip along a lonely New Hampshire road. The Hills initially recalled seeing a strange craft maneuvering in the night sky with abrupt changes of direction. As it closed in on their car, they looked at it with binoculars, and Barney saw humanoid figures in the lighted windows of the craft. At this point the Hills drove off quickly. Apart from hearing some strange beeping sounds, they recalled no further incidents as they returned to their home.

After the encounter, Betty Hill was troubled by weird dreams of being taken aboard the UFO by aliens, and Barney began to suffer from ulcers and other symptoms of stress. Between December 14, 1963, and June 27, 1964, they underwent hypnotherapy with psychiatrist Benjamin Simon, and there emerged a detailed story of alien abduction similar to that of Michael and Janet. This case was also investigated by Walter Webb, and it was written up in the book *The Interrupted Journey,* by John Fuller.[3]

Under hypnosis, Barney Hill described his captors in terms that would later become almost hackneyed. They had grayish, almost metallic-looking skin, no hair, large, slanted eyes that seemed to wrap around the sides of the head, two slits for nostrils, and a mouth that was a horizontal line. He said that they would speak to one another by making a "mumumumming sound," whereas their leader communicated with him mentally.[4] He also said that the leader exerted long-range mental control over him: "It was as if I knew the leader was elsewhere, but his effectiveness was there with me."[5]

In the years since 1964, many UFO abduction accounts have come to light. To get an idea of how often these experiences occur, I note that the British UFO researcher Jenny Randles knew of 32 continental European abduction accounts in 1988.[6] In 1981, Budd Hopkins, an American researcher who has become well known for his studies of UFO abductions, said he had personally investigated 19 abduction cases since beginning his UFO research in 1976.[7]

Hopkins went on to say that a total of about 500 abduction cases had been studied as of 1981. He based this estimate on 300 cases from HUMCAT, a catalogue of humanoid reports compiled by Ted Bloecher and David Webb, plus cases investigated by Dr. James Harder, an engineer, and Dr. Leo Sprinkle, a psychologist.[8] Jacques Vallee gave a comparable figure in 1990, saying, "At this writing over 600 abductees have been interrogated by UFO researchers, sometimes assisted by clinical psychologists."[9]

General Characteristics of UFO Abductions

Although the story of Michael and Janet seems quite strange, it has a number of features that come up again and again. Here is a list of some of these features, more or less in order as they appear in the story:

1. The UFO is often (but not always) described as a domed disc with various flashing or pulsating lights.

2. Unusual high-pitched sounds are often heard, especially in the beginning of the experience.

3. Witnesses sometimes see alien beings looking out of windows in their craft. The beings often seem to be exerting some kind of hypnotic influence on the people watching them.

4. These beings are often small, with big heads and eyes, and vestigial-looking mouths, noses, and ears. The term "Grays" is often applied to this racial type. Sometimes, however, UFO entities are said to have handsome human features, and in some cases humans and "Gray" entities seem to be working together in UFOs.

5. The entities often communicate with human witnesses telepathically. However, they are often said to communicate with one another by incomprehensible sounds.
6. They often say that they have visited this earth in the distant past and have returned because of our atomic testing.
7. They generally assure the witnesses that they will not be harmed.
8. There is typically a loss of memory of parts of the experience and a consequent time gap. This has become well known as the phenomenon of "missing time."
9. The entry of the witness into the UFO often involves a beam of light, but the precise mode of entry is not remembered.
10. Sometimes the witnesses report seeing the earth or other planets from outer space while on board the UFO.
11. Sometimes the UFO is taken into a larger "mothership."
12. These experiences often involve great fear. In this case, Janet experienced some fear, but it was not as extreme as in many cases.
13. The witnesses generally say that at a certain point they felt very calm, due to feelings of reassurance emanating from their captors.
14. The witness is typically subjected to a "medical" examination while lying on a table. This examination involves elaborate machines, and the witness's body is often probed, poked, scraped, and injected with fluids.
15. As we saw in this case, with the reference to "spawning consciousness," the examination often has something to do with reproduction, and ova or sperm samples may be taken.
16. The witnesses often see panels with many TV-like screens.
17. After the examination, there is often a sort of tour of the ship. The witness is taken to various rooms and sees various incomprehensible things.
18. Witnesses often are floated through the air on beams of light, and they sometimes report floating *through* walls.
19. Sometimes convocations of alien entities in large rooms are described.

20. Sometimes the witnesses are shown strange, surreal landscapes.

21. There are often experiences that seem hallucinatory or visionary. In this case Michael's fall into the globe faceted with TV screens is an example. At the same time, many aspects of these experiences seem to involve normal sense perception in strange circumstances.

22. The witnesses often report extreme exhaustion after the experience.

Recurring Small Details

In addition to these features, there are many small details that come up repeatedly in abduction accounts. For example, although witnesses frequently don't remember how they entered a UFO (point 9), there are cases where the witness remembered entering through a door, and there are many references to doors within UFOs. Almost invariably, the witnesses say that these doors vanish seamlessly when they are closed.

An example would be the abduction story recounted in *The Andreasson Affair,* by Raymond Fowler. In this story, a New England housewife named Betty Andreasson was visited in her home in 1967 by beings of the "Gray" type, floated by them *through* the closed door of her house (point 18), and then taken into a UFO parked in her yard. She entered the UFO and went from room to room within it by passing through doors in the normal way, and she commented that these doors could not be seen when they were closed.[10] Another example is the story of the Brazilian farmer Villas Boas, who reported seeing seamless doorways in a UFO in 1957 (see page 138).

In 1950, Frank Scully published a highly controversial book, *Behind the Flying Saucers,* about the recovery of a crashed flying disc near Aztec, New Mexico, in the late 1940s. In the book he mentioned that the door leading into the disc could not be seen when it was closed.[11] Many UFO researchers have rejected Scully's story as a hoax, but it has been vigorously defended by William Steinman.[12] Without taking sides on this issue, I note that Steinman introduced some additional testimony regarding seamless doorways. He maintained that Baron Nicholas von Poppen, an expatriate Estonian and expert photographer, was called by military authorities to the Aztec crash site to photograph the downed UFO. Von Poppen is said to have described what he saw to George C. Tyler in 1949. In his description, he said, "The door was so finely machined that when closed it left no indication that it was there."[13]

What is happening here? Did someone make up the seamless door story in 1949, or perhaps borrow it from some science fiction story? Did Betty Andreasson, a fundamentalist Christian housewife, obtain the idea from UFO literature and weave it into her own tale, perhaps unconsciously? Was this also done by the many other witnesses who mention seamless doorways, including the supposedly ignorant Brazilian farmer, Villas Boas? Or were independent witnesses actually observing seamless doors in UFOs?

Here is another example involving Scully's book. The Scully crashed disc supposedly had an outer ring of metal that revolved around a nonrotating, central cabin. As Scully put it in an article in *Variety* in 1949, "Its center remained at rest, but it had an outer edge that revolved at terrific speed." [14] This idea also came up in the testimony of Betty Hill regarding her UFO encounter. Under hypnosis, she said, "But there was this kind of rim that went around the craft. And I don't know why, but I had the idea that this rim was movable, that it would spin around the perimeter, maybe. Like a huge gyroscope of some kind." [15]

Some students of the UFO phenomenon have argued that the content of UFO close encounter experiences has been taken from various forms of fiction, including science fiction movies. According to this idea, fiction writers first invented certain themes. These themes were spread widely by the media, and they later emerged in UFO abduction stories told by people who are psychologically susceptible to accept fantasy as reality. These arguments were recently reviewed by the psychologist Kenneth Ring, who concluded that fiction has unquestionably influenced UFO abduction narratives. Ring, however, felt that "There is more to abduction stories than their science fiction parentage," [16] and he suggested that these stories may involve an objectively real "imaginal realm" which is somehow the cumulative product of imaginative thought.[17]

The repeated appearance of certain details in UFO narratives could be attributed to fictional ideas transmitted from one person to another by ordinary means of communication. However, in many cases these details are so obscure that it is hard to see how the UFO experiencer could have known of them. A striking example is the illustration in *The Andreasson Affair* showing an apparatus that Betty Andreasson supposedly saw on the underside of a parked UFO during one of her abduction experiences.[18] This apparatus consists of three glass balls held from above by supports reminiscent of the legs of an old-fashioned piano stool.

Researcher William Moore noted that in "Scully's Scrapbook" in *Variety* magazine of November 23, 1949, Scully described the landing gear of flying saucers as consisting of three "wheels like the glass balls once common to the legs of piano stools." [19] The piano stool theme is certainly very obscure, and it is hard to see how Betty Andreasson could have come across it unless she was researching the writings of Frank Scully prior to reporting her abduction memories. Ring's imaginal realm idea can be seen as a way in which the transmission from Scully to Andreasson could take place by nonordinary means of communication. The same could be said of the hypothesis that actual alien beings make use of human cultural materials when staging UFO encounters.

The 1955 movie *This Island Earth* gives another example of possible fictional antecedents for a UFO incident. In the movie, a classical disk-shaped UFO is seen lifting a small airplane by means of a green beam of light that emerges from a hole in the bottom of the UFO. There are shots of the hero and heroine in the cockpit of the airplane staring with astonishment as the cockpit is bathed in green light and the plane is lifted up. This movie scene is strongly reminiscent of the Mansfield helicopter incident in 1973, in which a UFO reportedly bathed the cockpit of an Army Reserve helicopter with a beam of green light and lifted the helicopter 1,800 feet, even though its controls were set for a dive (see pages 354–55).[20]

One could suppose that the UFO encounter of 1973 was simply a story inspired by the 1955 movie and perhaps triggered in the crewmen's minds by some natural phenomenon such as a meteor. But the four crew members consciously remembered the incident and their story was confirmed by seven ground witnesses.[21] It does seem strange that a scene from an old movie could exert such a strong irrational effect on the minds of so many people. Is an imaginal realm involved here? Another possible explanation is that the 1955 movie may have incorporated material based on real UFO incidents. This, of course, would put the lifting of objects by UFO light beams back to 1955 or earlier.

Injury and Disease

In many cases, symptoms of physical injury or disease are connected with UFO encounters, including abductions. For example, Barney Hill apparently developed ulcers as a result of anxiety caused by his experience. He also developed warts around his groin that were perhaps caused by an instrument he remembered being placed over his genitals while on board the UFO.[22] This experience and a "pregnancy test" administered to his wife, Betty, by the UFO entities are also illustrations of point 15.

According to another report, in November 1975, a young man experienced a bizarre encounter in the Catskill Forest Preserve involving an oval, semiluminous craft, an attack by robotlike figures, and a period of missing time. About a week after this event, he began to develop a series of straight raised welts running from his navel toward his groin in a converging pattern.[23] In the Villas Boas case, the witness was exhausted, nauseated, and unable to eat or sleep normally after his abduction experience. Subsequently, he developed an unusual chronic skin disease.[24] He also experienced severe headaches and burning and watering of the eyes.

Irritation of the eyes is apparently common in UFO close-encounter cases, since UFOs often employ dazzling beams of light. A number of examples of CE3 cases with eye injuries are cited in an article, "The Medical Evidence in UFO Cases," by John Schuessler.[25] According to Budd Hopkins, abduction witnesses frequently report eye irritation caused by brilliant lights seen within UFOs.[26]

At the same time, there are reports of remarkable healings connected with UFO close encounters. Some of these appear to be of a mystical nature (see pages 164–65). Others are attributed to medical interventions that seem to make use of recognizable high technology.

An example of the latter is a case reported by the psychologist Edith Fiore. One of her patients said that he was born with a malformed blood vessel in his brain that was likely to burst. Physicians had told his mother that he could only be expected to live for a few years and that he would be retarded. However, he is now in his late 40s and is normal. It turns out that under hypnosis by Fiore, he remembered "ETs doing a treatment on him which he saw on a screen, in which he saw some blood vessels which appeared to be on the outside of his brain."[27]

Fiore said that she has run across some 200 CE4 reports in the course of hypnotic regressions performed for purposes of psychotherapy. Of these, some fifty percent involved cures of life-threatening illnesses, such as cancers, or painful conditions, such as migraines.[28] Of course, one can suggest that people imagine these ET cures because they need to explain natural cures occurring for unknown reasons. But Western culture provides familiar mystical explanations of unusual cures (such as the grace of Jesus). So why should someone try to explain mysterious cures by invoking even more mysterious ETs?

The evidence that many UFO encounters tend to be accompanied by physical effects—injurious or beneficial—gives support to the hypothesis that these encounters are physically real. This is especially true in cases where the physical effect can be connected with recollections of specific events occurring within a UFO.

At the same time, however, it is known that states of mind can produce remarkable effects on the body. A well-known example would be the stigmata that have appeared on the bodies of certain Catholic monks and nuns who were meditating on the crucifixion of Christ. Some stigmata have been said to closely resemble actual nail wounds, but whether their cause was "natural" or "supernatural," they were clearly not produced by nailing. Could it be that UFO abductions take place on a mental level and involve unusual effects of mind over matter?

The issue of whether or not UFO abductions are physically real turns out to be very complex. Comparisons with Vedic material can help shed light on this issue, and I will discuss it further in Chapter 10 after some of this material has been introduced. For the moment, I would suggest that some UFO close encounters seem to involve gross physical phenomena and others seem to involve the action of subtle energies connected with the mind.

Anatomy of a Hallucination?

The UFO literature suggests that the humanoid beings involved in abductions are, with some notable exceptions, remarkably uniform in appearance and behavior. For example, the British researcher Jenny Randles pointed out that, on the basis of her data, two basic types of abducting entities can be identified.

Her first group, which she calls the "small, ugly beings," correspond to the "Gray" type that I have already mentioned. According to her, "These are between 3½ and five feet tall, have large, pear-shaped heads, big, round eyes and slit noses and mouths; they are usually hairless and often wear greenish uniforms; skin is sometimes said to be gray or wrinkled."[29] She pointed out that there is a staggering lack of variety in the descriptions of these beings.

Her second category is "the tall, thin ones." These are typically six feet tall or more, and they are often said to have Scandinavian features, including pale skin and blond hair. Their eyes are generally said to be oriental or catlike and are often blue or pink. They are often said to be strangely beautiful, and their appearance is more humanlike than that of the "Gray" beings.

The division between short and tall humanoids also shows up statistically in the table of 164 humanoid reports published by the UFO investigators Coral and Jim Lorenzen in 1976.[30] As I pointed out in Chapter 2 (page 69), the Lorenzens mentioned small and large humanoids, and (oddly enough) the size of the reported humanoids seems to correlate with the size of the accompanying UFO.

The Lorenzens also mentioned four other types of entities, which they spoke of as large and small robots and large and small monsters (recall the Cisco Grove robot story, pages 93–94). In their table, the large and small humanoids predominate, as we can see by counting the number of reports featuring entities in the different categories. (In their table there is only one case featuring two different entity types, but such cases are fairly common in the UFO literature.)

Type	Number
Large humanoid	60
Small humanoid	81
Large monster	3
Small monster	4
Large robot	1
Small robot	3

The Lorenzens also listed various bodily features of the humanoids. It turns out that among large humanoids, eight had notably large eyes, and ten were said to have normal eyes, a roughly equal division. But among small humanoids, eighteen were reported to have large eyes, and two were said to have small eyes. This is consistent with the standard description of the "Gray" entities as having large eyes.

In recent years a number of nonhumanoid physical types have been reported. One is the so-called Reptilians, who are described as erect, lizard-like creatures with catlike eyes.[31] These beings are typically said to sexually attack their human victims, and they are described as grotesque and repulsive. Another, even stranger, type is said to resemble an insect such as a grasshopper or preying mantis and is sometimes said to play a supervisory role in some abduction experiences. Reports of these beings seem to be significantly rarer than humanoid reports, although this may be an artifact of biased reporting.

Do the consistent anatomical patterns in the humanoid beings reflect the bodily structure of actual living organisms, or do they reflect the anatomy of some kind of hallucination? Although there may be other possible explanations for the humanoids, let us consider these two for the moment.

One could develop the hallucination idea as follows: For some reason, abduction stories featuring certain types of beings were first created by human imagination. These beings are humanlike because it is natural for people to imagine human forms. The stories are spread by normal means of communication. When people report vivid and shocking encounters with these beings, this may be due to a psychological process that incorporates the stories they have heard into a seemingly real experience. This could be called the folklore theory.

Randles cites a study of 200 abduction cases by a student of folklore named Thomas E. Bullard.[32] Bullard argued that if UFO stories spread as a kind of folklore, then they should show features expected of folklore. For example, the stories should show a degree of variation typical of products of human imagination. They should vary from one geographical region to another, and they should show the influence of highly publicized cases.

Randles summed up Bullard's conclusions as follows: Although American cases show markedly fewer instances of the tall beings than non-American cases, abduction cases from different parts of the world nonetheless tend to be highly uniform. Well-publicized cases seem to have no detectable impact on abduction reports. Also, abduction stories are highly stereotyped, and they show a much smaller range of variation than is found, for example, in science fiction. It seems that abduction stories don't follow the patterns expected of folklore.[33]

It might be argued, however, that abduction experiences tend to be highly uniform because of a psychological process that picks out certain ideas and intensifies them. However, abduction reports have become prominent since the early 1960s. Why have these particular stories become invested with psychological potency in recent years and not before?

Another drawback of this idea is that many of the uniform features that show up repeatedly in abduction accounts do not seem to be psychologically significant. For example, what would be the psychological significance of seamless doors in UFOs, or slitlike mouths in short humanoids? Why should people speak of bright lights inside the UFOs, and why should they imagine that entities would communicate telepathically? A good psychological theory of these features would relate them convincingly to known psychological principles.

One can still argue that certain key features of abduction reports are psychologically determined. The others are arbitrary creations of the imagination that are, so to speak, carried along for the ride by the psychologically compelling story elements. But if these elements carry no great significance for people, then why don't they vary greatly from one account to another on the basis of individual whimsy?

On the Evolution of Humanoids

If the reported humanoids are not products of psychology and folklore, perhaps they are real living beings. If this is so, then the commonly reported features in abduction stories might be due to the fact that these beings have certain physical and cultural traits. Some of them might really have slit-mouths, and seamless doors might be part of their technology.

This hypothesis is not proven by the close-encounter reports, but as far as I am aware, it remains a definite possibility. However, it immediately gives rise to the question of where and how the humanoids originate.

For the moment, let us restrict ourselves to the idea that the humanoids have physical bodies that arose by processes of neo-Darwinian evolution. Some scientists, such as Carl Sagan, have argued that intelligent life may have evolved on other planets within our galaxy. Others have argued that if the dinosaurs hadn't died out, evolution might have generated a large-brained, bipedal dinosaur that would look similar to a human being and have comparable intelligence. On the basis of these considerations, it has been suggested that humanoids might have evolved on another planet.[34]

However, the prominent evolutionist Theodosius Dobzhansky rejected this idea, and he explained why by discussing a thought experiment. He said, "Suppose that by some utterly unlikely chance there is another planet somewhere on which there arose animals and vertebrates and mammals like those which lived on earth during the Eocene period. Must manlike creatures develop also on this imaginary planet?"[35]

Dobzhansky estimated that changes in some 50,000 genes would be required for the development of modern humans from Eocene ancestors of some 55 million years ago.[36] These changes include mutations and other kinds of genetic alterations. Since each of these changes is simply one out of a large number of alternatives, the probability would be virtually zero that the changes would occur and be selected in the same sequence as they were in human evolutionary history. Small deviations in the sequence of changes could throw the evolution of prehuman creatures off the track to humankind. Deviations in the evolution of other plants and animals in the prehuman environment could also derail human evolution, and so could deviations of the planet's climate from the climatic history of the earth.

Dobzhansky therefore felt that the chances were vanishingly small that anything humanlike would evolve on his hypothetical planet. To get something that we would regard as similar to humanity, evolution would have to keep on an earthlike track for most of the 55 million years since the Eocene. Otherwise, the tree-climbing Eocene prehumans probably would have either gone extinct or evolved into some unfamiliar mammalian form.

The famous evolutionist George Gaylord Simpson came to similar conclusions. He defined a humanoid in very broad terms as "a natural, living organism with intelligence comparable to man's in quantity and quality, hence with the possibility of rational communication with us."[37] He argued that the evolution of such a being is contingent on a vast number of special circumstances and that it is extremely unlikely for equivalent circumstances to arise on another planet. These circumstances include the chemical conditions required for the production of living cells, the environmental conditions prevailing during millions of years of evolution on earth, and the many mutations required to produce complex organisms. He concluded, "I therefore think it extremely unlikely that anything enough like us for real communication of thought exists anywhere in our accessible universe."[38]

If this understanding of evolution is correct, then the existence of UFO humanoids as real, humanlike beings poses a challenge to current evolutionary theory. We can argue that evolution was bound to produce something, and humankind happens to be one of the things it produced on the earth. Thus the existence of humans on the earth presents no problem. But for evolution to produce something close to humankind independently on two different planets in this galaxy is highly unlikely.

Of course, one could argue that UFO humanoids only superficially appear to resemble humans. One hypothesis is that the humanoids people see are simply simulations of the human form that are being manipulated by an unknown agency. The agency is perhaps using these simulations to communicate with us. Or perhaps the motives of the agency are completely incomprehensible to us. The drawback of this simulation theory is that it can be applied to any data whatsoever, and it simply leaves us with a mystery. I suggest that it should be left as a last resort.

Another hypothesis is that UFO humanoids actually have certain humanlike features, but on a fundamental level they are totally different from us. This may well be partly true, but some of the extremely humanlike traits that these beings are reported to have make me doubtful that it is wholly true.

For example, the humanoids are often reported to wear clothing.[39] They are frequently said to exhibit human emotions such as friendliness, anger, humor, and fear. They communicate verbally with human beings by telepathy, and they are sometimes said to communicate with one another using what seems to be a spoken language.[40] With the exception of some of the reported Reptilians and Insectoids, the physical form of UFO entities is remarkably humanlike. This can be contrasted with the many weird alien forms portrayed in science fiction movies.

Are the humanoids just crude simulations of humans, or are they fundamentally similar to humans? The latter alternative is certainly a possible hypothesis. This hypothesis has the virtue of simplicity, and it makes a strong, specific statement. But it also asserts the existence of the sort of being that Simpson thought would not evolve anywhere within the accessible universe.

Let me sum up my observations in this section and the preceding one. The folklore-plus-psychology hypothesis can explain why UFO entities should seem very humanlike in many ways (we naturally imagine humans), but it cannot readily explain the consistently appearing strange features of these entities. The extraterrestrial-evolution hypothesis can explain these strange features as the physical and technological traits of an alien race (or races). However, this theory is not compatible with the many highly humanlike features of the reported entities.

What we are left with at this point is that the humanoids might be beings similar to ourselves but with a nonevolutionary origin. Or they might be illusions or outward manifestations of something incomprehensible. However, we are just beginning at this point. There are other considerations that may help shed light on the nature of the humanoids, and I will introduce these gradually in the course of this book. The final hypothesis that I shall present depends on these considerations.

Seduction and Genetics

Having said this, I now turn to one of the most disturbing aspects of the UFO abduction phenomenon. In case after case, there are reports of sexual interactions between human abductees and UFO entities. These seem to fall into two categories: (1) experiments with human reproduction involving medical manipulations and (2) direct sexual relationships between abductees and their captors. I will begin by briefly reviewing what has been written about these matters, and then make some observations.

The earliest known example falling in category (2) is the abduction story of the Brazilian farmer Antonio Villas Boas. This incident was investigated by Dr. Olavo T. Fontes, M.D., a few weeks after it occurred in October of 1957, and an English translation of Fontes's write-up was later published by a British UFO researcher, Gordon Creighton.[41] I will give a brief summary based on Creighton's publication.

When the reported incidents occurred, Villas Boas was 23 years old and lived with his family on a farm near the town of Francisco de Sales, in the state of Minas Gerais. He was intelligent but poorly educated and was engaged in work on the family farm.

On the 5th and 14th of October, Villas Boas observed strange lights that maneuvered around the farm at night and at one point mysteriously vanished. The main episode, however, occurred on the night of the 15th. He was alone in the fields, plowing with his tractor at 1 a.m. to avoid the blazing heat of the sun. He saw a red star that came rapidly toward him, growing in size until it appeared as a luminous egg-shaped object. The object halted about 50 meters above his tractor and illuminated the area as though it were broad daylight. It paused for a couple of minutes, then slowly landed. It was rounded in shape with small purplish lights and a large red headlight. It had three supporting legs and a glowing, rotating cupola on the top.

Villas Boas tried to escape in the tractor, but the engine died and the lights went out. He tried to run, but a short individual in strange clothing grabbed him by the arm. He pushed the attacker down but was grabbed by three others and dragged, struggling, toward the machine.

After he was dragged into the machine, the outer door closed, and its outline became invisible. His five captors conversed with sounds resembling the barking of dogs. They stripped him naked, washed him with some kind of liquid, and took him into another room. Again, when the door closed, it was no longer visible, even in outline. An apparatus was used to take two blood samples, and he was left alone in the room. Then nauseating smoke was blown into the room through tubes, and he vomited.

The men wore tight-fitting gray garments that appeared to be uniforms, and helmets that hid their facial features. Above their eyes, which seemed light in color, the helmets were twice the size of a normal head. The trousers were form-fitting and were joined without a break to the shoes, which were bulky and curved up in the front.

After some time, a naked, good-looking woman came into the room and seduced him. She seemed to be a normal human being in every respect, although her face was somewhat unusual, with nearly white blonde hair, large slanted eyes, very high cheekbones, thin lips, and a narrow pointed chin. After the woman left, his clothes were returned. After dressing and waiting for some time (and trying unsuccessfully to snatch an object as a proof of his experience), he was given a short tour around the outside of the craft and then dismissed.

At this point, the whirling cupola began to spin faster and change from greenish to red; the craft lifted into the air, and then shot off, like a bullet, to the south. Villas Boas estimated that he was in the craft for 4 hours and 15 minutes. In response to questioning, he said that he had not been under any telepathic influence during the experience.

As I mentioned in the previous section, Villas Boas suffered from a number of ailments after this episode. These were noted by Dr. Fontes, who also observed that Boas had two marks on the chin corresponding to the places where he said blood samples were extracted. Fontes also observed that Boas seemed psychologically normal, with good intelligence, and with no tendency toward mysticism.

Since Fontes thought the story obviously couldn't be true, he believed that Boas must be a liar of great imagination, with the ability to remember an imaginary story in great detail and recount it without slipping up. Fontes suppressed the story until sending it to Gordon Creighton in April of 1966. The story had first been published in English by *Flying Saucer Review* in January 1965, on the basis of an interview with Boas by Dr. Walter Buhler of Rio de Janeiro in 1962.

In this story, there are many details that come up repeatedly in UFO close-encounter cases. These include (1) lights of unknown source that move around and then abruptly vanish, (2) brilliant beams of light projected by a glowing UFO, (3) the stopping of the tractor engine, (4) doors that are invisible when closed, (5) helmeted form-fitting uniforms with shoes attached to the pant legs, and (6) heads with large slanted eyes, thin lips, and pointed chins. It would seem that if Villas Boas made up the story, then he was not merely imaginative but also well-versed in UFO lore.

The Villas Boas story is a bit unusual in that Boas was not paralyzed or mentally controlled during the experience. However, there have been cases in which men have reported being forced, while in a state of paralysis, to engage in sexual relations with strange-looking alien women. These experiences were particularly distressing and revolting to the men forced to undergo them, and the women involved resembled crosses between humans and the "Gray" beings. Three cases of this type are described in detail in Budd Hopkins's 1987 book, *Intruders*.[42] It is curious that the being described by Villas Boas back in 1957 also seemed to share human and "Gray" features.

Budd Hopkins is also well known for his discussion of cases in which a woman is somehow impregnated and the embryo is then removed by UFO entities. These cases sometimes involve "presentation" scenes, in which the woman is later shown a half-human, half-alien child that seems to be her offspring. An example would be Hopkins's case of "Kathie Davis."[43]

It has been argued that these stories are created by Hopkins in the minds of his subjects through the process of hypnosis. For example, Ann Strieber, the wife of abductee Whitley Strieber, pointed out that some 2,500 people wrote letters in response to Strieber's book *Communion,* describing their own UFO experiences. About these letters, she said:

> We have literally no letters that mention the Budd Hopkins's taking-the-fetus scenario. None of them, except for the ones that have been either heavily influenced by his book, *Intruders*, or—in most cases—hypnotized by him. There are only a few [of these] letters, but it just struck me that the ones who have been hypnotized by him—and it's only a few [of the] people who wrote—followed the scenario exactly. It's like they're religious converts.[44]

Curiously, Whitley Strieber himself accepted the taking-the-fetus scenario. He said to the journalist Ed Conroy, "I think that the visitors literally are us, and I wouldn't be surprised if some of these beings that I've seen that look like half-visitor and half-human aren't the progeny—they are what happens when one of the pure neonates is somehow crossed with a fully mature human being, and the fetus is then removed. . . ."[45] Strieber's abduction experience, as reported in *Communion,* also involved strange sexual interactions with his "visitors."

One should certainly be cautious about accepting testimony obtained through hypnosis (see pages 147–53). However, it should also be recognized that the taking-the-fetus scenario has been described by investigators other than Hopkins. David Jacobs, an associate professor of history at Temple University, has written a book presenting several abduction cases in which this scenario appears.[46] The scenario also emerges in testimony given under hypnosis by Betty Andreasson and reported in Raymond Fowler's book *Watchers*.[47] Likewise, Jenny Randles in England spoke of a "young woman who claims repeated and at least partially conscious memories of being taken into a room by small beings who have then impregnated her. Later the foetus has been removed."[48] In this case, hypnosis was not employed.

Randles cited about seven additional cases with strong sexual or gynecological features. In one, a "Mrs. Verona" recalled without hypnosis being raped in a 1973 UFO encounter in England. In this case the "entities" looked human, but they were equipped with a domed, disc-shaped craft and a metallic, robotlike "retrieval device," which they used to capture their victim.[49]

In another case, witnesses in Venezuela in 1965 were told by seven-foot-tall, blond-haired, large-eyed beings that they were "studying the possibility of interbreeding with you to create a new species."[50] In a case in 1978, a Brazilian man described under hypnosis a seduction episode very similar to that of Villas Boas.[51] In another Brazilian case, in 1979, the well-known concert pianist Luli Oswald was given a complete gynecological examination by "Gray" beings who said they came from "a small galaxy near Neptune."[52] And in a case investigated by Dr. Hans Holzer in New York State in 1968, small hairless entities used a long needle to take ova samples from a woman and told her she had been chosen to give them a baby.[53]

It would seem that many UFO abductions do have a strong sexual component. This feature is consistent with the hypothesis that the abduction experiences are expressions of human psychology. However, the question remains why people should choose UFO stories, with their many strange but recurring details, as a medium for sexual fantasy.

Ostensibly, the medical examinations in UFO abductions look like scientific studies of humans conducted by visitors from another planet. Indeed, it is often reported that UFO entities tell people this. For example, William Herrmann of South Carolina reported being abducted in 1978 by humanoids who said they came from the constellation Reticulum and have been abducting humans for research purposes.[54] But messages from UFO entities tend to be untrustworthy or self-contradictory, as we can see from the story of beings coming from a "small galaxy near Neptune."

Jacques Vallee has argued that physical examinations performed in UFOs could not be strictly scientific in nature, since the human body can be examined and tissue samples can be taken without resorting to the traumatic methods experienced by abductees.[55] This can be done by human doctors, and it certainly should be child's play for high-tech ufonauts. The sexual component of UFO abductions also suggests that they involve something other than objective scientific research. Genetic experiments could be carried out using sperm and ova collected without any awareness on the part of the individual, and direct sexual activity with abductees is certainly not required.

If UFO humanoids are actual beings that evolved on another planet, the reasoning of Dobzhansky and Simpson indicates that it is highly unlikely that they would be genetically compatible with humans or any other earthly life forms. One reason for this is that the earthly genetic code table could be set up in many ways, and if life were to arise independently somewhere else, then it is highly improbable that it would make use of the same code table. Even if the beings outwardly resembled us, their molecular machinery would surely be completely different from ours.

One could argue that high-tech ufonauts could easily overcome these difficulties and produce bodies genetically compatible with humans. But in that case, why produce semihuman forms and try to get them to mate with humans? Why not just create perfect human bodies using high-tech methods? One can say that the motives of the entities are incomprehensible. This may be true, but I suggest that we can always turn to the hypothesis of unknown agencies with incomprehensible motives as a last resort. A better strategy would be to seek comprehensibility first and see how far we can get.

One comprehensible hypothesis is that the humanoids are real beings endowed with a sexual psychology similar to our own. According to this hypothesis, sexual manipulation of humans by these beings is at least partly due to their own sexual motives. This hypothesis can be kept general, and it allows for various interpretations of the controversial cases in which women were apparently shown their half-alien offspring. For example, they may literally have had such offspring, or this may be an illusion created in the abductees' minds by the entities.

This hypothesis suggests that the humanoids have not evolved independently from humans in the Darwinian fashion. This is certainly implied if they are close enough to humans genetically to make cross-breeding attempts worthwhile. It is also indicated if the humanoids lack genetic compatibility with humans but nonetheless have a recognizable sexual psychology.

The key term here is psychology. On a hypothetical planet where technological capacity evolves, would recognizable, humanlike psychology also be likely to evolve? I suspect that Dobzhansky would say no. Independent evolution of humans and the humanoids is unlikely, and coevolution of both forms on the earth is ruled out by current evolutionary scenarios. If the humanoids are real beings, then it would seem that something non-Darwinian must be involved in their origins or in our own.

The Element of Fear

One commonly reported feature of UFO abductions is intense fear. This fear typically arises from the helplessness an abductee feels when put into a state of temporary paralysis by the UFO operators. A striking illustration of this is found in the experience of 16-year-old "David Oldham," as related by Budd Hopkins. In September of 1966 David was in a car with teenage friends, who were driving around aimlessly, looking for something to do. At one point the driver stopped the car on a side road, and the boys saw a large orange light hovering over nearby trees. David recalled wishing that he could talk to the others about the light but feeling that his mind was somehow blocked. His next conscious memory was of driving to a nightclub and going inside with his friends.

Under hypnosis, David recalled that on seeing the light, he got out of the car and began to walk toward it. Then he felt paralyzed and encountered beings who took him into the UFO. He responded to this with extreme terror:

> WHAT IS IT? WHAT . . . WHAT IS IT? (Very agitated breathing) What is it? Why . . . why . . . getting numb . . . all over . . . getting numb. . . . Oh! Oh! Oh! Can't move . . . can't move. Oh! Oh! What's going on? Can't move. Oh! What . . . what do you . . . want?[56]

Fear and other emotional reactions are sometimes mixed up in very complex ways in abduction cases. For example, overwhelming fear is one of the main themes of the books *Communion* and *Transformation*, in which the popular author Whitley Strieber described his encounters with humanoid beings he calls "visitors." Although he seemed preoccupied with fear, Strieber also stressed the idea of developing a positive relationship with these beings. Apparently, a similar attitude was expressed by many people who wrote to him about their experiences with UFO entities. His wife Anne said, "you get a lot of letters where people say 'They [the aliens] felt like family,' or 'I've always felt I didn't belong on this Earth; when I was little I would look up in the sky and I would tell my mother I came on a spaceship.'"[57]

Budd Hopkins, who takes a more uniformly negative view of alien entities, made some remarks on Strieber's possible motives for taking a positive view of his visitors:

> Strieber called me up early on, saying the aliens had told him to change the title of his book from "Body Terror" to "Communion," from a title that suggested they were scary to one suggesting that they had to be much nicer. Then he called me up a month later, very upset, saying it was extremely important that I change the title of my book from "Intruders" to something more palatable—or things might go

> against me. He was saying this not from the point of view of the publishing world, but that "they" didn't like it, he said.[58]

In many cases, people experiencing UFO abductions tend to find their captors uncaring and devoid of compassion, and they often say they felt as though they were being treated like experimental animals. Some people have seen ironic justice in this, in view of our own well-known cruelty toward defenseless humans and animals. Interestingly enough, the UFO entities themselves have been frequently said to deliver scathing comments about human motives and behavior.

However, not all abduction witnesses describe their experiences in negative or fearful terms. In some cases, people have reported meeting UFO entities initially during a fearful abduction scene and then developing a friendly relationship in the course of subsequent meetings. Two examples of this are the stories of William Herrmann[59] and Filiberto Cardenas.[60] These accounts represent a cross between UFO abductions and the so-called contactee cases, in which a person claims to have entered into a voluntary, friendly relationship with alien beings. The stories of Herrmann and Cardenas will be discussed along with the contactee phenomenon in Chapter 5.

Another example of a positive response to a UFO abduction was provided by John Salter.[61] Salter is a professor of sociology at the University of North Dakota and a social justice activist who was heavily involved in the civil rights movement in Mississippi in the 1960s. He said that on March 20, 1988, he was traveling with his son John III from North Dakota to Mississippi, where he was due to deliver a paper on civil rights. At that time, he had no interest in UFOs and had read virtually nothing on the subject.

He pointed out that for some reason he had chosen a route through Wisconsin that went far out of his way on small country roads. At a certain point along this route, both he and his son experienced amnesia for a considerable section of the trip, extending from late afternoon to about 7:45 p.m., with a short stretch of clear memory at about 6:30 p.m. After the point where their memory resumed, they drove on for some time, stayed overnight, and then resumed driving in the morning. At about 10:14 a.m., they saw a shimmering, silvery saucerlike form that swooped overhead and vanished at high speed. Both father and son felt that this had something to do with their experiences of the previous day.

In late June of 1988, the elder Salter began to spontaneously remember what had happened during his period of amnesia, and he pointed out that his son had similar recollections beginning in November of 1988. Salter said that he deliberately refrained from telling his son the content of his own recollections until his son's memories of the encounter surfaced.

He recalled that he had pulled off onto a narrow and rough road that went into a wooded area. On parking, he and his son were met by two or three small humanoid figures and one taller humanoid. The smaller figures were four to four and a half feet tall, with large heads and conspicuously large, slanted eyes. The taller one looked "more human." These beings led them to a parked UFO, where they were given medical examinations. The elder Salter said that they inserted some kind of implant up his right nostril and also gave him several injections. Then he and his son were returned to their vehicle.

Although this is a typical UFO abduction account, both Salter and his son felt strongly that their encounter was positive and beneficial. Salter also noted that his physical health had markedly improved after the encounter, and he attributed this to the treatments administered by the humanoids. That this reaction is not limited to Salter alone is indicated by a letter written to him by the folklorist Thomas Bullard:

> In my earlier studies a pattern of unfriendliness, coldness and exploitation came to the fore and persuaded me that these beings were up to no good. Since then I have received several letters from people who regarded their experience in the same positive light as you do. Some people certainly come away with negative feelings. That is understandable, even reasonable when taking the experience of kidnap at face value. Yet the more I learn, the more I realize that the experience has a less obvious positive side. Some abductees feel a deep and abiding affection for their captors, and sense a reciprocity of those feelings.[62]

Missing Time

The Salter case displayed the common feature of "missing time," in which people find a mysterious gap in their memory of UFO-related experiences. The Buff Ledge case is another example of this. In these two cases, we see that people responded differently to this period of amnesia. Thus Michael and Janet in the Buff Ledge case were not able to recall their experiences on the UFO without the aid of hypnosis, and even then, Janet's recall was less complete than Michael's. The Salters were able to recall their meeting with the humanoids spontaneously, but the elder Salter's recall began several months before his son's. This suggests that UFO-induced memory loss may, like other forms of amnesia, be partly due to psychological mechanisms within the individual.

Another possibility is that memory loss may be deliberately induced by UFO entities to hide their operations. In some cases, the entities are said to give the witness threatening posthypnotic suggestions to the effect that "you will die if you remember these events." Hopkins has given several examples suggesting this, including the stories of "Steven Kilburn,"[63] "Dr. Geis,"[64] and "Kathie Davis."[65] Barney Hill likewise remembered being told by his captors that "you have to forget it, you will forget it, and it can only cause great harm that can be meted out to you if you do *not* forget."[66]

In other cases, the entities are said to simply tell the witness that he or she will not remember. In one encounter occurring north of Los Angeles in 1956, they even seem to have convinced one of the witnesses, a woman named Emily, that she should not talk because nobody would really care to know about her experience.[67] Although Emily seemed to remember, she would not talk, and she appeared to side with the UFO entities against the people who were interrogating her.

There are cases where the witness notes missing time in connection with a UFO sighting, even though no abduction is known to have taken place. Here is an example dating back to September 1963, in England. Paul, aged 21, was driving toward the village of Little Houghton at 2 a.m. Suddenly he found himself on foot and soaking wet just outside Bedford at 7 a.m. His last conscious memory was of seeing a brilliant white light in the sky heading toward his windshield. A friend drove him back along highway A428 to search for his car, and they found it locked up in the middle of a rain-soaked field with no tracks showing how it got there. Paul had the keys in his pocket.[68]

In this case, hypnosis was not used. In the absence of further information, one might suspect that Paul suffered from petit mal epilepsy or some kind of psychological fugue state. However, the story of the car in the field, if true, adds a mysterious element to the story. How did it get into the field without leaving tracks unless, perhaps, it was lifted there through the air?

In another British case, in January of 1974, Jeff and Jane, aged 20, were out driving at 9:30 p.m. They experienced being followed by a green light in the sky. At a certain point, this seemed to have gone away, but when they stopped the car and got out, green and blue beams shone down from a black oval above. They drove off in panic but suddenly found themselves in another town at 1:30 a.m., with no memory of how they got there. Then after a few more minutes of driving they found themselves in another town 20 miles away with another memory gap. It was now 3:30 a.m.[69] Here again, hypnosis was not used.

Here is an example of a similar missing time case in which an abduction story did turn up when hypnosis was used to probe for possible lost memories. On June 11, 1976, in Romans, France, Helene Giuliana, a maid in the house of the mayor of Hostun, saw a big orange glow in the sky while returning home from a movie. At this point, her Renault stopped functioning. The light suddenly vanished, and she drove home—with about four hours of missing time. Under hypnosis, she reported being carried into a room by "small figures with big eyes and ugly faces, clamped on a table and examined, particularly around her abdomen."[70]

The Role of Hypnosis

Thus far, many of the abduction accounts I have presented have emerged under hypnosis. These cases might suggest that the process of hypnosis somehow conjures up abduction fantasies from the minds of UFO witnesses. There is certainly a great deal of support for this viewpoint. For example, after reviewing the scientific literature on hypnosis, the Council on Scientific Affairs (CSA) of the American Medical Association concluded that "recollections obtained during hypnosis can involve confabulations and pseudomemories and not only fail to be more accurate, but actually appear to be less reliable than nonhypnotic recall."[71]

This condemnation of hypnosis as a means of recovering lost memories must certainly be taken seriously. At the same time, there appears to be evidence indicating that hypnosis can enable people to genuinely recover lost memories, including memories of UFO abductions. What is going on here?

One important point is that most scientific studies of memory under hypnosis deal with items such as nonsense syllables or passages of poetry that carry little emotional significance for the experimental subject. This can be contrasted with memories of UFO abductions which deal with highly traumatic experiences. Thus the CSA report stated that "With respect to cases where there is a preexisting psychopathology and/or extreme emotional trauma, the current experimental literature is not definitive."[72] Of course, this is not surprising. It would be unethical to carry out experiments in which people are subjected to extreme emotional trauma.

Another point is that UFO abductees often recall being told by the abducting entities not to remember their abduction. In some cases they are told that they will experience great harm if they remember. This suggests that the role of hypnotic regression in UFO abduction cases may be to counteract the effects of both trauma and previous hypnotic commands to forget. It might be possible to scientifically investigate the effectiveness of hypnosis in reversing amnesia induced by previous hypnosis. However, to be truly relevant, such investigations would have to duplicate the traumatic character of UFO abductions.

According to the CSA report, hypnosis often "results in more information being reported, but these recollections contain both accurate and inaccurate details."[73] If this is so, does it also apply to UFO abduction accounts? Or do recovered memories of UFO abductions contain only inaccurate details?

In the Buff Ledge case, Webb reported that under hypnosis Janet verified 70 percent of Michael's descriptions of what occurred on the dock and 68 percent of his on-board descriptions (page 123). This means that she failed to verify some 30 percent of his descriptions. One could argue that this 30 percent consists of false memories and the 70 percent consists of accurate memories.

In the Buff Ledge case, the presence of two witnesses who never had a chance to discuss their experiences makes it possible to distinguish between false memories and memories that are quite likely to be genuine. In cases with only one witness, it is not possible to do this. Unfortunately, multiple witness cases with lack of communication between the witnesses are rare.

It is difficult to estimate what percentage of testimony retrieved under hypnosis is accurate and what percentage is inaccurate. This presumably varies from case to case and depends on such factors as the psychology of the witnesses and the methods used by the hypnotist. However, if a substantial portion of the details remembered in some UFO abduction cases are accurate, then we must conclude that UFO abductions are not mere fantasies. At the same time, we have to be cautious about assigning great weight to the details of particular cases.

In *The Allagash Abductions,* Raymond Fowler reported on a UFO abduction case in which four sane, responsible witnesses testified under hypnosis about a single UFO abduction.[74] This case is similar to the Buff Ledge case in that (1) the witnesses did not consciously remember the abduction before their hypnosis sessions, (2) they did consciously remember a UFO close encounter (of type CE-III) connected with the abduction, and (3) they were strongly encouraged not to communicate with one another about the results of their hypnosis sessions until the sessions were completed for all four witnesses.

The four Allagash abduction accounts broadly agree with one another, even though they differ in some details. Thus this case seems to add further support to the idea that UFO abduction memories recovered under hypnosis can contain both accurate and inaccurate material. The common features of the four abduction accounts suggest that the witnesses did undergo an extraordinary experience.

The psychologist John Carpenter has reported a case in which consistent information was obtained separately from two respectable, middle-aged women who reported a close encounter of type CE-III that involved missing time.[75] The women were given the MMPI test and found to be free of psychopathology or psychological problems. They were also found to have a low to moderate level of fantasy proneness. They testified that they had no interest in UFOs before their experience and they had not read any books on UFOs. In separate hypnosis sessions they produced abduction accounts that agreed in some 40 points, although there were also some disagreements. This also seems to be a case in which genuine memories were recovered under hypnosis.

One explanation for consistent narratives produced by two witnesses is that these may be due to *folie a deux.* This is a disorder in which a dominant, psychotic person induces another person to share his or her delusions. *Folie a deux* is ruled out in this case because the two witnesses were acquaintances who did not have a close personal relationship. Also, neither witness was psychotic. This is also true of the Buff Ledge case.

It is sometimes argued that hypnotists produce consistent UFO narratives by asking leading questions. This idea was explored by Alvin Lawson, a professor of English at the California State University at Long Beach, in a paper entitled "What Can We Learn from Hypnosis of Imaginary Abductions?"[76] Lawson hypnotized eight subjects and asked questions like, "Imagine you are seeing some entities or beings aboard the UFO. Describe them," and then, "You are undergoing some kind of physical examination. Describe what is happening to you." The eight subjects were supposedly unknowledgeable regarding UFOs, yet they produced abduction stories in response to such questions. To explain this, Lawson later published the theory that abduction scenes involving beings with big heads and spindly bodies are based on memories of the trauma of birth and one's bodily shape as a fetus.

However, Lawson's work has been widely criticized. Jenny Randles pointed out that he was definitely asking leading questions, and his data base of eight people was too small. In addition, his subjects' eight imaginary abduction stories featured six types of alien entities, including four types that almost never appear in abduction accounts. On top of this, his fetal memory theory is baseless, since people never see themselves or others as fetuses at the time of birth.[77]

Although Lawson's work was flawed, it does show how leading questions can influence a person's testimony under hypnosis. However, it should not be concluded that hypnotic regression carried out by UFO investigators will automatically invoke UFO abduction accounts. Randles noted that there are British cases in which hypnosis was used in an effort to uncover a UFO abduction but no abduction scenario emerged, and she pointed out that Budd Hopkins has cases in America where the same thing happened. In fact, Hopkins mentioned that out of 79 cases in one study, in 20 an abduction was recalled with the aid of hypnosis, in 11 hypnosis was used but no abduction was recalled, and in five cases an abduction was recalled without the aid of hypnosis. The remaining 43 cases had not been fully evaluated at the time of the report.[78]

Randles also noted a study by Dr. Thomas Bullard of over 200 abduction cases, in which one-third of the subjects had full conscious memory of the abduction experience. An analysis of these cases showed that abduction experiences recalled under hypnosis were essentially the same as those recalled without hypnosis. The most notable difference between the two sets of cases was that medical examinations were mentioned twice as often in hypnosis cases as in cases with direct recall.[79] Of course, this might be expected if the medical examinations are more likely to be blocked by amnesia than other, less traumatic aspects of abduction experiences.

The findings mentioned by Randles are paralleled by those of David Webb in a study of 300 abduction reports from HUMCAT, a data base of UFO encounters involving humanoids.[80] Out of these 300 reports, Webb found that 140 satisfied the following five reliability criteria: (1) an abduction was clearly indicated, (2) the case was reasonably well investigated, (3) there was no evidence of a hoax or of witness psychopathology, (4) there was enough data to assess the general scenario and degree of hypnosis used, and (5) the case did not involve crashed UFOs.

Webb divided these cases into three categories: those in which on-board abduction information was obtained (I) mainly with, (II) partly with, and (III) entirely without the use of hypnosis. The last category was divided into two subcategories, IIIa, in which hypnosis was used but added nothing new, and IIIb, in which no hypnosis was used. At the time of his report, he had reviewed 117 of the 140 cases and obtained the following results:

Category	No. of Reports	% of Total
I	61	52
II	11	9
IIIa	8	7
IIIb	37	32

Thus it turns out that in 39% of the cases, no on-board abduction information was obtained using hypnosis. Webb pointed out that the reports in categories I and III were remarkably similar in content.

Another point to make about hypnosis as a tool for retrieving lost memories is that there are instances in which information obtained under hypnosis is independently corroborated. An example would be the story of "Steven Kilburn" presented by Budd Hopkins. Under hypnosis, Steven (whose real name is Michael Bershad[81]) revealed a typical abduction scenario, including a physical examination by the standard "Gray" entities. This included what seemed like a neurological examination.

Steven described this examination to a neurosurgeon named Paul Cooper. Here is Dr. Cooper's reaction, as related by Hopkins:

> Steven is a remarkable young man. He's extremely bright, an excellent observer, and totally believable. . . . Everything he told me about what they did to him and how his body reacted accorded exactly with what should have happened if they stimulated the different nerves he said they touched. I tried to mislead him. . . . And he has no particular knowledge of the nervous system. He'd have to have known a great deal to make it all up, and I'm certain he's not the type to lie. He's a very decent guy, and I'm really impressed with him.[82]

It seems from this that under hypnosis there emerged specialized anatomical information that Steven had never consciously studied. Assuming that he wasn't lying about his medical education, one could hypothesize that he once read a neuroanatomy textbook, remembered it only on a subconscious level, and incorporated material from it into his abduction story. This sort of thing is actually known to occur, and it is called cryptomnesia. For example, there are cases where an apparent previous life that has been recalled under hypnosis has been traced to a book that the subject read but had forgotten.[83]

Although cryptomnesia is a possible explanation of Steven's testimony about his neurological examination, it seems to be an unlikely one. It is one thing to remember information from a medical textbook, and it is quite another to convert that knowledge into an accurate description of how the body would behave during an examination. That might well require practical medical training that would not readily be forgotten. Thus, if Dr. Cooper's testimony is genuine, it definitely seems to add weight to Steven's abduction story.

In summary, it appears that typical abduction experiences are sometimes remembered with or without hypnosis. Since the experiences remembered with the aid of hypnosis tend to be much the same as those remembered without it, it would appear that the process of hypnosis itself is not a major cause of abduction accounts. If abductions recalled without hypnosis might be factual, then so might abductions recalled with hypnosis.

I should note, however, that very doubtful stories can emerge under hypnosis. For example, the psychologist Edith Fiore recounted the story of a man named "Dan" whom she regressed hypnotically, hoping to recover memories of possible close encounters. Dan proceeded to describe a previous life as a cold-blooded soldier on an interstellar spaceship. His job was to make "drops" on designated planets and wipe out target cities with "force beams" without asking questions. The soldier and his compatriots were fully human, living a life reminiscent of "Star Trek." The soldier was supposedly "retired" by being mentally transferred into the body of a child in Washington State, displacing the child's original mind. This child then grew up as Dan.[84]

In this case, it seems doubtful that the story was produced as a result of leading questions, for the hypnotist had no idea that such a story would emerge. But in contrast to UFO abduction accounts, Dan's story seems very similar to familiar science fiction stories, and it also casts Dan in an ego-building role as a tough, self-reliant soldier. Dan had read a lot of sci-fi,[85] and it is possible that he had incorporated science fiction themes into a subconscious fantasy. Hypnosis, it seems, is an imperfect and poorly understood tool that can produce useful results but cannot be fully trusted.

Psychological Evaluation of Abductees

To many people, it is natural to attribute the strange stories of UFO abductees to some kind of mental aberration. To test this idea, a number of psychological studies have been made of abductees, and I will discuss some of them in this section.

It turns out that abductees as a group have proven to be free of overt psychopathology (although there are inevitable exceptions to this rule). Psychologists have reacted to this finding in a variety of ways. Some have tried to remain noncommital, while others have accepted the reality of abduction by aliens. Others have argued that the abduction phenomenon requires far-reaching reformulations of our basic ideas of reality. Still others have tried to preserve those ideas by arguing that although abductees are sane, they are nonetheless prone to accept fantasy as reality.

A pioneering study was undertaken in 1981 by two UFO researchers, Ted Bloecher and Budd Hopkins, and a psychologist, Dr. Aphrodite Clamar.[86] They selected five men and four women who had reported UFO abduction experiences involving missing time, encounters with aliens, on-board physical examinations, and so on. A psychologist, Dr.

Elizabeth Slater, was asked to evaluate them in order to determine their comparative psychological strengths and weaknesses. She was not told that the nine people had anything to do with UFOs, and they were instructed not to disclose this to her.

The subjects were a college instructor (photography), an electronics expert, an actor and tennis instructor, a corporation lawyer, a commercial artist, a business executive, the director of a chemistry laboratory, a salesman and audio technician, and a secretary. They were administered the MMPI, the Wechsler Adult Intelligence Scale, the TAT (Thematic Apperception Test), the Rorschach test, and the projective drawings test.

Slater concluded that the nine subjects were "quite heterogeneous" in personality, but they tended to share the following traits:

1. relatively high intelligence with concomitant richness of inner life.
2. relative weakness in the sense of identity, especially sexual identity.
3. concomitant vulnerability in the interpersonal realm.
4. a certain orientation toward alertness which is manifest . . . in a certain perceptual sophistication and awareness or in interpersonal hypervigilance and caution.[87]

She also found that the nine subjects tended to be anxious, sometimes overwhelmingly so. They tended to suffer from low self-esteem and a feeling of vulnerability to insult and injury. They tended to be wary and cautious, but Slater described this as oversensitivity rather than paranoia.

After learning of the subjects' UFO histories, Slater was flabbergasted. After reading Hopkins's book *Missing Time* and meeting with Clamar, Hopkins, and Ted Bloecher, she said in her final report:

> The first and most critical question is whether our subjects' reported experiences could be accounted for strictly on the basis of psychopathology, i.e., mental disorder. *The answer is a firm no.* In broad terms, if the reported abductions were *confabulated fantasy productions,* based on what we know about psychological disorders, they could have only come from *pathological liars, paranoid schizophrenics, and severely disturbed and extraordinarily rare hysteroid characters subject to fugue states and/ or multiple personality shifts.* [Slater's italics.][88]

Thus, the subjects' abduction experiences could not be explained psychologically. However, Slater observed that their anxiety and insecurity could be readily accounted for by real UFO abductions:

> Certainly such an unexpected, random, and literally otherworldly experience as UFO abduction, during which the individual has absolutely no control over the outcome, constitutes a trauma of major proportions. Hypothetically, its psychological impact might be analogous to what one sees in crime victims or victims of natural disasters, as it would constitute an event during which an individual is overwhelmed by external circumstances in an extreme manner.[89]

Slater pointed out that an analogy can be drawn between UFO abductees and rape victims. She concluded that although the study did not prove the reality of UFO abductions, it showed that the subjects' psychological problems could be explained in terms of such experiences and not the other way around.

Another psychological study of UFO abductees has been carried out by Rima Laibow, M.D., a psychiatrist in Dobbs Ferry, New York, and a graduate of Albert Einstein College of Medicine in New York City. On the basis of personal work with 11 abductees and familiarity with 65 cases, she made observations similar to Slater's. She expected to find psychosis in people reporting such bizarre experiences, but instead she found only the anxiety that such experiences would be expected to produce.

In the abductees she saw PTSD, or posttraumatic stress disorder, which is normally thought to be produced only by event-level trauma. (This term refers to traumas that actually occur physically, as opposed to fantasies generated within the mind.) She also observed that fantasies should vary widely from individual to individual, whereas UFO abduction stories tend to be very similar.[90]

Dr. June Parnell, a professional counselor at the University of Wyoming, wrote a 110-page paper entitled "Personality characteristics on the MMPI, 16PF and ACL tests, of persons who claim UFO experiences," which was published by the university in 1986. She applied these tests to 225 witnesses reporting all types of UFO encounters. She described people reporting exotic contacts and abductions as ". . . having a high level of psychic energy, being self-sufficient, resourceful, and preferring their own decisions . . . [with] above-average intelligence, assertiveness, a tendency to be experimenting thinkers, a tendency toward a reserved attitude, and a tendency toward defensiveness. There was also a high level of the following traits in these deep-encounter witnesses: "being suspicious or distrustful . . . creative and imaginative . . . .'"[91] This description is quite similar to Slater's, and Parnell concluded that UFO experiencers tend to be mentally healthy.

Kenneth Ring, a psychology professor known for his near death investigations, published an extensive study comparing near death experiencers with people having UFO experiences. Ring observed that while some researchers agree with Parnell's positive interpretation of her findings, other "psychiatrically oriented commentators" argue that UFO experiencers may be suffering from some form of mental illness.[92] Ring's own conclusion was that UFO experiencers are characterized by a tendency to dissociate which may have been brought about by abuse in childhood.[93] He stressed, however, that dissociative tendencies can be considered to be mentally normal and must be distinguished from dissociative disorders, which are pathological. Ring's theory is that both near death experiencers and UFO experiencers are able to enter an altered state of consciousness through dissociation. This enables them to enter an objectively real "imaginal realm" or alternate reality in which their unusual experiences take place.

Ring's ideas have led him to affirm a third state of consciousness which is distinct from both sanity and insanity as they are customarily defined. Philosophically, he has adopted a position reminiscent of idealism, in which matter and imagination are not considered to be separate categories. To Ring, this allows for remarkable possibilities. He suggests that in the future, "Veils will be lifted from the face of the non-physical, and we ourselves will become diaphanous beings, with bodies of light—*if* the speculations to be offered are a true reading of our future condition and experiential possibilities."[94]

Dr. John Mack, a professor of psychiatry at the Harvard Medical School, has also adopted radical new views of reality as a result of a study of UFO abductees. He accepts the experiences of the abductees as being essentially real and concludes that "To acknowledge that the universe (or universes) contains other beings that have been able to enter our world and affect us as powerfully as the alien entities seem able to do would require an expansion of our notions of reality that all too radically undermines the Western scientific and philosophical ideology."[95] Mack argues that the conceptual expansion caused by abduction experiences often causes abductees to undergo spiritual growth and thus it tends to have a positive effect.

Dr. Richard Boylan, a clinical psychologist, has published a study of abductees which presents UFO abductions as essentially positive experiences brought about by real alien beings. Boylan introduces the controversial idea that "when experiencers are properly debriefed and psychologically treated, they do not exhibit PTSD, *unless* the close encounter has caused a resurfacing of preexisting *human-caused* PTSD-level trauma."[96] For example, memories of childhood sexual abuse could cause a negative reaction to a UFO abduction. Boylan also argues that some traumatic UFO abductions are actually pseudo-abductions perpetrated by human agents (see page 47).

A more scientifically conservative approach to UFO experiences was taken by a team of psychologists from Carleton University in Canada led by Nicholas Spanos.[97] Spanos and his colleagues administered a battery of psychological tests to four groups of people: 15 intense UFO experiencers (including abductees), 20 nonintense UFO experiencers, 74 college psychology students, and 53 people from the general community. One of their most important conclusions was that UFO experiencers tend to be mentally healthy:

> The most important findings indicate that neither of the UFO groups scored lower on any measures of psychological health than either of the comparison groups. Moreover, both UFO groups attained higher psychological health scores than either one or both of the comparison groups on five of the psychological health variables.[98]

Spanos and his colleagues also tried to evaluate the idea that even though UFO experiencers tend to be sane, they nonetheless have tendencies towards fantasy and false experience that might account for their strange experiences. Here I use the phrase "false experience" to refer to impressions of unreal events that might arise due to hallucination, suggestibility, brain instability, or irrational (but not pathological) thought processes. Clearly this concept is an attempt to straddle a fine line between sanity and insanity.

The following table sums up some of the test results that address this issue.[99]

Psychological Variable	UFO experiencers		Control groups	
	Intense	Nonintense	General	Students
Paranormal experience	42.2	34.4	40.9	40.0
MMPI Schizophrenia	12.1	9.2	19.3	19.6
Perceptual Aberration	5.6	3.0	7.6	6.3
Magical Ideation	9.0	8.3	10.9	8.7
Temporal lobe lability	32.8	34.2	37.3	38.6
Fantasy Proneness	22.4	21.6	25.3	23.6

The six psychological variables in the table are supposed to assess the tendency of a person to have false experiences or indulge in fantasy. We can see that the intense UFO experiencers tend to score higher than the nonintense UFO experiencers on five out of the six variables. However, both UFO groups tended to score lower on the six variables than the control groups. The authors conclude:

> Subjects in the two UFO groups failed to differ from subjects in the comparison groups on any of the imaginal propensity measures, the temporal lobe lability index, the paranormal experiences index, or the hypnotizability measures. These findings clearly contradict the hypothesis that UFO reports—even intense UFO reports characterized by such seemingly bizarre experiences as missing time and communication with aliens—occur primarily in individuals who are highly fantasy prone, given to paranormal beliefs, or unusually suggestible.[100]

Nonetheless, Spanos and his colleagues could not accept that UFO experiences might be caused by some agency external to the experiencers. Their final conclusion was that "beliefs in alien visitation and flying saucers serve as templates against which people shape ambiguous external information, diffuse physical sensations, and vivid imaginings into alien encounters that are experienced as real events." [101] Of course, this does not account for the fact that many intense UFO experiencers maintain that they had no belief in UFOs before their experiences.

The Paranormal Factor

The test data reported by Spanos and his colleagues indicate that the intense UFO experiencers scored higher on the Paranormal Experience scale than the nonintense UFO experiencers or controls. This scale is supposed to measure belief that one has had paranormal experiences, and it turns out that many UFO abductees do report experiencing unusual psychic phenomena.

Spanos and his colleagues followed conventional scientific wisdom by implicitly treating paranormal experiences as false. Certainly there are many instances in which such experiences turn out to be delusory or fraudulent. However, it is possible that some paranormal experiences may be both genuine and closely connected with the UFO phenomenon. I will therefore discuss the connection between UFO close encounters and psychic phenomena in this section. But before doing this, I should say a few words indicating why reports of psychic phenomena should be taken seriously. I will do this by presenting strong evidence for some extremely controversial psychic phenomena.

The world of spirit mediums is famous for fraud, and one can find extensive accounts of this in books such as *The Psychic Mafia,* by the confessed psychic confidence man M. Lamar Keene.[102] However, it might be a mistake to dismiss the phenomena connected with spirit mediums as entirely bogus. Many cases could be cited showing significant evidence for the reality of these phenomena, and here I will briefly summarize one of them. This case is described in greater detail in a book entitled *The Limits of Influence,* by Stephen Braude, a professor of philosophy at the University of Maryland.[103]

In the early part of the twentieth century, there was a spirit medium named Eusapia Palladino, who became known for producing such things as supernormal movements of objects and ectoplasmic emanations. She was studied by a number of distinguished scientists, and she was also caught in fraud. The people investigating her agreed that she would cheat if she got the opportunity, but some argued that strange phenomena had been observed in her presence that she could not have produced by cheating.

In an attempt to resolve this question, the Society for Psychical Research in England put together a "Fraud Squad" consisting of:

1. The Hon. Everard Feilding, who claimed to be a complete skeptic regarding spiritualistic mediums and who had detected many of them in acts of fraud.
2. Hereward Carrington, an amateur conjurer who wrote *The Physical Phenomena of Spiritualism,* three-fourths of which was devoted to an analysis of fraudulent mediumship.

3. W. W. Baggally, a skilled conjurer, who "claimed to have investigated almost every medium in Britain since Home without finding one who was genuine."[104] (Daniel D. Home was a famous nineteenth-century medium.)

These investigators rented three adjoining hotel rooms in Naples, Italy, in November of 1908, and séances with Palladino were held in the central room. The room was illuminated by electric ceiling lights. Before each séance, the investigators would carefully examine the room, and they would set up a curtain, called the "cabinet," across one corner. Behind the curtain there was a small table surrounded by the walls, floor, and ceiling, with no doors or windows (and presumably no trapdoor). The table and curtain were carefully checked for hidden devices.

After these preparations, one investigator would go downstairs and escort Eusapia Palladino up to the room alone. The room would be locked, and Palladino would sit at a table in front of the curtain, accompanied by the investigators. Two of the men would sit on either side of 54-year-old Eusapia, holding onto her arms and legs, and observing her carefully. In front of the séance table was a table occupied by the stenographer Albert Meeson, who was a stranger to Eusapia and who wrote down whatever the investigators spoke to him.[105]

Feilding explained the strategy of the investigators as follows: "We felt that if, in a reasonable number of experiments, persons specially versed in conjuring tricks and already forewarned concerning, and familiar with, the particular tricks to be expected, were unable to discover them, it would not be presumptuous to claim as a probable consequence that some other agency must be involved."[106]

Here is an excerpt from Feilding's account of the séances. He began by stressing his skepticism, which was based on many observations of fraud. In the séances with Eusapia, however, he witnessed phenomena that he was unable to explain on the basis of fraud, and his reactions to this are interesting:

> The first séance with Eusapia, accordingly, provoked chiefly a feeling of surprise; the second, of irritation—irritation at finding oneself confronted with a foolish but apparently insoluble problem. The third sïance, at which a trumpery trick was detected, came as a sort of relief. At the fourth, where the control of the medium was

withdrawn from ourselves [due to the presence of "guest"sitters], my baffled intelligence sought to evade the responsibility of meeting facts by harbouring grotesque doubts as to the competency of the eminent professors, who took our places, to observe things properly; while at the fifth, where this course was no longer possible, as I was constantly controlling the medium myself, the mental gymnastics involved in seriously facing the necessity of concluding in favour of what was manifestly absurd, produced a kind of intellectual fatigue.

After the sixth, for the first time I find that my mind, from which the stream of events had hitherto run off like rain from a macintosh, is at last beginning to be capable of absorbing them. For the first time I have the absolute conviction that our observation is not mistaken. I realize, as an appreciable fact in life, that from an empty cabinet I have seen hands and heads come forth, that from behind the curtain of that empty cabinet I have been seized by living fingers, the existence and position of the very nails of which could be felt. I have seen this extraordinary woman sitting visible outside the curtain, held hand and foot by my colleagues, immobile, except for an occasional straining of a limb, while some entity within the curtain has over and over again pressed my hand in a position clearly beyond her reach. . . .[107]

A more detailed account of the heads and hands is given by the following passage, in which Feilding contemplated the possibility of producing the strange phenomena by means of an apparatus:

It would be an interesting problem to set before a manufacturer of conjuring machines to devise an apparatus capable of producing alternatively a black flat profile face, a square face on a long neck, and a 'cello like face on a warty nobbly body two feet long; also, a white hand with moveable fingers having nails, capable of reaching high above the medium's head, or patting, pinching and pulling hair, and of so vigorously grasping B. by the coat as almost to upset him into the cabinet. Our manufacturer must so construct the apparatus that it can be actuated unseen by a somewhat stout and elderly lady clad in a tight plain gown, who sits outside the curtain held visibly hand and foot, in such a way as to escape the observation of two practical conjurers clinging about her and on the look-out for its operation.[108]

One way of looking at this testimony is that it constitutes good evidence that some unknown agency actually produced the weird phenomena that Feilding described. But one can also say that we have no proof that Feilding and his colleagues were not lying or that they were not duped by Eusapia Palladino. The status of evidence for the paranormal is similar to that of UFO evidence. In the ultimate issue, incontrovertible proof is impossible. But there are cases of strong testimony for paranormal phenomena, and these tend to become more persuasive as they accumulate and display recurring law-like patterns.

With this background on psychic phenomena, I will now turn to the main topic of this section, beginning with the case of Betty and Barney Hill. The psychiatrist Berthold Schwarz observed that after the Hills' close encounter on a lonely New Hampshire road, they began to experience poltergeist phenomena in their home. Betty would find her coats unaccountably dumped on the living room floor, even though she had left them in the closet. Clocks would stop and start mysteriously, or their time settings would change. Water faucets would turn on when nobody was there, and electrical appliances would break down and then work perfectly without repair.[109]

The German word "poltergeist" literally means "noisy ghost" and is used to refer to disturbances in which objects move or behave strangely without any obvious physical cause. In recent years, parapsychologists wishing to avoid the word "ghost" have coined the term RSPK, or recurrent spontaneous psychokinesis, for these disturbances. This change is motivated by their hypothesis that poltergeist effects may be produced by some kind of energetic emanation from a target person. In this case, Betty Hill would be the likely target individual, and one could speculate that her UFO experience changed her energy balance and triggered the poltergeist effects.

Poltergeist phenomena have been known for centuries, and they include the kind of events reported by Betty Hill, as well as spontaneous fires, objects flying through the air, and spontaneously moving furniture. Frequently, the target person in a poltergeist case is emotionally disturbed or chronically ill.[110]

This doesn't seem to be the case with Betty Hill, but Schwarz pointed out that she had a past history of paranormal experiences. When she was in high school, she had many accurate precognitive dreams, including two in which she foresaw the deaths of school friends in automobile accidents. Many of her family members were also psychic, including her maternal grandmother and an adopted daughter. Her sister, Janet, reportedly lived in a house that was haunted by a child ghost named Hannah, whose name was revealed by a psychic and later confirmed by old records.[111] Apparently, Barney Hill and his family had no past history of psychic phenomena.[112]

Karla Turner, a college instructor with a Ph.D. in Old English Studies, began to investigate UFO abductions in an effort to understand abduction experiences involving herself and her family.[113] In a recent lecture, she observed that in a group of 21 UFO abductees, 16 reported heightened psychic abilities, 16 reported unexplained noises in their homes (such as footsteps on the roof), 16 reported unexplained electrical disturbances (such as TVs and lights turning on or off mysteriously), 16 reported the overnight appearance of strange marks on their bodies (such as punctures, bruises, and claw marks), and 12 reported poltergeist phenomena (such as the unexplained appearance or disappearance of household objects). She also pointed out that all 21 abductees reported hearing unexplained voices (often calling their names) and unexplained sounds (such as beeps and whistles).[114] Nearly all of these reported events fall broadly into the category of psychic phenomena.

Phenomena of this kind are often mentioned in published UFO accounts. For example, Ed Walters, a Florida businessman well-known for his UFO photographs, has recently published a book which includes accounts of his own UFO abduction and of poltergeist activities in his home.[115] Likewise, in Raymond Fowler's Allagash case, four men recalled a common UFO abduction under hypnosis (page 149). Two of them, the twin brothers Jim and Jack Weiner, had a history of childhood poltergeist experiences.[116]

The psychic phenomena reported in connection with UFO accounts tend to fall into the following two categories:

1. Psychic phenomena not directly connected with UFOs, such as poltergeist activities in the home. These may begin abruptly after a UFO encounter, or it may be that a close-encounter witness has a long history of paranormal experiences. Matters are complicated by the fact that UFO close-encounter witnesses often turn out to have a long history of UFO encounters going back to early childhood.
2. Psychic phenomena that typically occur during UFO encounters, including telepathic communication, levitation, passing of matter

> through matter, and mysterious healing. UFO investigators some times attribute these phenomena to high technology on the part of UFO entities. This may be correct, but it remains a fact that these phenomena have also been studied in the domain of psychical research without reference to UFOs. Of course, one should not rule out the possibility that psychic phenomena not connected with UFOs may also involve some kind of high technology.

We have already seen reports of UFO encounters involving telepathic communication, levitation, and the passing of bodies through walls. An example of mysterious healing is provided by the case of "Dr. X," a French medical doctor. This case was studied originally by Aime Michel in France and was recounted by Jacques Vallee.[117] According to Vallee, an important aspect of the case is that an astrophysicist, a psychiatrist, and a physiologist were able to gain rapid access to Dr. X and were able to monitor ongoing developments.

The doctor testified that on November 1, 1968, he was awakened by calls from his 14-month-old baby shortly before 4:00 a.m. On opening a window, he saw two hovering disc-shaped objects that were silvery-white on top and bright red underneath. After moving closer for some time, the two discs merged into a single disc, which directed a beam of white light at the doctor. The disc then vanished with a sort of explosion, leaving a cloud that dissipated slowly.

The doctor said that he had received a serious leg injury while chopping wood three days before. After the departure of the mysterious object(s), the swelling and pain from this injury suddenly vanished, and during subsequent days he also noticed the disappearance of all the chronic aftereffects of serious injuries he had received in the Algerian war. A few days after the encounter, Dr. X and his child each developed a strange, reddish, triangular mark on the abdomen, and this mark recurred in successive years.

During a two-year period following this incident, there was no recurrence of symptoms associated with either the war injuries or the leg wound. However, strange paranormal phenomena began to take place around the doctor and his family, including poltergeist activity and unexplained disturbances in electrical circuits. According to Jacques Vallee, "Coincidences of a telepathic nature are frequently reported, and the doctor has allegedly, on at least one occasion, experienced levitation without being able to control it."[118]

Other developments were even more bizarre. Dr. X recounted that he began to have mysterious meetings with a strange, nameless man he called "Mr. Bied." The doctor would hear a whistling noise inside his head and would feel guided to walk or drive to a certain location. There he would meet the strange man, who would discuss his UFO experience with him and instruct him on paranormal matters. Mr. Bied caused him "to experience teleportation and time travel, including a distressing episode with alternative landscapes on a road that "does not exist."' The stranger also once visited Dr. X at his home "accompanied by a three-foot-tall humanoid with mummified skin, who remained motionless while his eyes quickly darted around the room." [119]

Vallee remarked that even though UFO cases generally seem quite strange, the reports of such cases are often edited by the suppression of particularly bizarre or incredible aspects. Nonetheless, strange events of the kind reported by Dr. X also turn up frequently in other UFO cases.

For example, Vallee mentioned a case in Lima, Peru, on December 9, 1968, in which a customs officer was hit in the face by a purple beam from a UFO and then discovered that he was cured of myopia and rheumatism.[120] In South Carolina on April 21, 1979, UFO abductee William Herrmann reported being visited in his trailer by two alien beings who seemed to materialize in the midst of a blue glow while he was on the phone with a UFO investigator.[121] Whitley Strieber has reported many paranormal effects connected with his visitor experiences, as well as visions of surreal landscapes. These include poltergeist phenomena, spontaneous levitation, and out-of-body experiences.[122]

The British UFO investigator Jenny Randles has given many examples of persons who reported both UFO encounters and psychic experiences. For example, Joyce Bowles experienced being abducted into an unknown room along with a man named Ted Pratt and having an extended meeting with three tall humanoid beings. Randles commented that Bowles had also suffered a poltergeist attack and "had a track record of psychic experience." [123]

Raymond Fowler pointed out that Betty Andreasson and some of her family members reported a number of strange psychic experiences that occurred before her 1967 UFO encounter. For example, her daughter Becky (who was involved in that encounter) described how she woke up in 1964 to see a glowing yellow-orange ball hovering outside her bedroom window and directing a beam of light at her. Shortly after this, Becky began to produce pages filled with strange symbols by automatic writing.[124] Automatic writing is a well-known psychic phenomenon, and it also shows up in the stories of many UFO witnesses (see Chapter 5).

Fowler also gave the story of Mrs. Rita Malley, who was driving along Route 34 to Ithaca, New York, in 1967 when her car was stopped by a humming, domed, disc-shaped object. She said that a bright light beamed down from the object. "Then I began to hear voices. They didn't sound like male or female voices, but were weird, the words broken and jerky . . . like a weird chorus of several voices. . . . The voices named someone I knew and said that at that moment, my friend's brother was involved in a terrible accident miles away." [125] As the lady found out the next day, this message was correct.

In this case, the UFO experience included a precognitive warning. There is an extensive literature on precognitive warnings in dreams and sudden flashes of intuition, and it is reported that these predictions very often turn out to be true.[126] These paranormal warnings and the other psychic phenomena that I have mentioned have been part of human life since time immemorial, even though they do not fit into the mechanistic paradigm of modern science.

In summary, there is evidence that psychic phenomena (or close imitations of them) are often reported in conjunction with UFO close-encounter cases. These phenomena seem to be connected with both the human subjects of these cases and the humanoid entities they meet.

One possible inference from this is that the UFO humanoids are not as alien to us as one might suppose. This is based on the argument that beings with humanlike form, humanlike emotions, and humanlike paranormal powers may well be related to human beings on a fundamental level.

Of course, there are other possibilities, and I will mention two. First, there is the extraterrestrial hypothesis (ETH), which maintains that UFO entities have evolved on a distant planet. It is difficult to discuss the evolution of psychic abilities, since evolutionists generally deny that such abilities exist. But suppose, for the sake of argument, that psychic abilities have evolved according to standard evolutionary principles. Then the reasoning of Dobzhansky and Simpson must apply and we would not expect them to evolve on another planet. If that is so, it is strange that visiting ETs seem to exhibit psychic phenomena and also seem to awaken psychic abilities in humans.

Statements of the Hypothesis that UFO Manifestations Are Given Physical Reality by the Power of the Human Mind

1. Jacques Vallee: "One could theorize that there exists a remarkable state of psychic functioning that alters the percipient's vision of physical reality and also generates actual traces and luminous phenomena, visible to other witnesses in their normal state." [127]

2. Berthold Schwarz: "Almost all the data associated with UFOs have their analogies in spontaneous psychic phenomena, or have been noted to occur in the séance room." The reported UFO electromagnetic effects may be an exception, but even they might be produced psychically. He quotes Eisenbud as saying, "There is a small solid core of parapsychological data indicating that both animate and inanimate entities can be created (presumably under mental auspices) not only piecemeal, as a sort of intrusion into a more ordinary reality, but as a completely coexisting reality." [128]

3. D. Scott Rogo on UFO abductions: "The abduction is a real physical event, but it reflects concerns or traumas buried within the subject's unconscious. It might be called an "objectified' dream— i.e., a system of symbolic imagery which suddenly erupts into the three-dimensional world." [129]

4. Lieutenant Colonel (Ret.) Thomas Bearden: "In June 1947 Kenneth Arnold, flying over the state of Washington—the state closest to the Soviet Union at the time— encountered flying saucers, which were simply female mandalas modulated by our science fiction/Buck Rogers national/cultural unconsciousness. . . . Make no mistake, these tulpoids are actual materializations, not hallucinations or fantasies." [130]

5. Hilary Evans: "The entity experience has a material basis that can be reasonably conceived of as a physical communication, fabricated by an autonomously operating part of the percipient's mind, either on its own or in liaison with an external agent; expressed in the same encoded-signal form as any other mental communication; presented to the conscious mind as a substitute for sensory input from the real world; and occasionally being given a temporary external expression utilizing some kind of quasi-material psi-substance." [131]

6. Jenny Randles used Rupert Sheldrake's morphogenetic field idea to formulate a theory of UFO abductions: "If something becomes accepted as real, then it gains more and more actual reality. It would not be stretching Sheldrake's hypothesis too far to regard abductions as becoming real, because of their repeated emphasis within society." [132]

However, if we posit beings similar to ourselves with more highly developed, inborn psychic powers, then we can formulate an explanation that accounts for a number of empirical observations. For beings with inborn psychic powers, it would be natural to interact with humans using such powers. Such interactions might stimulate latent psychic abilities in human beings, thus accounting for poltergeist effects that follow UFO encounters. It is also conceivable that humans with "psychic track records" might be especially compatible with the alien entities, thus explaining why people who report UFO encounters often turn out to have such track records.

This brings us to the second hypothesis: UFO entities and all their paraphernalia are psychic projections of the human mind. This idea is related to Kenneth Ring's concept of an objectively real imaginal realm that is somehow connected with human thought (see page 156). Such an imaginal realm would generate both psychic and UFO experiences in sensitive people, and it might also create actual physical effects.

One could hypothesize that the imaginal realm is a product of individual human thoughts. This idea is quite popular and could be called the psychic hypothesis. The table on page 167 lists six different formulations of this hypothesis by popular UFO writers. (These are not necessarily the only UFO theories advanced by these writers.)

One could also propose that the imaginal realm exists as a collective "mind at large" that is independent of individual minds and may even be the underlying cause of the human mind.

To me, it seems doubtful that ordinary humans have the power to call into being flying objects that can reflect radar, chase jet planes, and interfere with automobiles. If human imagination has so much power, then why don't typical sci-fi movie monsters materialize in American cities? If an imaginal realm is the cause of UFO phenomena, then it is much more powerful than individual human minds. Since UFO phenomena only partially follow popular culture, such an imaginal realm must also be independent of the contents of our individual minds.

An independently acting imaginal realm that can produce physical effects is the equivalent of a transphysical world that interacts with the world of matter. Such a world might have its own inhabitants which are capable of visiting our world. If that transphysical world is the source of the physical world, then its humanoid inhabitants might be closely related to the human inhabitants of this world. In this way, the concept of an imaginal realm can be related to the idea that the UFO humanoids may be objectively real beings that have psychic powers and are fundamentally similar to ourselves.

Here I should note that psychic phenomena and UFO phenomena are both reported to violate the known laws of physics. Thus, when I say that the UFO entities may be similar to us, I do not mean to suggest that they are simply molecular machines, made of matter as we ordinarily think of it. Human beings may also be something more than molecular machines.

5

Contactees, Channels, and Communications

In this survey of the reported UFO phenomena, there are three additional topics that we will need to discuss: contactees, channeling, and UFO-related communications and doctrines. These have been part of the UFO scene since at least the late nineteen forties, and so it is difficult to disregard them. Yet they also involve a great deal of material that sounds highly implausible, and a natural response is to dismiss it out of hand.

However, there are good reasons for giving this material a careful appraisal. Some of it is clearly fraudulent, and we should be aware of this. Some of it may also be genuine, and it may shed important light on the nature of the UFO phenomenon. The unifying theme of the three topics in this chapter is that of detailed communication between human beings and ostensible nonhuman beings. There is great opportunity for fraud here because the idea of such communication appeals deeply to the human mind and therefore allows for many forms of exploitation and self-delusion. At the same time, detailed communications, when genuine, can reveal a great deal of useful information about the communicators.

I will begin with the topic of UFO contactees. In recent years, the study of UFO abductions has become respectable for many ufologists, even though it is largely ignored or rejected by the scientific community. However, there is another kind of reported close-encounter case that tends to be rejected even by established UFO researchers. This is the so-called contactee case, in which a person known as a contactee meets with beings from other worlds on a friendly basis. Contactees may claim to have been selected by these beings to carry their message to humankind, and sometimes they claim to have been taken on visits to other planets in spaceships.

The bad reputation of the contactees became established in the early 1950s, when a number of people began to publicly promote extraordinary extraterrestrial contact stories that were backed up by very little evidence. UFO researcher Richard Hall sarcastically characterized the typical early contactee as "a technician tinkerer, typically male, 40 to 60 years old, from a troubled or disrupted childhood, poorly educated and needing a ghost writer for his book." [1]

Typically, these men presented themselves as specially chosen prophets, and in some cases they tried to impress the public by falsely adopting grandiose titles and academic degrees. They could easily be dismissed as alienated incompetents who took to dishonest means to earn some money, make a name for themselves, or overcome some psychological imbalance.

Here is a brief list of some of the prominent contactees of the 1950s and 60s:

1. "Professor" George Adamski was an amateur philosopher and hamburger stand grill cook who claimed to have met Orthon, a man from Venus, on November 20, 1952, near Desert Center, California. He took pictures of Venusian spaceships that are widely regarded as hoaxes.[2]
2. Truman Bethurum claimed to have met a space woman named Aura Rhanes from the planet Clarion, which is on the other side of the sun. He was widely denounced as a charlatan, and Dr. Edward U. Condon went to the trouble of proving in his Condon Report that Clarion could not exist.[3]
3. Daniel Fry wrote that he was "recognized by many as the best-informed scientist in the world on the subject of space and space travel." He claimed to have met Space People who were descendants of the ancient civilization of Lemuria.[4]
4. Howard Menger served in the U.S. Army and subsequently ran an advertising and sign-painting business. He claimed to have had many contacts with Space People from Venus, Mars, Jupiter, and Saturn, starting at the age of ten. He said that he had lived on Venus in a previous life.[5]
5. George Van Tassel hosted the Giant Rock Space Conventions, which were highly popular between 1954 and 1970. He claimed to be in contact with the "Council of Seven Lights," which rules this solar system.[6]
6. Orfeo Angelucci was an uneducated but intelligent Italian-American who had UFO experiences of a highly mystical, religious nature His experiences were interpreted by the psychologist Carl Jung as sincerely reported products of subconscious mental processes.[7]

Jung, by the way, is known for his interpretation of UFO experiences as psychological projections. However, it is not so well known that he thought some UFOs were real objects. He wrote: "So far as I know it remains an established fact, supported by numerous observations, that Ufos have not only been seen visually but have also been picked up on the radar screen and have left traces on the photographic plate. . . . It boils down to nothing less than this: that either psychic projections throw back a radar echo, or else the appearance of real objects affords an opportunity for mythological projections."[8]

Although Jung treated Angelucci in a sympathetic fashion, most authors who write about the contactees simply dismiss them with contempt and ridicule. I suspect that such blanket dismissals may be naive and unjust in some cases, since real-life stories often turn out to be more complex than one might expect. Nonetheless, many of these men probably were trying to gain money or followers by promoting false claims. Such promotional activity is unfortunately still going on today, and so-called contactees are running mail-order businesses advertising such items as:

1. A Nuclear Receptor, based on extraterrestrial technology, that absorbs negative energy and transforms it into harmonious frequencies ($100).
2. "Readings designed to attune Light-Workers and Star People to their Individual Mission and Purpose for being in Earth Embodiment" ($125).
3. A videotape by a 36,000-year-old Space Commander that explains "the brutal, inter-universal war that has been going on between the Universa Federation and the Negitarian Confederation for thousands of years" ($19.95).

I have not investigated these particular ads, and thus I cannot insist that they are phony. But the overall impression conveyed by this material—and there is a vast flood of it—is that much of it consists of "con games" intended to separate fools from their money. This does not enhance the credibility of persons who claim to make contact with unknown beings. However, the fact that some people promote bogus stories does not imply that all claims of otherworldly contact are false. Each unusual claim has to be evaluated on its own merits.

The Adamski Case

Some contact stories may be deliberate hoaxes created from the beginning for commercial motives. In other cases, a person may try to exploit a genuine experience and later make dishonest additions to his story to enhance its commercial value. Or he may sincerely believe his story, become emotionally committed to it, and then try to embroider it to make it more convincing. In such cases we would expect to see evidence for a real experience overlaid by evidence reflecting increasing dishonesty and self-delusion.

The case of George Adamski may be an example of this. There is a sympathetic discussion of Adamski written by Lou Zinsstag, a niece of the psychologist Carl Jung. Zinsstag argued that some of Adamski's early contact experiences were genuine but that later on he somehow became misled and falsely claimed such things as having made a trip to Saturn to attend "the Twelve Counsellors' Meeting of our Sun System."[9] She also pointed out that he had a mail-order scheme in which people would send him a photo, their birthdate, and $5.00, and he would give them a reading telling them which planet they came from.[10]

Interestingly enough, Zinsstag said she had personally seen evidence indicating that Adamski possessed unusual psychic powers, and she noted that he had resorted to trance mediumship to contact space beings.[11] She speculated that in later years he may have been led astray by disinformation, implanted either by inimical psychic communicators or by inimical humans with expertise in hypnotism.[12]

Adamski's own developing egotism may also have played a role. Although Zinsstag was a leading member of Adamski's group in Europe, she later rejected him, saying, "he now wants co-workers who implicitly believe in him like in God. This is something I can't do." [13]

A similar appraisal of Adamski was made by the UFO researcher Ray Stanford. William Mendez, who investigated the famous Pascagoula case, was told by Stanford that he had known Adamski. Stanford said that "Adamski made up all that Venus stuff (and more) but at one time he really did have a flying saucer experience and *that,* in Adamski's mind, justified making up stories about them." [14]

From Abductee to Contactee

If abduction and contactee stories could be cleanly separated into two distinct categories, then one could simplify things by leaving aside the contactee stories. Unfortunately, however, this is not possible. Practically any feature that is found in contactee stories can also be found in some abduction accounts, and there seems to be a continuum of scenarios ranging from typical abductions at one end to contactee cases at the other.

For example, Betty Andreasson reported an encounter in which she was taken on board a UFO by "Gray" aliens in 1967 and subjected to a harrowing physical examination. She also recalled having striking religious experiences during this encounter, and she reported being told by the aliens that they had chosen her to "show the world." [15] Many features of her story are typical of UFO abductions, but the feature of being chosen as a messenger or prophet is reminiscent of contactee accounts.

The story of William Herrmann also shares features typical of both abduction and contactee cases. This case was studied by the UFO researcher Wendelle Stevens, and I will present information taken from his written account and from a videotape he produced featuring interviews with Herrmann.[16]

Unlike contactees who extensively exploit their stories, Herrmann seemed to regard his UFO experiences simply as impediments to his normal life. He is a fundamentalist Christian, and his involvement with UFOs has apparently created serious difficulties for him with his fellow church members. In discussions of his experiences, he has mainly expressed bewilderment and a desire to understand what happened to him. He has also insisted that he did not believe in UFOs or have any interest in them prior to his UFO experiences.

Herrmann reported that he was abducted on March 18, 1978, near Charleston, South Carolina, by beings identifying themselves as Reticulans. He described these beings as short, hairless, and large-headed, with slitlike mouths and small noses, and he said that they brought him on board their craft by hitting him with a blue beam of light. He then became bewildered, and his next clear memory was of lying on a table in the presence of three of the beings. After being given a tour of the ship and seeing all kinds of incomprehensible machinery, he was returned to the earth in a state of terror fifteen miles from his starting point.[17] At that point, he lost all memory of his experiences on board the UFO, and he recovered these memories later with the aid of hypnosis.

Thus far Herrmann's story followed the standard abduction scenario. He testified, however, that the Reticulans later began to transmit complex messages to him through automatic writing, and they also completely unblocked his memory of the abduction.[18] Subsequently, he entered into a friendly relation with these beings, and he was taken on board their craft voluntarily.[19] He recalled without hypnosis that they took him on a ride down to the Rio Salado in Argentina, and then back north to Florida, where they showed him the Manned Space Complex. This part of his story is typical of contactee cases.

The reception of messages through automatic writing is an example of a process, popularly known as channeling, in which a person writes or speaks material that he does not recognize as coming from his own mind. Such material is often thought to emanate from some other being, who is acting as an information transmitter, but it may actually originate in the mind of the channeler.

Channeled communications have frequently taken place in contactee cases, and thus the automatic writing produced by Herrmann is a link between his case and cases of this kind. I will discuss the content of some of the messages he produced later in this chapter (see pages 180–85).

Another report combining features of abduction and contactee cases involved Filiberto Cardenas, a Cuban immigrant living in Hialeah, Florida. This case was investigated by a lawyer and UFO investigator named Virgilio Sanchez-Ocejo, and my information on the case comes from his account.[20]

In Cuba, Cardenas had studied physical therapy and had become an electrocardiac technician. He joined the Cuban Army, found himself on the wrong side of Fidel Castro's Communist revolution, and wound up spending nine years in prison. On being freed from prison, he emigrated to the United States, where he worked at various jobs and ran a gift shop and later a gas station.

On the evening of January 3, 1979, Cardenas, his friend Fernando Marti, and Marti's wife and 13-year-old daughter were driving around on the outskirts of Hialeah, looking for a pig they could buy for a roast. They were unsuccessful, and on the way home their car engine quit.

The two men testified that the lights and starter wouldn't work, and so they got out and began to look under the hood. At this point, they suddenly saw red and violet alternating lights reflecting off the engine and heard a sound "like many bees." The car began to shake, the light turned to a brilliant white, and Fernando began to crawl further under the hood for protection. Meanwhile, Filiberto felt paralyzed, and he began to rise into the air, shouting, "Don't take me. Don't take me."

Fernando saw him rising up, and by the time he got out from under the hood, all he could see was a "bulky object that ascended and then moved away." [21]

The next thing Filiberto remembered was being nearly run over by a car on the Tamiami Trail about 16 kilometers from where he had been lifted up. The police were sufficiently puzzled by the story that they listed the Type of Offense as "close encounter of the third kind" in their official report.[22]

Under hypnosis, Filiberto initially refused to say what happened during the abduction because "they told me not to say anything." [23] Later he recounted a strange and elaborate story that began when he awoke to find himself sitting, paralyzed, in the presence of a robotlike being and two small men in tight-fitting suits.

One of the men tried to speak to Filiberto in German, English, and finally Spanish, and he rotated a button on the side of his chest each time he switched languages. Filiberto was given an examination that he said left 108 marks on his body, and then he was taken into the presence of an individual who was seated on a high throne and who wore a cape and a chain from which hung a triangular stone. This personage spoke to him at length, both telepathically and in perfect Spanish, and showed him many remarkable scenes displayed on the walls.[24]

Filiberto said that the alien beings looked quite human. They had elongated eyes with eyelashes, small flattened noses, long lipless mouths, and light beards. They also wore a symbol on the right of their chest, consisting of a serpent on a lazy X.[25]

The story becomes even more extraordinary: The beings proceeded to take Filiberto to an undersea base, traveling beneath the sea at high speed through a tunnel of "firmed water" that seemed to open in front of the craft so that water did not touch it. At the base, he met a human who was working with the aliens, and he was led through what seemed like a city. He was again paralyzed and examined, and a semen sample was taken. Afterwards, another caped, enthroned figure gave him instructions illustrated with images on banks of TVs. After many similar experiences that seemed to go on for many days, he was dropped off near the Tamiami Trail after a lapse of about two hours in local earth time.[26]

This could be called a story of not enough missing time. The story is certainly difficult to believe, but there is no need to suppose that it must be completely true or completely false. It is possible, for example, that Filiberto Cardenas was actually carried off into the sky as Marti testified. But the experiences he related under hypnosis may have been partially generated by his own mind. Or they may have been projected into his mind by the agency that carried him off.

As in the Herrmann case, there was a second meeting with the aliens. On this occasion, Filiberto and his wife Iris voluntarily walked up a ramp into the alien ship and had a friendly conversation with its nearly human occupants. They were subsequently able to directly remember this experience, with no need for hypnosis.[27] This kind of voluntary meeting on an alien ship is typical of contactee stories, but for two witnesses to participate in such a meeting is unusual.

Although the Cardenas case has many features typical of contactee cases, it also has many standard features of UFO abduction reports. These include stories of implants that could not be detected medically, stories of crosses between aliens and humans, and stories of psychic phenomena that followed the abduction.[28] Of course, it also includes the dramatic abduction itself, which was confirmed by three eyewitnesses.

The Quality of UFO Communications

I will refer to information people have reported receiving from UFO entities as "UFO communications." As we saw in the Cardenas case, this information may be presented during an abduction in the form verbal statements by the entities and elaborate visual imagery. After an abduction, information may also be transmitted through automatic writing or unusually vivid, forced dreams that are attributed to the entities. If we make a survey of UFO communications, we find that they tend to contain a large proportion of misleading or completely false information, mixed with material that may be true. This tendency towards falsehood is consistent with a hypothesis of alien deception, and in this section I will illustrate it with a number of examples.

One example involves a story related to UFO investigator Jacques Vallee by a woman he called Helen.[29] Helen was traveling with three friends from Lompoc, California, to Los Angeles in the summer of 1968. While driving in a flat, open area at about 3:00 a.m., all four persons saw a white light in the sky that moved in an erratic fashion and approached their car. As it approached, they saw that it was a white, glowing object with a width of about six freeway lanes. It swooped over the car and projected four funnel-shaped lights over the bodies of the four witnesses. This caused them to separate from their bodies and float out of the car, which apparently continued down the road. Vallee said that he got in touch with two of the other witnesses, and they independently confirmed this part of the story.

Under hypnosis, Helen recalled being taken on board the UFO and meeting a man dressed in white who showed her an amazing motor. She became determined to build this motor. In fact, this became the central concern of her life, and she initially approached Vallee in order to enlist his help in building it. But Vallee pointed out that the motor, as the woman described it, is completely unworkable.

A similar story involves the UFO witness named Sara Shaw, whose case was investigated by Ann Druffel and the parapsychologist D. Scott Rogo.[30] The story began with a frightening experience involving missing time in a lonely cabin in Tujunga Canyon near Los Angeles. After this experience, Sara became interested in medical knowledge and took a job in a hospital. While working there, she saw a method of curing cancer in a sudden revelation, which seemed to her to come from some source outside herself. As in the case of Helen and the motor, Sara became determined to reveal this cure to the world.

When her experience in Tujunga Canyon was investigated using hypnosis, a classical UFO abduction scenario emerged. In addition, Sara reported that she was told about the cancer cure while on the UFO. Unfortunately, the cancer cure, which involves injecting vinegar into cancerous tumors, is an old folk remedy for cancer that doesn't work.

Now one might argue that the cancer cure really came to Sara through partially forgotten knowledge of the folk remedy and that the UFO story was simply a creation of her subconscious mind under hypnosis. However, this does not explain her preoccupation with this cure and the fact that her fascination with medical knowledge began immediately after her Tujunga Canyon experience.

It is also curious that Sara's story involved seeking out a physician to tell about the cure. At a certain point, she realized intuitively that a certain Dr. Allini was the right physician. When she approached him with the cure, it turned out that he was indeed receptive to studying it. In fact, he said he had already heard about it from a local man who claimed to have received it from UFO entities.[31] So we seem to have two independent stories in which the same ineffective cancer cure was allegedly promoted by beings from UFOs.

Now why should beings flying about in high-tech vehicles cause people to develop overriding interests in impossible motors and ineffective cures? Whatever the reason, there is evidence suggesting that such beings sometimes give quite elaborate presentations of nonsensical information. An example of this is the abduction case of William Herrmann (see pages 175–76).

Technical Gibberish

Herrmann claimed that in 1979 the beings contacting him identified their point of origin as the stars Zeta1 and Zeta2 Reticuli.[32] Apparently, they did this by transmitting information to Herrmann through the medium of automatic writing.

At that time, these stars were quite famous in UFO circles as a result of the celebrated star map reported by Betty Hill. The Hill abduction occurred on September 19, 1961, and Betty Hill recalled dreaming of a star map on a wall in the UFO. This map supposedly included the aliens' home star. Betty Hill first drew this star map in 1964 while under hypnosis. In 1966, Marjorie Fish, an intellectually brilliant grade-school teacher, began an effort to model the star patterns in the vicinity of the earth and identify the pattern represented by Betty Hill's map. In the early autumn of 1972, she concluded that the home base on the map must be Zeta1 or Zeta2 Reticuli. This finding was written up in the July 1973 issue of *Saga* and the January 1974 issue of *Pursuit*.[33] It was discussed in *Astronomy* magazine in the issue of December 1974.[34]

How did Herrmann come to mention Zeta Reticuli? He was a born-again Christian and an auto mechanic, and he swore that he had no interest in UFOs before his close-encounter experiences. If this is true, it is unlikely that he had heard of Betty Hill's star map. However, it is possible that he may have heard of the star map in conversations with UFO investigators after his abduction in March of 1978. One can then hypothesize that the information could have emerged from his unconscious mind during his automatic writing.

However, Herrmann's messages from the Reticulans do have a number of strange features that are hard to account for by the hypothesis that they are entirely produced by his mind. Here is an extract from one of the messages:

Reticulan Technology

Propulsion Evolutionary-Hypothesis:

> A combination of gravity equilibrium manipulation by electro-magnetic energy-mass conversion within a unified field of positive and negative particle beam fusions . . . using kinetic energy and harnessed static electricity a conversion takes place that increases the energy flow into the electro-magnetic wave cohesive force chamber . . . thus resulting in action/reaction basis of fluctuation. The manipulation effect is maintained by continual increase and decrease of the electromagnetic wave MPS (manipulation per sequence).[35]

Wendelle Stevens observed that this kind of statement is completely out of character for Herrmann, and it is not to be expected from a person of his educational background. Possibly it involves more than Herrmann's own unconscious mind.

At the same time, the message does not look like a genuine communication of technical information. If one wanted to transmit technical knowledge using this kind of vocabulary, the only rational way to do it would be to make step-by-step definitions of terms that would be intelligible to the intended audience, and that is not what we see here. The message is reminiscent of Helen's impossible motor or Sara's unworkable cancer cure. It would appear that, for some reason, gibberish was being transmitted to Herrmann. One can also postulate that a meaningful message was being garbled in the process of being transmitted through Herrmann's mind. But intelligent beings responsible for the transmission would be expected to know about such distortion and be able to correct for it.

Some of this gibberish makes use of technical knowledge of the kind one could look up in various reference books. For example, in one communication there are complicated mathematical formulas and a reference to "ORBITAL ECCENTRICITY: 0.0167."[36] In fact, according to astronomy textbooks, the eccentricity of the earth's orbit is 0.0167.[37]

Unless Herrmann studied astronomy, it seems unlikely that he would have run across this datum. In his writing and in videotaped interviews, he comes across as a sincere person who would be unlikely to consult textbooks to fabricate a phony story. At the same time, it seems doubtful that alien space travelers would use this particular three-significant-digit eccentricity figure in their calculations. We are thus left with the alternatives of fraud on Herrmann's part or fraud on the part of beings communicating with him.

Some of the technical points in the messages clearly were not taken from current textbooks. For example, in the mid 1970s, astronomers maintained that Zeta1 and Zeta2 Reticuli are 36.6 light-years from the earth.[38] In contrast, the communications to Herrmann repeatedly mention a distance of 32 light-years.

Another curious point is that Herrmann's communications from the Reticulans refer repeatedly to an organization they call "the Network." Now, the word *reticulum* means "network" in Latin. It would seem that whoever created Herrmann's Reticulan story may have been indulging in bilingual puns. This seems a bit strange either for a South Carolina auto mechanic or for aliens from another planet.

There is a curious story behind the phrase "MPS (manipulation per sequence)" in the Reticulan message quoted above. This phrase showed up in notes dated 9/10/85 on a telephone interview between the UFO investigator James McCampbell and the physicist Paul Bennewitz.[39] Bennewitz was investigating UFO activities in the Albuquerque, New Mexico, area, and some have said that he was driven off the deep end by UFO disinformation spread by government agents (see pages 110–15).

In the notes presented by McCampbell, Bennewitz referred to the term MPS, saying that for alien craft "the MPS (manipulations per sequence/second) changes its frequency on a periodic basis."[40] Compare this to Herrmann's reference to "continual increase and decrease of the electromagnetic wave MPS (manipulation per sequence)."

There are other apparent coincidences between the statements of Bennewitz and Herrmann. Herrmann said he observed the Reticulan vehicles moving in triangular patterns, and he said that the Reticulans explained to him during his abduction that they did this to avoid harmful effects from U.S. military radar.[41] Bennewitz maintained that he had photographed UFOs as they flew in triangular or square patterns, making acute-angled turns in a 20th of a second. He also said that high-powered radar can interfere with these UFOs.[42] He said that some of the aliens might come from Zeta Reticuli, and he mentioned that they came from distances "up to and larger than 32 light-years away."[43] Like Herrmann, he also said that they have a federation called "The Network."[44]

This information seems to solidly link Herrmann with the material ascribed to Bennewitz. Some possible explanations of this are: (1) Bennewitz or some disinformer in contact with him copied from Herrmann (whose statements preceded Bennewitz's), (2) Bennewitz and Herrmann were both victims of the same group of disinformers, or (3) there is some connection between Herrmann's abductors and the aliens discussed by Bennewitz. It is difficult to say which alternative is correct.

Throughout this chapter, one hypothesis that has always been in the background is that material in ostensible communications from UFO entities is really being transmitted through human society by ordinary means. Let us look more closely at the idea of radar-induced UFO crashes from this point of view.

The history and possible genesis of the radar-induced-crash story is a bit difficult to unravel. An article in the Kansas City newspaper *The Wyandotte Echo* of January 6, 1950, gave a version of the radar interference explanation of saucer crashes. The article said that "since they seem to invariably crash near radar installations, it is surmised that they are attracted by radar, or possibly radar waves interfere with their control systems." [45] According to William Moore,[46] the story in *The Wyandotte Echo* can be traced to friends of Silas Newton, who was the source of information for Frank Scully's controversial book *Behind the Flying Saucers*. This book came out in 1950 and discussed a UFO crash that supposedly took place in 1948.

The radar story also appeared in a memo supposedly sent to the director of the FBI from Guy Hottel on March 22, 1950. This memo gave a somewhat artificial sounding description of three 50-foot-diameter flying saucers, each containing three humanoid bodies, that had been recovered by the Air Force in New Mexico. It went on to say:

> According to Mr. . . . informant, the saucers were found in New Mexico due to the fact that the government has a very high-powered radar setup in that area and it is believed that the radar interferes with the controlling mechanism of the saucers.
>
> No further evaluation was attempted by SA (Deleted) concerning the above.[47]

Moore also claimed that this memo can be traced back to *The Wyandotte Echo*, although this seems doubtful since the newspaper article speaks of two flying saucers with two bodies apiece. [48] In any event, it seems that the radar-induced-crash story goes back as far as 1950.

During Herrmann's first UFO abduction in 1979, his captors reportedly told him that some of their space ships were sensitive to radar. Apparently, some of their ships went out of control and crashed because radar interference damaged their on-board computers. The entities told Herrmann that this had last happened about 30 years previous to the date of his abduction.[49] Since the abduction occurred in 1979, this means that the last crash was in 1949. This ties in Herrmann's radar story with the stories relating to UFO crashes in about 1948.

Did Herrmann's story come to him from these earlier stories by ordinary means of communication? The story is definitely obscure, and we would have to suppose that Herrmann heard it from some irresponsible UFO investigator (or some such person) and then falsely incorporated it into his own account. Or we would have to suppose that, contrary to his testimony, Herrmann really had done considerable reading of UFO literature.

The radar-induced-crash idea came up in another close-encounter story. On December 3, 1967, a police officer named Herb Schirmer saw a strange, lighted object on the road ahead of him at 2:30 a.m. When he flashed his high beams at it, it took off, and Schirmer reported seeing a flying saucer. This came to the attention of the Condon committee, and arrangements were made to hypnotize Schirmer. Hypnosis revealed a complex experience in which Schirmer was met in his car by beings who took him aboard the UFO. There the beings told him many bizarre-sounding things, including that their craft operated by reverse electromagnetism, that they drew power from water reservoirs, and that their ships had been knocked out of the air by radar.[50] Jacques Vallee, interestingly enough, regards this as disinformation by the UFO beings.

To make things worse, Schirmer said that the beings were wearing coveralls with an emblem of a winged serpent.[51] Likewise, William Herrmann reported that the beings he saw had a metallic figure worked into the fabric on the left breast of their one-piece jumpsuits. It was an image of a winged serpent.[52] Filiberto Cardenas and his wife reported seeing an emblem of a serpent on a lazy X on the right breast of the jump suits worn by their captors.[53] Also, the psychologist John Carpenter mentioned the case of a 29-year-old abductee who had not read any UFO books but who recalled captors of the "Gray" type wearing "a tight-fitting uniform with an emblem on the chest depicting a winged serpent."[54]

Are serpent emblem stories also circulating and being incorporated into the UFO tales of supposedly honest witnesses? Why would anyone want to adopt these pointless stories and lie about them? It could be argued that people hear them, forget them, and then bring them forth from their subconscious minds. But why do such arbitrary stories have such an effect on the subconscious mind that they can override people's power of discrimination between imagination and reality?

Another radar story comes up in the case of Staff Sgt. Charles L. Moody of the U.S. Air Force, who had a UFO close encounter on August 13, 1975, near Alamogordo, New Mexico (page 220). Over a period of two months, Moody gradually remembered a typical abduction by beings of the classical "Gray" type. Among other things, he reported being told by these beings that radar interferes with their navigational devices.[55]

It is impossible to know for sure how these stories are being transmitted. Some of them may be literally true, but those that are false are not necessarily due to manmade lies and delusions. Vallee's option is also a possibility. For example, it is conceivable that a manmade radar story from 1950 might have been transmitted to Herrmann, Schirmer, and Moody by actual nonhuman beings, perhaps as part of their own disinformation plan. This would be consistent with the use of constants from astronomy textbooks in communications to Herrmann.

One might ask why humanoid beings would want to spread disinformation about themselves. One possible answer is that the function of disinformation is to make a subject unbelievable. If "they" want to conceal their activities, then the spreading of ridiculous stories about themselves is a very practical way to achieve this.

To sum up this subsection, one can always dismiss Herrmann as a fraud or a suggestible victim of human manipulators. But there is also the possibility that he was telling a genuine story of his experiences. Perhaps Herrmann did have an encounter with strange beings riding UFOs. If so, it appears that these beings may have presented him with nonsensical communications using material—some of it very obscure—that was borrowed from earthly culture.

The Theory of Genetic Intervention

The material I have just discussed gives us some idea of the quality and level of truthfulness of UFO communications. In this section and the next, I would like to discuss some of the particular themes that come up repeatedly in this material. I will begin by considering the story that extraterrestrials created modern humans by mating with existing primitive people on the earth. The theory that humankind was created by some kind of extraterrestrial genetic manipulation comes up repeatedly in UFO-related communications, and it is related to stories of present-day genetic manipulation of humans by UFO entities.

According to some scholars, the genetic intervention theory can be found in ancient Hebrew and Sumerian texts. For example, the geologist Christian O'Brien argued that these texts describe a race of beings called Shining Ones—his translation of the Hebrew word Elohim.[56] These beings created modern humans from earlier human forms by genetic manipulation. Some of these beings, called Watchers, mated with humans, and this was considered a crime by the leaders of the Shining Ones. O'Brien argued that the Shining Ones were superior but mortal beings of unknown origin.

The Israeli scholar Zecharia Sitchin used ancient Sumerian and Babylonian texts to argue that modern human beings were created by space travelers called the Nefilim, who then mated with them and quarreled over what to do with them. According to Sitchin, the Nefilim created humans by genetically modifying *Homo erectus*.[57]

O'Brien and Sitchin based their ideas on ancient Near Eastern texts, but in 1950 Pope Pius XII arrived at a very similar idea in what appeared to be an attempt to reconcile evolution with the Bible. He decreed that it is acceptable to Catholics that the human body evolved from other living matter already in existence. But he held that Catholics must believe that present humans are descended from Adam and Eve, since otherwise there could be no doctrine of original sin. This implies that Adam and Eve resulted from divine intervention, but that all other organisms, including ape-men, evolved in the Darwinian fashion.[58]

The genetic intervention theory has shown up in the stories circulating about UFOs and the U.S. Government. The American UFO researcher Linda Howe claimed that a version of this theory was part of a "presidential briefing paper" shown to her by AFOSI agent Richard Doty. (AFOSI stands for Air Force Office of Special Investigations.) According to this paper, the extraterrestrials in contact with the U.S. Government have come to the Earth at various times to manipulate

DNA in existing terrestrial primates. Supposedly, this was done 25,000, 15,000, 5,000, and 2,500 years ago. In addition, two thousand years ago extraterrestrials created a being that was "placed on this earth to teach mankind about love and non-violence."[59]

Doty has publicly denied showing such a document to Linda Howe, and the genetic manipulation story belongs to the realm of rumors and disinformation.[60] But where did the story originate?

The genetic intervention theory also surfaces through a variety of UFO-related channeled communications. This might indicate that the idea has a strong grip on people's minds and therefore tends to emerge from the subconscious. Or it may be that it is actually being communicated to channelers from a source external to themselves. A combination of these possibilities could also be true.

Here is an example in which the theory emerges from a channeled communication. The medium Carla Rueckert produced elaborate trance communications that purportedly came from the Ra entity, a "social memory complex" that had visited the earth in spaceships in the days of ancient Egypt. Ra's story of human origins can be summed up as follows:

> War on Mars caused that planet to become inhospitable and its human population died. The "Yahweh" group produced humans of modern type on the earth 75,000 years ago by cloning genetic material from the dead Martians. The first modern humans on earth appeared at this time; half were from the Martians, half developed from barely erect native bipeds, and a quarter came from other planets.[61]

The Ra communications stated that the Yahweh group was a task force of advanced extraterrestrials. The genetic intervention theory presented here is similar to others we have seen, but there are differences, such as the reference to Martians. This is typical of UFO and channeled communications. Certain themes come up again and again, but the stories all tend to differ in detail.

The genetic intervention theory dates back at least to the early 1950s. At that time Ralph M. Holland, an engineer living in Cayahoga Falls, Ohio, said he was in touch with humanlike space travelers that he called Etherians. These beings claimed to live on an etheric plane of existence. According to Holland, they told him that they had created the human race in the following manner:

> When these groups first came to the physical plane of your planet, they found that their physical bodies were not entirely suited to the environment. In an effort to improve the situation, they began, by selective breeding and cross-breeding, to develop a better adapted physical body. The final choice was the ancestor race of the present Adamic races, which was a cross between the Elder Races themselves and a certain manlike animal native to your planet.[62]

Linda Howe has presented an abduction case in which the alien intervention theory shows up. This case involved a New Jersey woman named A. Allen, who is black and American Indian. She remembered under hypnosis having an encounter with a 7-foot-tall male being who had eyes with vertical slit pupils.[63] Howe stated that this woman "believes that *Homo sapiens* were originally created to be someone else's work force on Earth to mine minerals and do physical labor for a tall race of beings who have been harvesting this planet for eons."[64]

When I asked her where the woman may have gotten these ideas, Howe said that they emerged during the hypnosis sessions probing her abduction. She said that the woman was not well educated, and she was not a reader of many books. However, the idea that humans were created as miners appears in Zecharia Sitchin's book *The 12th Planet*.[65] The miner detail may link Allen's story with Sitchin's book.

My last example of the genetic intervention theory is found in Raymond Fowler's book *The Watchers*. In that book, Fowler asked, "Was Cro-Magnon Man placed on earth intact or was he the result of a genetic transformation of Neanderthal Man by alien beings?"[66] He based this speculative idea on the evidence he has gathered for a genetic element in UFO abduction reports, combined with the well-known idea that Cro-Magnon Man abruptly replaced the Neanderthals. Fowler also noted the Biblical passages about the "Sons of God" who found the daughters of man to be fair and who mated with them, producing "mighty men, men of renown."

Fowlér called the extraterrestrials the "Watchers" and speculated that they have been concerned with the human race from its very beginning. He got this term from the contactee Betty Andreasson, who was extensively studied by Fowler, and who said that her abductors referred to themselves as "Watchers."[67] The genetic manipulation of humans is a prominent theme in Andreasson's abduction accounts.

All of these versions of the genetic intervention theory share common elements that are found in human cultural traditions—in this case traditions recorded in Biblical apocrypha and Sumerian mythology. It is puzzling that this theory keeps coming up in alien contact stories, but here are some possible reasons for this: (1) This happens because the genetic intervention theory has a strange psychological attraction that induces people to imagine being told about it by alien beings. (2) It happens because a cabal of sinister disinformers is spreading the theory. (3) It happens because UFO entities are taking the genetic intervention theory from human culture and using it for their own program of conditioning human society. (4) The theory is a true picture of our origins, and the entities are presenting it to us as such.

There is some support for option (1). The genetic intervention theory is a neat compromise solution to the conflict between Darwinian evolution theory and Biblical creationism. For this reason it could have intellectual appeal to many people. But this still doesn't explain why people should say they experience hearing about it from alien beings.

Option (2) also doesn't explain why people should report such experiences. However, this is explained by options (3) and (4). Option (4) cannot be strictly true, since there are many differing versions of the genetic intervention theory, and they cannot all be correct. This leaves us with (3).

The case of Betty Andreasson provides another example of a UFO-related experience that is consistent with option (3). When hypnosis was used to probe her 1967 abduction, Betty recalled being taken in a UFO to a tunnel bored through solid rock. This tunnel led to a strange landscape with a view of an ocean, a distant city, and a pyramid surmounted by an "Egyptian head." She was conducted by two entities along an elevated track to a place where she saw a vivid enactment of the Egyptian myth of the Phoenix, a giant bird that consumes itself with fire and is then resurrected from the ashes.[68]

This experience had strong religious overtones, and Betty Andreasson is a fundamentalist Christian. However, the Phoenix story is not used by modern fundamentalists, even though it was used by early Christians. Thus the Phoenix motif may have been chosen by visiting entities, rather than by Andreasson's conscious or unconscious mind. In support of this idea, Fowler pointed out that the enactment of the Phoenix story recalled by Andreasson involved small details that are part of the original Egyptian myth but are not widely known (such as the fact that a worm emerged from the ashes, rather than a young bird).

Disasters and More Genetics

A common theme in UFO communications is that human beings are in danger of some terrible disaster caused either by nature or by their own actions. This theme tends to be intertwined with the theme of genetic manipulation. In this section, I will discuss these themes with the aim of gaining some more insight into the motives behind UFO communications and their possible sources.

Disasters involving the earth's atmosphere are mentioned repeatedly in UFO communications. For example, Whitley Strieber said he was shown "graphic depictions of the death of the atmosphere, not to mention the entire planet simply exploding."[69] William Herrmann said that his Reticulan contacts informed him that the earth's magnetic field was decaying and that radiation from space would soon wreak havoc on living organisms.[70]

The Ra communicator, in a more philosophical vein, spoke of a coming transition of the earth in which it would no longer be inhabitable by grossly embodied beings of the "third density." This involves a crisis attended by ruptures in the earth's "outer garment"—which presumably is the atmosphere.[71] In 1953, a medium named Mark Probert made the following statement in a trance communication on UFOs and their occupants: "Your present danger, mitigated for a time by the Guardians, lies in the progressive breakdown of the upper ethers, i.e., of the ionosphere."[72]

As we see in other UFO communications, these statements about the atmosphere have a surreal quality. They seem to be expressed more in dream symbolism than in objective scientific language. Probert's statement sounds the most realistic, although the ozone layer, which is now thought to be breaking down, lies below the ionosphere. The statement was made in 1953, well before the time of the controversies about the ozone layer in the early 1970s. The other statements about the atmosphere were, of course, made during or after this period.

Some feel that channeled communications such as Probert's are dubious and shouldn't be mentioned. Nonetheless, they may have an important bearing on communications received during UFO encounters because (1) they are often similar in content and (2) channeling takes place in some UFO contact cases. It is perhaps significant that many of the things mentioned in current UFO communications were also being mentioned in channeled communications in the early 1950s.

Two other examples of this are the genetic intervention theory and the idea that radar can cause UFOs to crash.

The dangers of man-made pollution and nuclear testing are frequently mentioned in UFO communications. For example, these topics came up in a close-encounter case that took place in May 1973 near Houston, Texas. The witness, Judy Doraty, was driving along with four family members. All five people remembered seeing a very bright light in the sky that paced their car. The family members remembered Judy pulling off the road and walking to the back of the car, then returning, getting back in, and complaining of thirst and nausea. She drove home with the light still following, and upon arriving they all watched it perform strange antics in the sky. They found that they had lost about an hour and fifteen minutes.

This time gap was filled in through hypnosis administered on March 3, 1980, by Dr. Leo Sprinkle, then Director of the Division of Counseling and Testing at the University of Wyoming. Sprinkle also tested the woman psychologically and said that she turned out to be completely normal. Under hypnosis, Judy Doraty related information communicated to her by entities on a UFO that she visited through an out-of-body experience. The entities were cutting up a calf, and they explained that they were doing this to monitor progressive pollution of the environment. The entities said that humans are going to destroy themselves through pollution, and they said that nuclear testing, including testing in outer space, is having very harmful effects on the earth.

Close-encounter witnesses often say they were warned about the dangers of man-made pollution, including pollution caused by nuclear tests. Of course, we all know that these dangers exist, and one explanation of Judy Doraty's testimony is that worries about pollution and nuclear testing were simply surfacing within her mind, perhaps as a result of being under hypnosis. However, there are curious features in this testimony that are corroborated by other UFO accounts. For example, the beings said that humanity's dangerous activities affect other unspecified beings:

> *Doraty:* It's like if we continue like we are now, it's going to involve not only us but possibly other . . . and they're trying to stop something that could cause a chain reaction. And maybe involving them. I don't know.

Sprinkle: Did they say what kind of chain reaction?
Doraty: No, only that it involves . . . we're not the only ones to be concerned.
Sprinkle: Do they say who else is involved?
Doraty: No.
Sprinkle: Do they talk about their origins, where they're from?
Doraty: That they're stationed here.
Sprinkle: On Earth?
Doraty: I don't know.[73]

This point also came up in a case studied by Dr. James Harder, a professor of civil engineering at UC Berkeley and a longtime UFO researcher. Pat Price, the main witness in the case, recalled sitting in front of a desk on board a UFO and conversing telepathically with the "leader." Here is part of the conversation, as recalled under hypnosis:

Price: Well, (pause) he drew me a circle, and he showed me some lines, and he told me "people can coexist—and not know it."
Harder: What kind of lines did he draw in the circle?
Price: Parallel lines.
Harder: What did he mean by this do you think?
Price: He said, "What we do, destructively, will affect them too." (sigh). I don't know what he was talking about—he just scared me.[74]

If our activities affect "them," one possible reason is that some of them might live here on the earth. This idea came up in the Doraty transcript, and it was mentioned explicitly in the testimony of Betty Andreasson. Here is a quotation from a hypnosis session in which alien entities seemed to be speaking through Betty Andreasson's voice, using her as a kind of channel or spirit medium. She spoke of many races of visiting beings that work cooperatively together and pointed out that some of these races live on this earth:

Interviewer: Betty, do they have enemies as we have enemies?
Betty: There is one planet that is an enemy, and also many men are enemies, only because they do not understand. . . .
Interviewer: Betty, are many of these clans or races visiting earth right now from many planets?
Betty: Yes . . . Seventy . . . races.
Interviewer: Do these races work together?

> *Betty:* Yes, except for the offensive one.
> *Interviewer:* They come from different planets, then? They don't come from the same planet? Is that correct?
> *Betty:* Some. Some come from realms where you cannot see their hiding place. Some come from the very earth. . . . Yes, there is a place on this very earth that you do not know of.[75]

The Andreasson testimony also presented the idea that pollution is going to cause serious harm to the human race. In one experience recalled under hypnosis, Betty had seen two fetuses taken from a woman abductee in a UFO. With horror she saw the aliens put long needles into the head and ears of one fetus and place it in a tank of liquid connected to a strange apparatus. The aliens gave her the following explanation for this:

> They're telling me they *have* to do this. And I'm saying, "*Why* do you have to do such a terrible thing?" And one of them is saying, "We *have* to because as time goes by, *Mankind will become sterile.* They will not be able to produce because of the pollutions of the lands and the waters and the air and the bacteria and the terrible things that are on the earth!"[76]

This statement ties in the pollution problem mentioned to Judy Doraty with the idea that an alien race is engaged in genetic experimentation with human beings. These two themes also come up in a report presented by Jenny Randles.[77] On February 5, 1978, in Medinaceli, Spain, a 33-year-old veterinarian named Julio was abducted, along with his dog, by tall, Nordic-looking entities. The entities examined Julio, taking samples of blood, gastric juices, and semen. They told him that their world is a dark spoiled place, and they want to study our wonderful life-filled world before we make a mess of things as they did. They also mentioned small, ugly entities who have a strange idea of biologically reprogramming humans.

The entities reported by Betty Andreasson seemingly correspond to these "small, ugly" beings concerned with genetic manipulations. However, she remembered an occasion in which she was taken to their world, and she described it as gray, dark, and hazy all the time.[78] This agrees with the statement of Julio's Nordic-looking entities, and one wonders if there are two dark worlds. Or perhaps there is a relationship between the two types of entities.

John R. Salter, who had an abduction experience on March 20, 1988 (see pages 144–45), also spoke of a dimly lit alien world. On January 9, 1990, he had a vivid dream in which he recalled being told that the beings who abducted him came from the Zeta Reticuli star system. On March 4, 1990, he had another vivid dream, which he perceived not as a recollection but as a direct telepathic communication from one of those beings. In this dream, he was visiting their world, and he noted that the light was very dim and the buildings were all white.[79] This kind of vivid dream is similar to a channeled communication in the sense that the subject receives information that he feels is coming from outside of himself. (Salter told me that he knew about Betty Hill and Zeta Reticuli before the dream of January 9, 1990, but he had only a slight acquaintance with the Hill case and no knowledge of the Zeta Reticuli story before his abduction experience in March of 1988.)

Another idea about the aliens' world was expressed by "Lucille Forman," a New York psychotherapist whose abduction experience was studied by Budd Hopkins. According to her testimony, her alien abductors, who were of the "Gray" type, come from a dying society that stresses intellectual development at the expense of emotional growth. Something has gone wrong with them genetically. Their children are dying prematurely, and they are engaged in a desperate struggle to survive. Hopkins tied this in with their interest in human genetics and reproduction.[80] There is also a tie-in with the dark world idea, although here the aliens' world is dark metaphorically, rather than literally.

Turning to another report cited by Jenny Randles, in Pudasjarvi, Finland, a woman named Aino Ivanoff was driving her car in the early hours of April 2, 1980. Suddenly, the car was surrounded by a mist, and she found herself in a room where she was examined on a table by small entities. These beings told her that war is evil and she should support peace groups. They also said they were unable to beget their own children.[81]

This fits in with Betty Andreasson's statement—on the other side of the Atlantic—that alien females cannot bear children and that human females are used to carry alien fetuses.[82] She gave this as the reason the aliens are concerned that human beings will destroy themselves: "The fetuses *become them*—like them. They said they're *Watchers* . . . and they keep seed from man and woman so the human *form* will not be lost."[83]

The italics in this quotation were supplied by Raymond Fowler. This is the statement that seems to tie in Andreasson's story with the old Hebrew story of the angels called Watchers that mated with human beings.

If we put together some of these communications, we get the picture of the "Gray" aliens as a parasitic race that depends on humans for reproduction and is concerned that humans may be on the verge of wiping themselves out. But then there are stories contradicting this. The communications reported by Lucille Forman and those from Julio's Nordic-looking entities do not exactly agree with this theory. Also inconsistent is the fact that abductions with a gynecological slant have been reported frequently only within recent decades. Why weren't the aliens' reproductive activities evident in the nineteenth century if they need humans in order to reproduce? Also, stories indicating that the aliens are from a distant star, such as Zeta Reticuli, are incompatible with the idea that they depend on earth humans for reproduction.

Conclusion

In conclusion, communications reported to come from UFO entities often contain certain standard themes. These range from disturbing statements regarding human genetics and origins to seeming trivia, such as the radar-induced-crash story. The communications often have a surreal quality, and they often contradict one another. Many seem to be a cross between disinformation and sheer nonsense, and many contain material found in human cultural traditions.

Hoaxes and delusions undoubtedly occur, and there is evidence suggesting that highly organized hoaxes have been perpetrated. There is also the possibility that sinister intelligence agents are spreading UFO disinformation. However, this does not mean that UFO contact stories are all products of human lies and delusions. We can always attribute them to these causes, but if we do so, I think we unnecessarily lower our estimate of apparently sane and responsible human witnesses. Ultimately, this lowers our estimate of our own ability to discriminate truth from illusion.

In the real world, we often encounter mixtures of truth and falsehood. It may be difficult to separate the true from the false, but I think it would be a mistake to dismiss a body of material just because it contains false elements. Indeed, we can turn things around and suggest that if a body of human testimony seemed to contain nothing false, then that would be contrary to human nature.

We can also turn around the theory that UFO tales are simply a phenomenon of folklore aided by deceit. It is possible that actual nonhuman beings are responsible for many reported UFO communications. These beings may be trying to condition people's thought processes by making use of themes taken, in some cases, from the people's own cultural traditions.

If this is so, then by manipulating traditional themes they add to the traditions. One can ask, To what extent are cultural traditions orchestrated by the intervention of various kinds of intelligent beings? Also, to what extent are cultural traditions true, and to what extent are they "disinformation" introduced—not by conniving priests and imaginative poets—but by transhuman sources? To what extent is this cultural orchestration good, and to what extent might it have bad effects?

PART 2

Vedic Parallels to UFO Phenomena

6

Transhuman Contact in Vedic Civilization

The reports of UFO sightings and contacts during the last 45 years have suggested to some that the human race is being contacted by intelligent beings that are not human but are surprisingly similar to ourselves. This similarity is so great in many cases that the term "alien" seems to be a misnomer. Yet these beings do seem to be aliens in the sense that there is alienation between them and us. The whole subject of UFO contacts and encounters is filled with secrecy and disinformation, and it appears that this cannot all be blamed on the U.S. Government. UFOs seem to behave in an elusive way, and communications with UFO entities are ambiguous and contradictory. They seem intended to influence human society from a distance, without establishing relationships based on clear mutual understanding.

There are no formal, socially recognized relationships between present-day human society and the beings responsible for the UFOs. In most countries, official scientific, academic, and governmental bodies do not acknowledge that such beings might exist and be in contact with human society. As a result, knowledge of UFOs is not regulated by established academic bodies, and the UFO field is a free-for-all in which serious researchers must cope with an outpouring of unscholarly or fraudulent material.

The UFO beings themselves seem to plan their contacts with people in such a way that there is very little tangible evidence that they really exist. These contacts involve phenomena that are very strange from the modern human perspective, but the ufonauts make little effort to reduce this strangeness. Close-encounter witnesses may have a history of encounters going back to childhood, but they are nonetheless given few explanations and practically no opportunities to introduce their otherworldly visitors to a wider circle of witnesses. Although many witnesses seem to be responsible people who have had genuinely unusual contact experiences, the information they have received during their contacts often seems absurd or contradictory, and it does not enhance their credibility.

Surprisingly, it may be that things have not always been this way. Among tribal societies, mystical contacts with higher beings have been standard since time immemorial, and these contacts are said to still go on today. The civilized societies of ancient times also claimed to be in contact with higher beings. In many cases, the available transhuman-contact material from these sources is placed in the category of religious doctrine, and it involves the unique experiences of a few mystically gifted individuals. However, there are reports of earthly human societies that have had regular diplomatic links with a hierarchy of extraterrestrial and higher-dimensional beings.

This is true, in particular, of the ancient Vedic society of India. Voluminous literature exists describing this society, and from it we can learn a great deal about how its people lived and how they interacted with a larger transhuman society. In this chapter, I will give a brief overview of the ancient Vedic world view. I will show that many features of the modern UFO phenomenon can be seen in Vedic accounts of encounters between humans and members of other humanlike races. I will also show how the social organization of the ancient Vedic people allowed for regular contact with higher beings.

To the best of my ability, I will present the Vedic material as it is understood by those who are immersed in the traditional Vedic outlook. This material may initially seem very strange to persons of Western cultural background, and some may feel reservations based on a religious or scientific perspective. However, the only scientific way to understand another culture is to try to enter into the actual world view of people who live in that culture. My advice would therefore be to suspend judgment and simply try to appreciate the Vedic material as it is. I discuss my approach to interpreting Vedic literature in greater detail in Appendix 1.

As I pointed out in the Introduction, we know that modern UFO accounts can seem very strange. So we shouldn't be surprised if the stories and traditions of people in regular contact with higher beings should also seem strange to us. They may help us attain a broader understanding of the strange universe which includes our own system of knowledge and culture as a small part in a much larger reality.

A Synopsis of the Vedic World View

The *Bhāgavata Purāṇa,* the *Mahābhārata,* and the *Rāmāyaṇa* are three important works in the Vedic tradition of India. They are well known as Hindu religious scriptures, but they should not be regarded simply as mythology or as presentations of some sectarian creed. Their real value lies in the fact that they reveal in detail a completely different way of seeing the world, and of living in it, which was followed for thousands of years by a highly developed human civilization.

From the viewpoint of modern Indologists, these works range in age from the 9th century A.D. for the *Bhāgavata Purāṇa* to the 5th or 6th century B.C. for the *Mahābhārata* and *Rāmāyaṇa.* However, Indologists agree that the existing texts incorporate material much older than the historical periods in which they believe these texts were written. The very word *purāṇa* means ancient, and according to native Indian tradition, all three texts date back to at least 3000 B.C.

Here I should make a technical remark about the use of the word "Vedic." Modern Western scholars insist that this word can be applied only to the four *Vedas: Ṛg, Yajur, Sāma,* and *Atharva.* However, in living Indian tradition this word applies to a much broader category of literature. This includes the *Purāṇas,* or ancient cosmological accounts, and the *Itihāsas,* or historical epics. The *Bhāgavata Purāṇa* is one of the 18 principal *Purāṇas,* and the *Mahābhārata* and *Rāmāyaṇa* are *Itihāsas.* I will therefore use the word Vedic to refer to these works as well as the four *Vedas.*

Vimānas

One important point to make about ancient Vedic society is that aerial vehicles, called *vimānas* in Sanskrit, were well known. They could be grossly physical machines, or they could be made of two other kinds of energy, which we can call subtle energy and transcendental energy. Humans of this earth generally did not manufacture such machines, although they did sometimes acquire them from more technically advanced beings.

There are ancient Indian accounts of manmade wooden vehicles that flew with wings in the manner of modern airplanes. Although these wooden vehicles were also called *vimānas,* most *vimānas* were not at all like airplanes. The more typical *vimānas* had flight characteristics resembling those reported for UFOs, and the beings associated with them were said to possess powers similar to those presently ascribed to UFO entities. An interesting example of a *vimāna* is the flying machine which Śālva, an ancient Indian king, acquired from Maya Dānava, an inhabitant of a planetary system called Talātala. The story of Śālva is presented later in this chapter, and additional information on *vimānas* is presented in Chapter 7.

Other Worlds

In Vedic society, it was understood that travel to other worlds is possible. This could involve travel to other star systems, travel into higher dimensions, or travel into higher-dimensional regions in another star system. It was also understood that it is possible to leave the material universe altogether and travel through a graded arrangement of transcendental realms.

The Vedic literature does not use geometric terms such as "higher dimensions" or "other planes" when referring to this kind of travel. Rather, the travel to other worlds is described functionally in terms of the experiences of the travelers, and it is necessary for the modern reader to deduce from the accounts that this travel involves more than motion through three-dimensional space. Since people of modern society are accustomed to thinking that travel is necessarily three-dimensional, I will use the term "higher-dimensional" to refer to Vedic accounts that cannot be understood in three-dimensional terms.

The objection might be raised that surely the ancient people of India had a very naive and unscientific understanding of stars and planets, and so it does not make sense to suppose that they might have actually been in contact with beings from such places. The answer is that the Vedic description of the universe sounds very strange and mythological to a person of Western background because it contains many ideas that are completely foreign to familiar Western conceptions. However, it also contains many ideas about the universe that are found in modern science.

For example, consider the following description of the travels of the hero Arjuna into the region of the stars:

> No sun shone there, or moon, or fire, but they shone with a light of their own acquired by their merits. Those lights that are seen as the

> stars look tiny like oil flames because of the distance, but they are very large. The Pāṇḍava saw them bright and beautiful, burning on their own hearths with a fire of their own. . . .
>
> Beholding those self luminous worlds, Phalguna, astonished, questioned Mātali in a friendly manner, and the other said to him, "Those are men of saintly deeds, ablaze on their own hearths, whom you saw there, my lord, looking like stars from earth below."[1]

This passage displays a mixture of familiar and unfamiliar elements. We expect that if we traveled among the stars we would be far away from the sun and moon, and we wouldn't see them. We also think that the stars are large, self-luminous worlds that seem small because of the distance. However, we don't expect to find them inhabited by "men of saintly deeds," and it seems strange to refer to the stars as men. It appears to be customary in Vedic texts to refer to a star as a person, and this person is normally the ruler of that star, or its predominating inhabitant.

The objection could also be raised that the earth was regarded as flat in ancient India. Actually, in Vedic literature two ideas of the earth are described. The earth is described as a globe 1,600 *yojanas* in diameter in the Sanskrit astronomical text *Sūrya-siddhānta*.[2] The *yojana* is a measure of distance, and it can be argued that this text uses about five miles per *yojana*. This would make the diameter of the earth about 8,000 miles, which agrees well with modern figures. The same text gives the diameter of the moon as 480 *yojanas,* or 2,400 miles. This can be compared with the modern figure of 2,160 miles.[3]

The earth is also described as a flat disk, called Bhū-maṇḍala, which is 500,000,000 *yojanas* in diameter. However, a careful study of Vedic texts shows that this "earth" actually corresponds to the plane of the ecliptic.[4] This is the plane determined, from a geocentric point of view, by the orbit of the sun around the earth. This plane is, of course, flat, and thus in one sense the Vedic literature does speak of a flat earth. One has to be alert to the fact that the term "earth," as used in Vedic texts, does not always refer to the small earth globe.

Higher-dimensional, inhabited realms are understood in Vedic thought to extend within the earth and on it, as well as through outer space. In particular, the flat "earth" of Bhū-maṇḍala is an inhabited realm that extends more or less through the plane of the solar system and is not directly visible or accessible to our gross senses. The general

Sanskrit term for such inhabited realms is *loka,* which is often translated as "planet" or "planetary system." There are fourteen grades of *lokas,* seven higher and seven lower. Bhū-maṇḍala or Bhū-loka is the lowest of the seven higher planetary systems.

The sun, the moon, and the planets Mercury, Venus, Mars, Jupiter, and Saturn are called *grahas,* and they are all regarded as being inhabited. (However, I have not come across any reference to Uranus, Neptune, or Pluto in Vedic texts.) Not surprisingly, the inhabitants of the sun are regarded as having bodies of fiery energy, and the bodies of the inhabitants of other planets are said to be built from types of energy suitable for the environments on those planets.

Humanoids

The *Purāṇas* speak of 400,000 humanlike races of beings living on various planets and of 8,000,000 other life forms, including plants and lower animals. Out of the 400,000 humanlike forms, human beings as we know them are said to be among the least powerful. This, of course, ties in with the picture that emerges from accounts of UFO encounters.

I have been using the word "humanoid" to refer to humanlike beings reported in UFO encounters, and I will also use it to refer to the Vedic humanlike races. UFO accounts often portray humanoids as looking strange or repulsive, but some have been described as beautiful. The Vedic humanoids also vary widely in appearance. Some of them, such as Gandharvas and Siddhas, are said to have very beautiful human forms. Others are said to be ugly, frightening, or deformed in appearance. One group is called the Kimpuruṣas. Here *kim* means "is it?," and *puruṣa* means "human."

Many of the Vedic humanoid races are said to naturally possess certain powers called *siddhis.* Humans of this earth can also potentially acquire these powers, and some people have greater abilities in this regard than others. Here is a list of some of these *siddhis.* Since they seem to be directly related to some of the powers attributed to UFO entities, I will discuss them in greater detail in later sections.

1. Mental communication and thought-reading. These are standard among Vedic humanoids, but normal speech through sound is also generally used.

2. Being able to see or hear at a great distance.
3. *Laghimā-siddhi:* levitation or antigravity. There is also a power of creating enormous weight.
4. *Aṇimā-* and *mahimā-siddhis:* the power to change the size of objects or living bodies without disrupting their structure.
5. *Prāpti-siddhi:* the power to move objects from one place to another, apparently without crossing the intervening space. This power is connected with the ability to travel into parallel, higher-dimensional realms.
6. The ability to move objects directly through the ether, without being impeded by gross physical obstacles. This type of travel is called *vihāyasa*. There is also a type of travel called *mano-java*, in which the body is directly transferred to a distant point by the action of the mind.
7. *Vaśitā-siddhi:* the power of long-distance hypnotic control. Vedic accounts point out that this power can be used to control people's thoughts from a distance.
8. *Antardhāna*, or invisibility.
9. The ability to assume different forms or to generate illusory bodily forms.
10. The power of entering within another person's body and controlling it. This is done using the subtle body (defined below).

Many different Vedic humanoid races are said to live in parallel, higher-dimensional realms within the earth, on its surface, and in its immediate vicinity. One striking feature of Vedic accounts is that different races such as Siddhas, Cāraṇas, Uragas, Guhyakas, and Vidyādharas are often described as living and working together cooperatively, even though they differ greatly in customs and appearance.

These beings are generally well endowed with the various *siddhis.* In the past, many of these humanoid types were to be found on the earth, either as visitors or as inhabitants. Indeed, large areas of the earth's surface have sometimes been controlled and populated by a variety of humanoid species. This is the basic setting of the *Rāmāyaṇa,* which tells how Lord Rāmacandra rescued his wife Sītā from the kingdom of Laṅkā, to which she had been taken by a Rākṣasa named Rāvaṇa. The Rākṣasas are one of the 400,000 humanoid races, and they were ruling Laṅkā at that time.

There is a wide range of life spans among the Vedic humanoid species. According to Vedic accounts, earthly human beings had much longer life spans thousands of years ago. For example, prior to about 5,000 years ago, the human life span is said to have been about 1,000 years. Typical life spans of humanoid beings living outside this earth are in the order of 10,000 years. There are also said to be beings called Devas, who are administrators of the universe and who live for hundreds of millions of years.

People in India still report encounters with humanoids of the classical Vedic type. Two examples of this are the case of the smallpox lady and the case of the Jaladevata in Appendix 2.

The Soul

One key feature of the Vedic world view is that living beings are souls dwelling within bodies. The soul is called the *ātmā,* or *jīvātmā,* and is endowed with the faculty of consciousness. The body consists of a gross body composed of the familiar physical elements and a subtle body made of the energies known as mind, intelligence, and false ego. These energies cannot normally be detected by our current scientific instruments, and thus the established scientific view is that they do not exist. However, according to the Vedic understanding, these energies naturally interact with gross matter, and when properly controlled they can exert a powerful influence on it.

The soul and subtle body are said to transmigrate from one gross body to another, and they can also travel temporarily outside of the gross body. The process of transmigration is regulated by universal laws, and there are humanoid beings involved in controlling this process. There is a natural process of evolution of consciousness, whereby souls gradually attain higher and higher types of bodies.

At the highest level of consciousness, it is possible for the soul to become free from the subtle body and attain liberation from the material world. The state of liberation, or *mukti,* involves transfer of the soul to a completely transcendental realm. Broadly speaking, there are two forms of liberation. These are (1) experience of Brahman, or transcendental oneness, and (2) experience of variegated activity in the service of the Supreme in the spiritual planets of Vaikuṇṭha.

According to Vedic philosophy, all manifestations emanate from the Supreme Being, who is known by many names, including Kṛṣṇa,

Govinda, Nārāyaṇa, and Viṣṇu. The individual souls are understood to be parts of the Supreme Being, and are compared to sparks within a great fire. They all share the qualities of the Supreme in a minute degree, and for this reason they are all closely related to one another. The liberated souls fully display these spiritual qualities, but those who are encased in material bodies tend to display perverted qualities due to the influence of the material energy.

UFO reports contain many references to the soul, to transmigration, and to out-of-body experiences. This is discussed in Chapter 10. There are also references to the experience of Brahman, and I will discuss this topic in Chapter 11.

The Cosmic Hierarchy

One idea that often comes up in UFO communications is that there is law and order within the cosmos. Various confederations of planets are mentioned, and these are often said to follow higher authorities who have greatly elevated states of consciousness and live in higher planes or vibrational states. I pointed out in Chapter 5 that these UFO communications do not seem to be very reliable. Nonetheless, it is interesting that the basic idea of a hierarchical universal government is a key element of the Vedic world view.

In the Vedic cosmic hierarchy, there is a graded series of higher planetary systems, each of which is inaccessible to the inhabitants of the systems below it. The topmost authority in the material universe is known as Brahmā, and he lives in the highest material planetary system, called Brahmaloka. Beneath Brahmaloka there are the planetary systems Tapoloka, Janaloka, and Maharloka, which are inhabited by sages (*ṛṣis*)who live as ascetics and cultivate knowledge and transcendental consciousness.

Beneath these planets, there is the realm of Svargaloka, which is predominated by the beings known as Devas. The Devas are organized in a military hierarchy. They engage in politics and warfare, and their battles with lower forces may sometimes have an impact on life on the earth. However, due to the extremely long life spans of the Devas, their social and political relationships tend to be stable.

Although the universe is completely under intelligent control, higher-level controllers such as the Devas and great sages do not generally intervene directly in the lives of subordinate beings, including earthly humans. Rather they make arrangements for these beings to transmigrate from body to body according to their work, and thereby allow for their gradual evolution in consciousness. They also make arrangements for the dissemination of spiritual teachings in various societies so as to guide embodied souls in the direction of higher spiritual development. According to the Vedic perspective, spiritual advancement should be the main goal of human life.

Above the cosmic hierarchy of the material world, there is a spiritual hierarchy predominated by the Supreme Being. Although this hierarchical system places a great distance between the Supreme Being and humans of this earth, the Vedic literature stresses that all spirit souls are intimately related with the Supreme and that the Supreme Being accompanies each soul as the Paramātmā, or Supersoul. Also, the Supreme Being descends personally as an *avatāra* on various material planets. The story of the *avatāra* known as Kṛṣṇa is the subject of the *Bhāgavata Purāṇa,* and the *Rāmāyaṇa* is the story of the *avatāra* known as Lord Rāma or Rāmacandra.

Self-Centered Elements

Among the different humanoid types, there are races who have an essentially self-centered outlook. These are distinguished from those who tend to be dedicated to the service of the Supreme Being and the cosmic hierarchy. Some are like celestial playboys who live in great opulence. Others are characterized by an alienated state of consciousness, and yet others are strongly inimical. The self-centered races tend to be greatly attracted to the exploitation of mystic powers and technology. This is illustrated by Maya Dānava, the being responsible for the construction of the *vimāna* of King Śālva mentioned above.

All of these different groups of beings are under the control of the universal hierarchy, and thus they are not able to act fully according to their own propensities. This would explain why we are not simply taken over by them. However, there are beings who actively rebel against the cosmic hierarchy and who do sometimes interfere strongly in earthly affairs.

The most famous rebels are the Asuras, who are close relatives of the Devas. The *Purāṇas* describe protracted wars in Svargaloka between the Devas and the Asuras, and the basic plot of the *Mahābhārata* has to do with an invasion of the earth by the Asuras. This is discussed in Chapter 10 in connection with harmful activities that have been attributed to UFOs.

Since the Devas are beings of a godly nature who hold administrative posts in the universal hierarchy, the word "demigod," taken from classical Greek and Roman mythology, is often used to refer to them. In contrast, the rebellious Asuras are often referred to as "demons," since they tend to be atheistic and to oppose the divine order.

Actually, the word "demon" acquired its negative connotations under the influence of Christianity. This word comes from "daemon," which in classical Roman times meant a being intermediate between the demigods and man. The Romans and Greeks thought that there were many types of beings in this category, and these were not all regarded as evil or "demonic." The Vedic literature also describes many races intermediate between the Devas and human beings, and these include the Vidyādharas, Uragas, and Rākṣasas.

The Rākṣasas are demonic and highly inimical to humans. The Vidyādharas and Uragas are essentially neutral—they cooperate with the universal hierarchy, but they have their own agendas to pursue, and they neither favor nor oppose the human race. They belong to a category of beings known as Upadevas, or almost-Devas.

Human Origins

According to the Vedic system of thought, the various species of living beings have been brought into existence by a process of creation and emanation. The spirit souls are all emanations of the Supreme, and so is the body of Brahmā, the first living being within the universe. Brahmā generated various bodily forms by direct mental action, and generations of descendants were produced from these forms by sexual reproduction. Unlike living species in our experience, these beings carried *bījas,* or seeds, for many different types of beings, and thus they could produce different types of offspring. (The bodies of these beings are composed of subtle forms of energy, and thus the *bījas* are not made of gross matter, like DNA.)

The different humanoid races were all produced in this manner, and thus they are all related by common ancestry. The humans of this earth, in particular, have descended from the Devas along several lines at different times, and thus they have a very complex celestial ancestry. The Vedic accounts clearly indicate that interbreeding can take place between different humanoid species. In particular, some of the heroes of the *Mahābhārata* were said to be descendants of a human mother and Deva fathers. This topic is discussed in greater detail in Chapter 8 (pages 276–77).

Contact

In the ancient Vedic civilization, contact with various nonhuman races was on a solid footing. Celestial *ṛṣis* and Devas would regularly visit the courts of great earthly kings. There were established diplomatic relationships and satisfying mutual understandings between leading members of human society and representatives of other societies in the cosmic hierarchy. This is illustrated by the description in the *Bhāgavata Purāṇa* of the Rājasūya sacrifice performed by King Yudhiṣṭhira, which took place according to traditional dating about 5,000 years ago in the city of Indraprastha, near present-day New Delhi. The winding up of this event is described as follows:

> The assembly officials, the priests and other excellent *brāhmaṇas* resoundingly vibrated Vedic mantras, while the demigods [Devas], divine sages [*ṛṣis*], Pitās and Gandharvas sang praises and rained down flowers. . . .
>
> The priests led the King through the execution of the final rituals of *patnī-saṁyāja* and *avabhṛthya.* Then they had him and Queen Draupadī sip water for purification and bathe in the Ganges. . . .
>
> Next the King put on new silken garments and adorned himself with fine jewelry. He then honored the priests, assembly officials, learned *brāhmaṇas* and other guests by presenting them with ornaments and clothing.
>
> In various ways King Yudhiṣṭhira, who had totally dedicated his life to Lord Nārāyaṇa, continuously honored his relatives, his immediate family, the other kings, his friends and well-wishers, and all others present as well. . . .
>
> Then the highly cultured priests, the great Vedic authorities who had served as sacrificial witnesses, the specially invited kings, the *brāhmaṇas, kṣatriyas, vaiśyas, śūdras,* demigods, sages, forefathers and mystic spirits, and the chief planetary rulers and their followers—all of them, having been worshiped by King Yudhiṣṭhira, took his permission and departed, O King, each for his own abode.[5]

The forefathers, or Pitās, are inhabitants of Pitṛloka, a planet connected with regulation of the transmigration of souls. The Gandharvas are a race of very beautiful beings who fall into the category of Upadevas, and the planetary rulers are prominent leaders of the Devas. The phrase "mystic spirits" refers to the Bhūtas, which are ghostly beings with a rather negative, alienated mentality. When it is stated that these various beings took permission from King Yudhiṣṭhira to depart for their abodes, this doesn't mean that he was their ruler. They were simply observing etiquette in their relation with the king.

Vedic Accounts of Close-Encounter Phemomena

There are many parallels between the Vedic view of reality, as described above, and the picture that emerges from UFO reports. Certainly the Vedic literature has not been influenced by UFO lore, since even the most recent dating of key Vedic texts places them at the beginning of the Middle Ages. However, it is possible that Vedic information may have influenced some alleged UFO communications. For example, some Vedic material appears in the works of the Theosophists and other Western mystical writers, although they all rework it in their own way. There are three ways in which some of this material might enter into UFO communications. The first is that people presenting false communications may use some of this material, which is circulating widely in popular circles. There is also the possibility of material surfacing from the unconscious mind and being woven into encounter stories told by sincere people. This is called cryptomnesia.

The third possibility is that UFO entities might take such material from popular human culture and weave it into messages given to people they contact. In Chapter 5, I argued that material from Western culture, such as the Egyptian myth of the Phoenix, sometimes shows up in UFO close-encounter cases. I also raised the question of whether or not nonhuman beings might be influencing human culture by introducing their own ideas into it. For example, it is conceivable that the myth of the Phoenix could have originated centuries ago in a nonhuman culture.

There is much material in Vedic texts that is practically unknown to Western people who do not have an explicit interest in Indian culture. Some of this material shows parallels with commonly reported features of the appearance and behavior of UFOs and UFO entities. For Western people to consciously or unconsciously fake these reported features using Vedic material, their interest in Vedic subject matter would have to be far greater than it is generally observed to be. Likewise, it seems implausible that UFO entities operating in Western countries would create extensive hoaxes based on Vedic texts. These parallels could therefore indicate a genuine relationship between the experiences of people living in Vedic times and modern experiences involving UFOs. In the remainder of this chapter, I will illustrate this with a number of examples.

The Aerial Bombardment of Dvārakā

A number of interesting parallels with UFO accounts can be found in the story of Śālva in the Tenth Canto of the *Bhāgavata Purāṇa.* Śālva was a king of this earth who developed an intense animosity toward Lord Kṛṣṇa and vowed to destroy Kṛṣṇa's city of Dvārakā. To do this, he acquired a remarkable *vimāna* by worshiping Lord Śiva. I will begin by quoting a description of the flight of Śālva's *vimāna,* which is referred to by the translator as an airplane:

> The airplane occupied by Śālva was very mysterious. It was so extraordinary that sometimes many airplanes would appear to be in the sky, and sometimes there were apparently none. Sometimes the plane was visible and sometimes not visible, and the warriors of the Yadu dynasty were puzzled about the whereabouts of the peculiar airplane. Sometimes they would see the airplane on the ground, sometimes flying in the sky, sometimes resting on the peak of a hill, and sometimes floating on the water. The wonderful airplane flew in the sky like a whirling firebrand— it was not steady even for a moment.[6]

It is significant that in his extensive writings, the translator of this passage, A. C. Bhaktivedanta Swami Prabhupāda, has never referred to UFOs or flying saucers. Yet the flight characteristics of this "airplane" resemble those of UFOs in many respects. The vehicle glows, and it moves in an irregular fashion, like a firebrand whirled by a dancer. It also appears and disappears. UFOs are well known for this kind of behavior, and they are also described as landing or hovering over water and then abruptly taking off.

As an example of this, consider the case of a UFO observed by Air Force personnel over the south-central U.S. on July 17, 1957. This case was summarized in the journal *Astronautics and Aeronautics* as follows:

> An Air Force RB-47, equipped with electronic countermeasures (ECM) gear and manned by six officers, was followed by an unidentified object for a distance of well over 700 mi. and for a time period of 1.5 hr., as it flew from Mississippi, through Louisiana and Texas and into Oklahoma. The object was, at various times, seen visually by the cockpit crew as an intensely luminous light, followed by ground-radar and detected on ECM monitoring gear aboard the RB-47. Of special interest in this case are several instances of simultaneous appearances and disappearances on all three of these physically distinct "channels," and rapidity of maneuvers beyond the prior experience of the air crew.[7]

One of the apparent disappearances of the object occurred as the RB-47 was about to fly over it. The pilot remarked that it seemed to blink out visually and simultaneously disappear from the scope of ECM monitor #2 (an electronic surveillance device). At the same time it disappeared from radar scopes at ADC site Utah. Moments later the object blinked on again visually and simultaneously appeared on the ECM monitor and ground radar. The observers on the RB-47 also noted that the UFO sometimes generated two signals with different bearings on their electronic monitoring equipment. Although we don't really know what the UFO was doing, this is reminiscent of the statement that Śālva's *vimāna* sometimes appeared to be in multiple forms.

How did Śālva acquire his remarkable vehicle? In view of the controversy regarding agreements between the U.S. Government and aliens, it is noteworthy that Śālva's *vimāna* was manufactured by a technological expert from another planet. Here is the story. (Paśupati and Umāpati are two names of Lord Śiva.)

> Having thus made his vow, the foolish King [Śālva] proceeded to worship Lord Paśupati as his deity by eating a handful of dust each day, and nothing more.
>
> The great Lord Umāpati is known as "he who is quickly pleased," yet only at the end of a year did he gratify Śālva, who had approached him for shelter, by offering him a choice of benedictions.
>
> Śālva chose a vehicle that could be destroyed by neither demigods [Devas], demons [Asuras], humans, Gandharvas, Uragas nor Rākṣasas, that could travel anywhere he wished to go, and that would terrify the Vṛṣṇis.

> Lord Śiva said, "So be it." On his order, Maya Dānava, who conquers his enemies' cities, constructed a flying iron city named Saubha and presented it to Śālva.
>
> This unassailable vehicle was filled with darkness and could go anywhere. Upon obtaining it, Śālva went to Dvārakā, remembering the Vṛṣṇis' enmity toward him.
>
> Śālva besieged the city with a large army, O best of the Bharatas, decimating the outlying parks and gardens, the mansions along with their observatories, towering gateways and surrounding walls, and also the public recreational areas. From his excellent airship he threw down a torrent of weapons, including stones, tree trunks, thunderbolts, snakes and hailstones. A fierce whirlwind arose and blanketed all directions with dust.
>
> Thus terribly tormented by the airship Saubha, Lord Kṛṣṇa's city had no peace, O King, just like the earth when it was attacked by the three aerial cities of the demons.[8]

We see from this account that Śālva did not engage engineers to manufacture his flying machine on the earth. As I will point out in Chapter 7, there are descriptions in Sanskrit of mechanical, airplane-like flying machines that are said to have been built by human beings. However, as far as I am aware, there are no accounts indicating that ordinary human beings ever built vehicles like Śālva's, which exhibited mystical modes of flight.

It is significant that Śālva dropped such things as snakes, stones, and tree trunks from his *vimāna.* There is no mention of bombs, and it would seem that even though Śālva possessed a remarkable flying machine, he did not have the kind of aerial weapons technology used in World War II. He did, however, have a quite different technology, which could be used to affect the weather and produce whirlwinds, thunderbolts, and hailstones.

In this story, as in many others, the manufacturer of the *vimāna* was the being named Maya Dānava. This person was the ruler of a kingdom of Dānavas situated in the planet known as Talātala. The Dānavas were a powerful group of humanoid beings who were known for their expertise in technology. The word *māyā* means the energy that makes up the material universe, and it also means the power of illusion. Maya Dānava was so named because he was an expert manipulator of *māyā.*

Umā, the wife of Lord Śiva, is also known as Māyā Devī, or the goddess in charge of the illusory energy. She is also the Mother Goddess who has been worshiped all over the world by many different names. Since Śiva is Umā's husband, he is the master of illusion and technology. Thus there is a natural connection between Lord Śiva, who Śālva approached to obtain his *vimāna,* and Maya Dānava, the master of illusion who manufactured it.

It is significant that Śālva asked for a vehicle that could not be destroyed by Devas, Asuras, Gandharvas, Uragas, or Rākṣasas. These are all powerful races of humanoid beings that were openly active on the earth or in its general environs in Śālva's time, and so naturally he wanted to be able to defend himself against them.

Śālva's vehicle is described as an iron city, and thus it must have been metallic in appearance and quite large. As we will see in Chapter 7, many Vedic *vimānas* are described as flying cities, and one is reminded of the very large "mother-ships" that are sometimes discussed in UFO reports. Also it is described as the "abode of darkness," or *tamo-dhāma.* Here "darkness" refers to the mode of ignorance, or illusion, that characterizes the material world in general and is particularly associated in Vedic literature with beings of negative character, such as the Asuras and Dānavas. It refers to a lack of spiritual insight, rather than to a lack of technical knowledge.

Invisibility and Sound-Seeking Arrows

The story of Śālva's *vimāna* contains a number of features that may give us some insight into the UFO phenomenon. I have already mentioned the power of the *vimāna* to become invisible. It is interesting to see how Kṛṣṇa, acting as a human warrior in the defense of Dvārakā, dealt with this invisibility. Here Kṛṣṇa is speaking to King Yudhiṣṭhira:

> I took my glittering bow, best of the Bhāratas, and cut with my arrows the heads of the Gods' enemies on the Saubha. I shot well-robed arrows, which looked like poisonous snakes, high flying and burning arrows, from my Śārṅga at King Śālva. Then the Saubha became invisible, O prosperer of Kuru's lineage, concealed by wizardry, and I was astounded. The bands of the Dānavas, with grimacing faces and disheveled heads, screeched out loud, as I held my ground, great

> king. I quickly laid on an arrow, which killed by seeking out sound, to kill them and the screeching subsided. All the Dānavas who had been screeching lay dead, killed by the blazing sunlike arrows that were triggered by sound.[9]

From this passage, we can see that even though Śālva was a human king, a contingent of grotesque-looking Dānava soldiers was present on his *vimāna*. This, of course, makes sense if we consider that Śālva obtained the craft from the leader of the Dānavas. There are many Vedic accounts of such alliances between human beings and other humanoid races. Although to modern historians it would seem doubtful that they ever existed, it is clear that the idea of such alliances was prominent in ancient India. And if they did exist then, one implication is that such alliances could also be made today.

The passage also shows that the bows and arrows used by the defenders of Dvārakā were not on a primitive or medieval level of technology. The bow was used as a launching device for many different kinds of arrows. These arrows are often described as "blazing" or "sunlike," and in this case they were endowed with some kind of guidance system that enabled them to find their targets by sound. Clearly, technological development does not have to be linear, so that all forms of technology higher than our own are further developments of the kind of technology that we have now.

The story of the sound-seeking arrows also shows that when Śālva's *vimāna* became invisible, it was still physically present, and sounds emanating from it could be heard. There are a number of UFO accounts that also have this feature. One example is given by the story of a man named Maurice Masse, which I will briefly summarize here.[10]

Masse was a lavender grower of the French Provensal village of Valensole. On the morning of July 1, 1965, at about 5:45 a.m., he was finishing a cigarette before starting work. Suddenly he heard a whistling noise and turned to see a machine shaped like a rugby football and the size of a Dauphine car. This was standing on six legs, with a central pivot stuck into the ground beneath it. He saw two boys near the object, but on approaching he found that they were not boys. At a distance of about 5 meters, one of the beings pointed a pencillike device at him, and he was paralyzed.

After some time, the beings returned to their machine, and Masse could see them looking at him from within the craft. At this point, the legs retracted, and with a thump from the central pivot, the machine floated silently away. At 20 meters it disappeared, but it left traces of its passage in the lavender field for 400 meters. It is said that newly planted lavender will not grow at the spot where the vehicle stood.

We can see from this description that the vehicle must have been physically present after it disappeared from sight at a distance of 20 meters. At least, this is a natural inference from the fact that the lavender crop was disturbed for a distance of up to 400 meters by the vehicle's passage. Thus its invisibility seems to have been similar to that of Śālva's *vimāna*. Both seem to have involved manipulation of light or the sense of vision so as to hide the aerial craft, which still betrayed its presence by sound or by disturbances of the air.

The power of invisibility was not limited to Śālva's *vimāna* as a whole. Śālva was also capable of personally becoming invisible and traveling in that state to another place. He was also able to project illusory forms:

> Lord Kṛṣṇa, in great anger, struck Śālva on the collarbone with His club so severely that Śālva began to bleed internally and tremble as if he were going to collapse from severe cold. Before Kṛṣṇa was able to strike him again, however, Śālva became invisible by his mystic power.
>
> Within a few moments, a mysterious, unknown man came before Lord Kṛṣṇa. Crying loudly, he bowed down at the Lord's lotus feet and said to Him, "Since You are the most beloved son of Your father, Vasudeva, Your mother, Devakī, has sent me to inform You of the unfortunate news that Śālva has arrested Your father and taken him away by force, just as a butcher mercilessly takes away an animal." When Lord Kṛṣṇa heard this unfortunate news from the unknown man, He at first became most perturbed, just like an ordinary human being.... While Śrī Kṛṣṇa was thinking like this, Śālva brought before Him in custody a man exactly resembling Vasudeva, His father. These were all creations of the mystic power of Śālva.
>
> Śālva addressed Kṛṣṇa, "You rascal, Kṛṣṇa! Look. This is Your father, who has begotten You and by whose mercy You are still living. Now just see how I kill Your father. If You have any strength, try to save him." The mystic juggler Śālva, speaking in this way before

> Lord Kṛṣṇa, immediately cut off the head of the false Vasudeva. Then without hesitation he took away the dead body and got into his airplane.[11]

Immediately after this, Kṛṣṇa realized that there actually was no body of Vasudeva. This was simply an illusion projected by Śālva using methods he had learned from Maya Dānava. Later in this chapter (pages 236–37), I will discuss cases in which UFO entities apparently caused people to see illusions—such as the vision of a beautiful deer in the woods—in order to manipulate their behavior. Examples of this can be multiplied in both Vedic and UFO literature.

The UFO literature also contains cases in which an individual being suddenly disappears and travels to another location. One reported instance of this took place at Nouatre, Indre-et-Loire, France, on September 30, 1954. At about 4:30 p.m., Georges Gatey, the head of a team of construction workers, encountered a strange-looking man standing in front of a large shining dome that floated about three feet above the ground. My concern here is with the way in which these odd apparitions disappeared:

> Suddenly, the strange man vanished, and I couldn't explain how he did, since he did not disappear from my field of vision by walking away, but vanished like an image one erases suddenly.
>
> Then I heard a strong whistling sound which drowned the noise of our excavators; the saucer rose by successive jerks, in a vertical direction, and then it too was erased in a sort of blue haze, as if by a miracle.[12]

Mr. Gatey, a pragmatic war veteran, maintained that he was not used to flights of fancy, and his story was corroborated by several of the construction workers.

Another story of a disappearing UFO involved Constable Charles Delk, an elected law-enforcement officer of Forrest County, Mississippi. At 8:15 p.m. on October 7, 1973, Delk was watching TV when the sheriff's dispatcher called to ask him to investigate a nearby UFO sighting. Delk skeptically declined to waste his time on this and returned to his program, but when called again he agreed to check out the complaint. Predictably, when he arrived at the location of the sighting, the UFO was gone.

However, on his way home Delk saw a glowing top-shaped object with flashing lights that floated slowly through the air. While keeping radio contact with his dispatcher, Delk described how the object hovered over an electrical power installation and emitted hissing torchlike jets. After Delk had followed it for several miles, it came closer, and his engine, car lights, and radio failed. The object departed, and after about fifteen minutes the car and radio resumed operation. Delk again caught up with the object and watched it slowly turn upside down. Then, while in full view, it abruptly disappeared. Delk was described as a pragmatic, elected law officer who had a solid reputation and who had nothing to gain by concocting wild stories.[13]

In summary, the story of Śālva's *vimāna* involves a flying machine with features similar to those reported in UFO sightings. It also involves persons who exhibit unusual powers and patterns of behavior that are typical of those reported in UFO close encounters. In a droll way, this was recognized by J. A. B. van Buitenen in the introduction to his translation of the *Mahābhārata*. Here are his remarks on Kṛṣṇa's battle with Śālva:

> Here we have an account of a hero who took these visiting astronauts for what they were: intruders and enemies. The aerial city is nothing but an armed camp with flame-throwers and thundering cannon, no doubt a spaceship. The name of the demons is also revealing: they were Nivātakavacas, "clad in airtight armor," which can hardly be anything but spacesuits. It is heartening to know that sometime in the hoary past a man stood up and destroyed the spaceship and aborted its mission with bow and arrow.[14]

The Nivātakavacas are a subgroup of the Dānavas. The word *nivāta* means "no air," and *kavaca* means "armor." Perhaps this does refer to spacesuits.

Levitation, or Laghimā-Siddhi

Let us return to the story of Maurice Masse and consider his description of the beings that he saw. He said that the creatures were less than four feet tall, wore close-fitting, gray-green clothes, and had pumpkin-like heads. They had high fleshy cheeks, large slanted eyes that went around the sides of the face, slit mouths without muscular lips, and sharply pointed chins. Their movement was described as "rising and falling in space like bubbles in a bottle without apparent support" or "sliding along bands of light."

References to strange beings that glide or float in the air are extremely common in UFO close-encounter reports. Another example is found in the story of Air Force Sergeant Charles L. Moody of the USAF Human Reliability Program, an elite group whose prospective members are carefully screened by psychiatrists for emotional disorders (page 185). He reported being abducted from his automobile outside of Alamogordo, New Mexico, in the early morning hours of August 13, 1975. He described his captors as dwarflike and said, "It's going to sound ridiculous and I hope nobody sends me a straitjacket, but these beings did not walk, they *glided*." [15]

The Vedic literature describes a mystic power called *laghimā-siddhi,* which enables a person to overcome the force of gravity. There are innumerable references to beings and objects that float weightlessly by this power, and it is commonly used by the Devas and related humanoid races. Thus, it is stated in one commentary on the *Bhāgavata Purāṇa:* "The residents of the upper planetary systems, beginning from Brahmaloka . . . down to Svargaloka . . . are so advanced in spiritual life that when they come to visit this or similar other lower planetary systems, they keep their weightlessness. This means that they can stand without touching the ground." [16]

It is said that *yogīs* can acquire the power of *laghimā-siddhi.* A description of how this can be done is given by Kṛṣṇa in the 11th Canto of the *Bhāgavata Purāṇa:*

> I exist within everything, and I am therefore the essence of the atomic constituents of material elements. By attaching his mind to Me in this form, the *yogī* may achieve the perfection called *laghimā,* by which he realizes the subtle atomic substance of time.[17]

It is curious that weightlessness should be connected with time. It is perhaps significant that in the general theory of relativity, gravitation is connected with transformations of space and time. It is also noteworthy that the idea of atomic particles was well known in Vedic times.

Disappearance and Reappearance

When Śālva disappeared after being struck by Kṛṣṇa, it is possible that he simply became invisible and walked away. However, there are also Vedic accounts in which a person disappears physically at one location and reappears somewhere else without crossing the intervening space in an ordinary way. According to the Vedic perspective, the ability to do this is simply a natural mystic power, or *siddhi,* which some beings, such as the Cāraṇas and Siddhas, inherit at birth, and which others can acquire by certain practical methods. As with ordinary bodily powers, this mystic power depends on the laws of nature and the gross and subtle organization of the body.

Here is one of many Vedic accounts in which this power is a standard element of the plot. When he was a child, the great sage Vyāsa had made a promise to his mother Satyavatī, saying, "Mother, if ever you need me, just set your mind on me, and I shall appear before you." Years later, an occasion arose in which Satyavatī needed to consult with Vyāsa. Her youngest son Vicitravīrya, who was king of the Kuru dynasty, had died without issue, and according to the law, his brother Vyāsa could beget children in his wives to continue the royal lineage. After this course of action was approved by the elder statesman, Bhīṣma, "Satyavatī fixed her mind on her son, who at the moment was reciting the *Vedas.* When the wise sage understood that his mother had set her mind on him, he appeared before her within a moment." [18]

In this case, it seems that Vyāsa disappeared from the place where he was reciting the *Vedas* and immediately appeared before his mother in a completely different location. The fact that he did this "within a moment" suggests that he traveled to his mother's location in some paranormal fashion. The telepathic communication between Vyāsa and his mother is also a standard feature of Vedic accounts.

The sage Vyāsa was a human being who was endowed with great mystic abilities as a result of being directly empowered by the Supreme Being. Such a person is known as a *śaktyāveśa-avatāra.* Vyāsa is famous as the compiler of the *Vedas,* and he is said to be still living in the Himalayas. In Vedic civilization, sages of his type served to link earthly human society with the celestial hierarchy.

There are many accounts in the UFO literature of beings who abruptly appear or disappear, and who seem to be able to travel invisibly in a mysterious way. In the case of William Herrmann in South Carolina (see pages 175–76 and 180–85), two alien beings of the short, large-headed type reportedly appeared in the midst of a blue glow in Herrmann's bedroom while he was out in the hall talking on the phone with UFO investigator John Fielding. Herrmann recognized one of them as a being he had met on a UFO during an earlier abduction. After a short telepathic conversation in which the beings said they could trust Fielding, they went back into Herrmann's bedroom and disappeared. Herrmann said that a short while earlier they had caused an inscribed metal object to appear in front of his eyes in the midst of a glowing ball of blue light.[19]

In a case in Altrincham, England, in late 1984, a man reportedly encountered a small, ugly-looking being in his bedroom on two occasions. He raised two interesting questions about these experiences: (1) He was nearsighted, and he had to strain to focus on the being, just as he would normally do with real objects. Why should a hallucinatory experience follow the rules of his poor eyesight? (2) The figure vanished in a flash without making any sound. Why didn't it leave a partial vacuum, and hence make a noise?[20]

A somewhat different case is the account given by a twenty-year-old college student of a being he saw in his bedroom when he was a child: "It was short and had big eyes . . . it also seemed to have an aura or some kind of glow around it. I don't think I could move while it was there. It seemed to speak to my brother, but I don't know what it said. It appeared to come out of the wall behind the dresser, and that's the way I think it left."[21] Of course, one could dismiss this story as a nightmare or hallucination. However, since such stories occur repeatedly and often involve beings of standardized appearance, it is also possible that the reported experiences are caused by visits from actual entities.

Bodily Travel Through Matter and Space

In these examples of appearance and disappearance, it would seem that physical objects are not merely moved invisibly through space. They are also moved through, or somehow around, solid matter. For example, the beings who visited William Herrmann seem to have entered his bedroom without coming in through the door of his trailer, either visibly or invisibly. It turns out that there is a Vedic process of travel, called *vihāyasa,* in which a physical object is moved directly through the ether to another location, without interacting with intervening gross matter. Here the word ether is used to translate the Sanskrit word *ākāśa*. *Ākāśa* is space, but it is considered to be a substance or plenum, rather than a void.

The story of the abduction of Aniruddha in the *Bhāgavata Purāṇa* contains an example of *vihāyasa* travel. A young princess named Ūṣā was living in the closely guarded inner quarters of her father's palace in the city of Śoṇitapura. One day, Ūṣā had a vivid dream about a beautiful young man who became her lover. She was certain that the person in her dream really existed, and she engaged her friend, the mystic *yoginī* Citralekhā, to find him for her:

> Citralekhā said, "I will remove your distress. If He is to be found anywhere in the three worlds, I will bring this future husband of yours who has stolen your heart. Please show me who He is."
>
> Saying this, Citralekhā proceeded to draw accurate pictures of various demigods, Gandharvas, Siddhas, Cāraṇas, Pannagas, Daityas, Vidyādharas, Yakṣas and humans.
>
> O King, among the humans, Citralekhā drew pictures of the Vṛṣṇis, including Śūrasena, Ānakadundubhi, Balarāma and Kṛṣṇa. When Ūṣā saw the picture of Pradyumna she became bashful, and when she saw Aniruddha's picture she bent her head down in embarrassment. Smiling, she exclaimed, "He's the one! It's Him!"
>
> Citralekhā, endowed with mystic powers, recognized Him [Aniruddha] as Kṛṣṇa's grandson. My dear King, she then traveled by the mystic skyway [*vihāyasa*] to Dvārakā, the city under Lord Kṛṣṇa's protection.
>
> There she found Pradyumna's son Aniruddha sleeping upon a fine bed. With her yogic power she took Him away to Śoṇitapura, where she presented her girlfriend Ūṣā with her beloved.[22]

The name Citralekhā means one who can make beautiful drawings. In typical Vedic fashion, Citralekhā considered that Ūṣā's lover might come from a wide variety of humanlike races. Once she identified Him as Aniruddha, she traveled directly through space to Aniruddha's bedroom, which was located in a palace in another city. Thus, from the point of view of a person in that bedroom, she appeared there out of nowhere, grabbed the sleeping Aniruddha without disturbing Him, and vanished from sight. She brought Him directly into the inner sanctum of Ūṣā's palace without having to use normal entrances and alert the guards assigned to protect Ūṣā's chastity. This story is similar to many UFO abduction accounts, with the exception that Aniruddha did not find the experience to be traumatic—at least not until Ūṣā's father found out what was happening.

There are many examples of UFO encounter stories that seem to involve this kind of mystical travel through space and matter. One example is the abduction case of Sara Shaw (see pages 179–80). She noted that her abductors entered her room through a closed window. In response to questions about how this was done, she said, "I feel really dumb, but it seems like they passed through the pane of glass without breaking it."[23]

In another case, the psychologist Edith Fiore was using hypnosis to explore the abduction experience of a person named Gloria. In the course of this, Gloria said, "I floated out of there. That's how I got to the sidewalk. Went right through the wall and I was on the sidewalk."[24]

In yet another case investigated by Edith Fiore, a witness named Fred recalled the following experience without the aid of hypnosis:

> During the night, I was awakened. I looked around and there was nobody in the room. The bedroom faced the street. It had a large window, an eight-by-six-foot window, with those old-style Venetian blinds. . . . The blinds were open. I was lying there and I started to move. This is the one thing about the whole experience that I have never been able to fully accept or understand. And I probably never will. I went through those Venetian blinds! I was absolutely terrified. I went through them. They didn't open. The window didn't open. I literally went through those blinds. To this day it astounds me! The next thing I saw was a sign saying Church Street.[25]

In a case investigated by Trevor Whitaker in the United Kingdom, an ambulance driver named Reg reported encountering strange visitors in his bedroom in February 1976. The visitors were two tall beings, with gray faces and large catlike eyes, who treated him like a specimen. He was told to lay prone on his bed and was paralyzed. He experienced floating up through the ceiling into the sky toward a hovering bathtub-shaped UFO. He underwent a medical examination on board, and he heard a number of Biblical references about the Alpha and the Omega through telepathy. He was told that "a thousand of your years are but a day to us," and he was informed that a wormlike being like him should not ask the visitors about their identity. He later found himself back in his bedroom with large memory gaps, but his experience was recalled without hypnosis.[26]

My last example is taken from the story of Betty Andreasson. In this case, hypnosis was used to aid the witness's memory. Here is part of the transcript of one of the hypnosis sessions:

> *Joseph Santangelo:* How did they get there, Betty?
> *Betty:* They came through the door.
> *Joseph Santangelo:* Did you open the door for them?
> *Betty:* No.
> *Joseph Santangelo:* Did *they* open the door?
> *Betty:* No. . . . They came in like follow-the-leader. . . . They are starting to come through the door now . . . right through the wood, one right after the other. It's amazing! Coming through! And I stood back a little. Was it real? And they are coming, one after another. . . . Now they are all inside.[27]

Andreasson said that later on two of the beings took positions in front of her and behind her, and they floated her through the door to a UFO parked outside.[28]

According to the Vedic literature, mystical travel can be done directly by an individual, using the powers of the mind. For example, in the *Bhāgavata Purāṇa,* Kṛṣṇa explains one mode of mystic travel, called *mano-java,* as follows:

> The *yogī* who completely absorbs his mind in Me, and who then makes use of the wind that follows the mind to absorb the material body in Me, obtains through the potency of meditation on Me the mystic perfection by which his body immediately follows his mind wherever it goes.[29]

In the case of Citralekhā and Aniruddha, or in the case of Vyāsa in the preceding section, it is clear that travel was accomplished by the direct potency of the individuals involved. Citralekhā was able not only to move her own body through matter, but she was also able to carry along the sleeping body of Aniruddha. This may be a parallel to the UFO accounts in which a person is carried through a solid wall by a humanoid being.

People have sometimes argued that high-tech machinery located in a UFO may be used to move bodies and transform them so that they can pass through matter without interference. There seem to be some UFO accounts that refer to such machinery. For example, Budd Hopkins recounted the story of how close-encounter witness Kathie Davis was positioned on a round platform within a UFO by her alien captors. Then the room seemed to shimmer, and she felt sudden pain in her chest. The next thing she knew, she was lying on the grass in her backyard, and she could see the departing UFO looking like a head-band with little lights on it.[30] A similar account was related by Jack Weiner, one of the four witnesses in the Allagash abduction case reported by Raymond Fowler.[31]

Teleportation of matter probably cannot be done by mechanisms obeying the known laws of physics, but it is perfectly possible that there exist principles of physics that have not yet been discovered. Apparent teleportation is often reported in poltergeist cases, where objects known to have been in another place are seen to appear suddenly in midair and follow strange trajectories.[32] These phenomena may also be related to Vedic *siddhis* such as *prāpti* and *mano-java*.

Since the *siddhis* are natural principles, it is possible that machines might be constructed that take advantage of them, and some *vimānas* and UFOs might operate on this basis. Thus, *laghimā-siddhi* could be used to make the craft weightless, and *mano-java* could be used to move it through the ether. Other vehicles might make use of more familiar mechanical or electromagnetic propulsion methods, or they might employ a combination of *siddhis* and more familiar principles.

The *siddhis* and the known physical principles are all features of nature, which in Vedic philosophy is regarded as a manifestation of divine potency. A deep understanding of both *siddhis* and the laws of physics would presumably allow for a unified understanding in which they appear as features of a larger whole.

The Abduction of Arjuna by Ulūpī

In the Villas Boas case (see pages 137–39), a human being was apparently abducted by ufonauts for the purpose of having sex with one of their female members. The story of Ūṣā and Aniruddha is not really comparable with this, since they entered into a relationship as equals and ultimately got married. However, there are Vedic accounts in which a human being is abducted for motives of lust by a member of another humanoid race.

This is illustrated by an account in the *Mahābhārata* involving the Pāṇḍava hero Arjuna. The story began when Arjuna was exiled for twelve months because of having intruded accidentally on his brother Yudhiṣṭhira and their common wife, Draupadī. Arjuna, known also as the son of Kuntī, went to visit Haridvāra along the Ganges River in the Himālayas. There he began to participate in sacrificial rites with a number of sages.

> While the son of Kuntī resided there among the *brāhmaṇas,* O Bharata, the sages brought to fruition many *agni-hotras,* the offering to the sacred fire. As the fires on both banks of the river were roused and brought to blaze, the offerings poured, and flowers offered in worship by learned, self-controlled sages, duly consecrated and fixed as great souls on the spiritual path, then, O king, the gateway of the Ganges shone with exceeding splendor.
>
> When his residence was thus crowded with divinity, the darling son of Pāṇḍu and Kuntī then went down into the Ganges water, to be consecrated for holy rite. Taking his ritual bath and worshiping his forefathers, Arjuna, happy to take his part in the rite of fire, was rising out of the water, O king, when he was pulled back in by Ulūpī, the virgin daughter of the serpent king, who could travel about at her will and was now within those waters. Holding onto him, she pulled him down into the land of the Nāgas, into her father's house.[33]

Ulūpī then proposed to Arjuna, arguing that she felt extreme desire for him and therefore he should be merciful and satisfy her. Arjuna did this in accordance with the code of the *kṣatriyas,* the Vedic warrior class. Thus, "The fiery hero Arjuna spent the night in the palace of the Nāga king, and when the sun rose he too rose up from Kauravya's abode."[34]

Kauravya is the name of the Nāga king. Note that when Ulūpī pulled Arjuna down, instead of finding himself on a rocky or sandy river bottom, he found himself in the Nāga kingdom. This is another example of mystic travel, but in this case the travelers entered into a parallel or higher-dimensional world. The Nāgas are a race of intelligent beings that are said to live either in the planetary system called Bila-svarga, or in parallel realities on the surface of the earth. Parallel realities are discussed in greater detail in the next section and in Chapter 8.

For the moment, recall that UFO entities have been reported to say that they share our world or that our activities directly affect their worlds (see pages 192–93). If some of them live in parallel worlds like those of the Nāgas, then this would make sense.

The Abduction of Duryodhana

Although sexual attraction seems to play a role in both modern and Vedic abduction stories, there may also be other motivating factors.

The abduction of King Duryodhana in the *Mahābhārata* is an example in which the underlying motives involved politics and military strategy.

King Duryodhana once had an encounter with some Gandharvas, who had cordoned off an area around a lake for recreational purposes and had blocked Duryodhana's army from entering. When Duryodhana tried to enter anyway, a fierce battle took place, and he was captured by the Gandharva forces. At this point, Arjuna, who was staying nearby, used his political connections with the Gandharvas to free Duryodhana. Arjuna and his brothers had been driven into exile by Duryodhana, but Arjuna intervened to save him from the Gandharvas on the grounds that he was a relative and a human being.

Duryodhana was humiliated at being saved by a person he had scorned and mistreated as an enemy, and he decided to give up everything and fast unto death. However, it seems that some other parties had long-standing plans for Duryodhana, and they weren't at all pleased by this turn of events:

> Thereupon the Daityas and Dānavas, hearing of his decision, the gruesome denizens of the nether world who had been defeated by the Gods, now, in the knowledge that Duryodhana would wreck their party, performed a sacrificial rite in order to summon him.[35]

With *mantras*, the Dānavas summoned a "wondrous woman with gaping mouth," and asked her to fetch Duryodhana. This woman was a *kṛtyā*, a type of demonic being, and she was able to transport the king by mystical travel: "Kṛtyā gave her promise and went forth and in a twinkling of the eye went to King Suyodhana [Duryodhana]. She took the king and entered the nether world and a little while after handed him over to the Dānavas."[36]

The "nether world" is not exactly the region beneath the surface of the earth. According to the Vedic literature, there are three regions known as Svarga, or heaven. These are delineated in relation to the ecliptic, or the orbital path of the sun against the background of fixed stars. There is Divya-svarga (divine heaven), the region of the heavens to the north of the ecliptic; Bhauma-svarga (earthly heaven), in roughly the plane of the ecliptic; and Bila-svarga (subterranean heaven), to the south of the ecliptic. The Bhauma-svarga is sometimes referred to as Bhū-maṇḍala, and it is the "flat earth" mentioned previously (see pages 203–4).

The "nether world" is Bila-svarga. It is "out there" in the heavens, but at the same time it can be reached through mystical travel by entering into the earth.[37] The nether regions can also be entered by taking the *pitṛ-yāna* path, which is said in the *Viṣṇu Purāṇa* to begin near the constellations Scorpio and Sagittarius and extend to the south in the direction of the star Agastya, or Canopus.[38] This is described in more detail in Chapter 7 (page 264).

Once Duryodhana was in the presence of the Dānavas, they explained to him that his birth on earth was arranged in advance as part of their plan. His great bodily strength and his near immunity to weapons were arranged by their manipulations. He therefore shouldn't spoil everything by taking his life. Dānavas and Daityas, taking birth as earthly heroes, would assist him in his battle with the Pāṇḍavas. The Dānavas also pointed out that they would use mind control to make sure that this battle had the desired outcome:

> Other Asuras will take possession of Bhīṣma, Droṇa, Kṛpa, and the others; and possessed by them they will fight your enemies ruthlessly. When they engage in battle, best of the Kurus, they will give no quarter to either sons or brothers, parents or relatives, students or kinsmen, the young or the old. Pitiless, possessed by the Dānavas, their inner souls overwhelmed, they will battle their relations and cast all love far off. Gleefully, their minds darkened, the tiger-like men, befuddled with ignorance by a fate set by the Ordainer, will say to one another, "You shall not escape from me with your life!" Standing firm on their manly might in the unleashing of manifold weapons, best of the Kurus, they will boastfully perpetrate a holocaust.[39]

If this wasn't enough, the Dānavas also explained that the hero Karṇa and the "sworn warriors" (a band of Rākṣasas) would slay Arjuna. After convincing Duryodhana that he would be victorious, the Dānavas arranged for his return:

> The same Kṛtyā brought the strong-armed man back when he was dismissed, to the very spot where he had been fasting unto death. Kṛtyā put the hero down, saluted him, and when the king had dismissed her, vanished then and there.
>
> After she was gone, King Duryodhana thought that it all had been a dream, Bhārata, and he was left with this thought: "I shall vanquish the Pāṇḍus in battle."[40]

This story from the *Mahābhārata* has a number of features that are also seen in UFO abduction accounts. These include:

1. A strange being takes Duryodhana bodily to another location, where he has a meeting with other strange beings.
2. Mystical or higher-dimensional transport is used.
3. The strange beings have human form, but look "gruesome." Certainly they are "aliens."
4. These beings have been guiding Duryodhana's life from the very beginning.
5. They designed his body so that he would be nearly impervious to weapons. Thus they apparently engaged in genetic manipulations, or something similar.
6. The aliens were planning to manipulate human beings through mind control.
7. After his interview, Duryodhana was returned to the spot where he was taken, and after setting him down, his captor disappeared.
8. After the experience, it seemed to be a dream.

There are accounts in the UFO literature that parallel the story of Duryodhana's paranormal transportation from one place to another. For example, Whitley Strieber said he once woke up to find one of his strange visitors beside his bed—a being of roughly humanoid form that seemed to be female and also insectlike. This visitor controlled his movements and proceeded to float him out of the bedroom: "If I walked, everything was normal. But if I stopped, I began to float along. I could feel her pushing from behind. . . . I had no control at all over my direction. I was not moving, I was being moved." [41]

At first, he found himself moving through the interior of his house, which looked perfectly normal. He grabbed his cat, Sadie, while passing by her so as to have some proof that he wasn't dreaming. But on reaching an outside door he entered into another state:

> We moved again, and this time I entered a profoundly different situation. No longer could I see normally. There was a glittering blackness before me. I could still feel Sadie in my arms, and I was very glad for her companionship. The next thing I knew I was standing in a room. It was an ordinary room. I was in front of a big, plainly designed desk.[42]

There were three other beings in the room: an ordinary-looking woman, a 6.5-foot-tall blond man in a jumpsuit, and a long-faced character with round, black eyes and a toupee, who "looked like something from another world wearing the clothes of the forties." [43]

Induced Paralysis and Long-Distance Hypnosis

One common feature of many UFO abduction reports is that the person being abducted is somehow paralyzed by his captors. This seems to involve a hypnotic power that abductees often associate with the eyes of their strange visitors. A classical example of this would be Barney Hill, who reported feeling overpowered by the gaze of an alien being that he saw through binoculars as it stared at him from the window of a hovering UFO (see pages 124–25).

The power to paralyze by glancing plays a role in a story in the *Mahābhārata*. Leaving aside various complexities of the plot, the story begins with Indra, the king of the Devas, being led to the top of King Mountain in the Himālayas by the goddess Gaṅgā:

> Indra followed her as she led the way, and he saw nearby on the top of King Mountain a beautiful and tender boy sitting on a throne, surrounded by young female companions and playing with dice. Indra, king of gods, said to him, "Know that this universe is mine, for the world is under my control. I am the lord." Indra spoke with anger, seeing the boy completely distracted with his dice.
>
> The boy, who was also a god, simply laughed and then slowly lifted his eyes toward Indra. As soon as the boy glanced at him, the king of gods was paralyzed and stood as stiff as a tree trunk.
>
> When the boy was finished with his game, he said to the weeping goddess, "Bring him close to me, and we shall see that pride does not again overwhelm him." [44]

The boy turned out to be Lord Śiva, who then punished Indra to cure him of false pride. In this story, the persons involved were all Devas. However, there is a mystic power of long-distance mind control, called *vaśitā-siddhi,* which is possessed by many humanoid races, and which can be acquired by human *yogīs*. This power has been described as follows:

> By this perfection one can bring anyone under his control. This is a kind of hypnotism which is almost irresistible. Sometimes it is found that a *yogī* who may have attained a little perfection in this *vaśitā* mystic power comes out among the people and speaks all sorts of nonsense, controls their minds, exploits them, takes their money, and then goes away.[45]

There is some evidence that even "ordinary" hypnosis can act at a distance. This could have important implications regarding the attainability of the *vaśitā-siddhi* by ordinary people, and it may also shed light on the nature of mind control and induced paralysis in UFO cases.

Here is a possible case of long-distance hypnotic suggestion that was reported by the psychical researcher F. H. W. Myers in the late 19th century.[46] The story begins at 9:00 p.m. on the evening of April 22, 1886. Four researchers, Ochorowicz, Marillier, Janet, and A. T. Myers, crept quietly through the deserted streets of Le Havre, France, and took up their stations outside the cottage of Madame B. They waited expectantly. "At 9.25," Ochorowicz wrote later, "I saw a shadow appearing at the garden gate: it was she. I hid behind the corner in order to be able to hear without being seen."[47]

At first, the woman paused at the gate and went back into the garden. Then at 9:30 she hurried out into the street and began to make her way unsteadily toward the house of Dr. Gibert. The four researchers tried to follow as unobtrusively as possible, and they could see that she was obviously in a somnambulistic state. Finally, she reached Gibert's house, entered, and hurried from room to room in an agitated state until she found him.

This was the planned outcome of an experiment in long-distance hypnotic influence. Madame B. was a person who could be easily hypnotized, and she was the subject of many experiments arranged by Professor Pierre Janet and Dr. Gibert, a prominent physician of Le Havre. In this research, they were joined by F. Myers of the London Society for Psychical Research, the physician Dr. A. T. Myers, Professor Ochorowicz of the University of Lvov, and M. Marillier of the French Psychological Society.

On this occasion, the plan was that Dr. Gibert would remain in his study and try to mentally summon Madame B. to leave her cottage and come to see him. The cottage was about a kilometer from his house, and neither Madame B. nor any of the other people living there had been informed that the experiment would take place. Gibert began issuing mental commands to summon her at 8:55 p.m., and within about half an hour she began her journey to his house. According to F. Myers, out of 25 similar experiments, 19 were equally successful.[48]

Experiments like this one by Dr. Gibert and his colleagues may seem unreliable. They were rather loosely organized and did not make use of the kind of strict laboratory protocols that we associate with accepted scientific work. However, many carefully organized experiments in distant hypnotic influence have been performed in laboratory settings.

For example, many experiments were performed by Professor Leonid Vasiliev of the University of Leningrad in the 1920s. In one series of experiments, a subject named Fedorova would regularly arrive at Vasiliev's laboratory at about 8 p.m. After about 20 minutes of rest and conversation, she would lie down on a bed in a darkened chamber. She was told to regularly squeeze a rubber balloon connected to an air tube as long as she was awake and to stop squeezing it when she began to fall asleep. The air tube was connected to an apparatus in the next room that kept a record of when she would fall asleep and wake up.

Once Fedorova was in the darkened room, she had no further contact with the experimenters. When she entered the room, the experimenter who had been talking with her would send a signal to another experimenter, called the sender, who was waiting two rooms away. The sender would then climb into a special lead-lined chamber and open a letter that had been prepared in advance and had not been read by the subject or the two experimenters. This letter would instruct the sender to do one of the following three things: (1) stay within the lead-lined chamber and mentally order the subject to go to sleep, (2) stand with his head outside the chamber and issue the same mental commands, and (3) stand with his head outside the chamber and make no mental commands.

In 29 runs of this experiment, the average time for the subject to go to sleep when no mental commands were given came out to 7 minutes and 24 seconds. In contrast, the average time when commands were given from inside the chamber was 4 minutes and 43 seconds. When the commands were issued outside the chamber, the average time was 4 minutes and 13 seconds.[49]

It seems that the subject was going to sleep faster when a person two rooms away was mentally ordering her to do so than when no orders were being given. Vasiliev performed many other carefully organized experiments of this kind, and he reported similar results.

These even included one successful experiment involving sending mental commands for sleeping and waking from Sebastopol to Leningrad, a distance of 1,700 kilometers.

The purpose of the chamber was to ascertain whether or not the long-range influence was transmitted by radio waves, which would be blocked by the lead. Vasiliev concluded from many experiments that radio was not involved, since the sender seemed to get the same results inside the chamber as outside.

This research is relevant to the topic of UFO abductions, since in case after case, the witness or witnesses will report being lured mentally to a certain location where a UFO close encounter takes place. The experience of Madame B., who was mentally directed to the home of Dr. Gibert, is a manmade example of this very phenomenon. The empirical evidence concerning long-distance hypnosis and the Vedic information about *vaśitā-siddhi* suggest that this kind of mental control may be a natural faculty of the minds of both human beings and related humanoid races.

Projection of Illusory Forms

The projection of illusory forms provides another parallel between the UFO phenomenon and the Vedic world view. There are many UFO accounts in which some kind of illusion is projected. In some cases, it seems that a UFO is disguised as an ordinary object, and in others the UFO entities seem to disguise themselves by assuming or mentally projecting unreal forms, including forms of animals. Here is an example, reported by Jacques Vallee, in which one man reported seeing a strange craft at close range, while his companion reported seeing only an ordinary bus.

On November 17, 1971, at 9:30 p.m., a Brazilian man named Paulo Gaetano was driving back from a business trip, accompanied by Mr. Elvio B. As they passed the town of Bananeiras, Paulo said the car was not "pulling" normally, but his companion's only response was that he was tired and wanted to sleep. According to Paulo, the engine stalled, and he pulled to the side of the road. A beam of red light seemingly caused the door to open. Several small beings then appeared, took him into a nearby craft, and subjected him to some kind of medical examination, which included taking a blood sample from his arm. He was also shown two panels depicting an atomic explosion and a plan of a nearby town. He could not recall how he and Elvio got back home.

Elvio told a different story:

> Near Bananeiras, Paulo had begun to show signs of nervousness, Elvio reported. He told him that there was a flying saucer accompanying them, when in fact what was following them was a *bus* which was keeping at a reasonable distance behind the car. [50]

Elvio saw the car pull off to the side of the road, and he remembered finding Paulo on the ground behind the parked car. But he did not remember seeing Paulo get out of the car, and he did not know what had happened to him. He took Paulo by bus to the nearby town of Itaperuna but could not explain why they had abandoned the car. The police in this town observed the cut on Paulo's arm, and they found the car parked by the highway.

If we accept this story as truthful, we must posit some kind of illusion to account for the reported events. One could suppose that either Paulo's or Elvio's experience was illusory, or that both were illusory. If we assume that Paulo saw an illusory UFO (and that Elvio may have seen a real bus), then we must explain why Elvio was so confused, why Paulo experienced an illusory abduction involving a cut on his arm, and how he received his actual cut. Of course, we can always attribute the confusion and the illusory abduction to an unknown agency within or outside the witnesses' minds. The cut could also have been made by this agency, or it may have been produced in some other way that the witnesses forgot in their confusion.

This is a complex explanation, but we can arrive at a simpler one if we assume that Paulo had a genuine experience. In this case, we can attribute Elvio's observation of the bus and his confusion to false perceptions induced by the UFO entities. This option has the advantage that it involves a simple illusion, the bus, rather than a complex one, namely the abduction experience. It also provides likely suspects for the perpetrators of the illusion, namely the UFO pilots, rather than attributing this to an unknown agency.

The psychologist Richard Boylan presented another story of a UFO disguising itself as a motor vehicle. In this case, the witness claimed to see the UFO transform into a car: "The light descended vertically to the level of the sagebrush tops, moved horizontally right until it lined up with the oncoming lane of U.S. 93, began moving toward me as a large single light at surface level on the highway, changed in an instant to two conventional headlight beams, then drove past me, 20 seconds later, as a late-model passenger car!"[51]

I turn next to an example in which a teenage girl on a picnic with her family reported a remarkable experience of seeing a beautiful deer in the woods—only to find through hypnosis that this was apparently a cover for a UFO close encounter. The witness, now a lawyer with a major corporation, was given the pseudonym Virginia Horton by Budd Hopkins in order to protect her identity. Here is her description of the deer, as recalled directly without the aid of hypnosis:

> Well, I just thought about it, but nothing came back to me except to remember again the sense of wonder that I had at the time at the beautiful, beautiful deer that I saw. You know, it was as though I had walked out of the woods and claimed that I saw a unicorn. There was that sense of excitement and wonder. . . . And the way I remember it is that the deer was looking at me and saying good-bye. The deer was saying good-bye telepathically to me. . . . It was as though I was talking to it and saying, "Well, don't leave yet," and then it just sort of dematerialized, disappeared.[52]

Upon returning to her family after seeing the deer, Virginia found that they had been worried by her absence, even though she felt that she hadn't been gone for a long time. Her mother also noticed that she had blood on her blouse, as though her nose had been bleeding. This piece of evidence was captured on a home movie made by Virginia's father.

Under hypnosis administered by the psychologist Aphrodite Clamar, Virginia revealed being lured into a strange craft parked in the woods:

> I'm walking through the woods. There's a very bright light. There's a ship just like they have in the movies. It's round. It's top-shaped roughly, but I can't tell exactly. . . . There's so much light you can't really see clearly. . . . And then I hear almost like a whisper, "Virginia . . . Virginia," and I think it was in my head that they were calling me.[53]

Virginia then proceeded to describe a highly complex encounter with typical big-eyed entities on the craft. This involved a procedure in which an instrument was inserted up her nose, apparently to take a tissue sample. This, of course, is consistent with the blood on her blouse. The deer episode occurred after she left the craft, perhaps to provide a natural excuse for her prolonged absence.

There are other instances of illusory animal forms that seem connected, directly or indirectly, with visitations by UFO entities. For example, Whitley Strieber pointed out that he has sometimes remembered his visitors as owls, a phenomenon he interprets as a "screen memory" generated by the mind to disguise their actual horrifying form.[54] Ed Walters, the main witness in the famous Gulf Breeze photo case, reported an encounter at age 17, which involved being followed by a sinister, abnormal-looking black dog during the day and being visited that night by a frightening, large-eyed, baldheaded being.[55]

In another case mentioned by Budd Hopkins, a friend he called Mary remembered seeing a beautiful hummingbird in 1950, when she was a child of about six. She tried to catch it in a jar and thought she had succeeded. But to her dismay, the jar turned out to be empty, and she discovered that her leg was mysteriously bleeding. Hopkins found on the back of her calf a hairline scar that she had been unaware of, and he noted that this scar was similar to those he had found in a number of UFO abduction cases.[56] She was reminded of this nearly forgotten incident on hearing Hopkins describe the Virginia Horton case at a meeting of friends.

To some researchers, the fact that illusions take place in UFO encounters, coupled with their general dreamlike quality, suggests that the encounters are entirely illusory. The fact that some encounters seem to take place in an out-of-body state also plays a role in this interpretation (see pages 340–43). These researchers suggest that the illusion might be due to some poorly understood feature of human psychology, or it might be due to some kind of astral agency.[57] Of course, the presence of physical evidence, such as cuts and bloodstains, suggests that a physically real agency may be involved. One could hypothesize that this agency consists of physically embodied beings who possess the power, through technology or natural endowment, to create illusions in people's minds.

The Vedic literature tends to support the latter view. Many different races of beings are said to have the ability to create illusory bodily forms, as well as illusory objects of various kinds. In some cases, the illusory forms seem to have physical substance. For example, there are descriptions of cascades of rocks produced on a battlefield—rocks that do real damage when they strike enemy soldiers. In other cases, the illusory form seems less substantial, since it ceases to exist when the being that generates it is incapacitated.

In the epic called the *Rāmāyaṇa,* there is a famous story involving the latter type of illusion. In this story, Lord Rāmacandra, the heir to the throne of Ayodhyā, had been banished as a result of political intrigue. He was living in the forest, accompanied only by His wife Sītā and His brother Lakṣmaṇa. Although He acted in the role of a human being, Lord Rāmacandra was actually an incarnation of Viṣṇu, the Supreme Godhead, who had appeared on earth to demonstrate the conduct of an ideal king.

At one point in the story, Rāvaṇa, the king of the Rākṣasas, became lustfully attracted to Sītā, and he made a plan to abduct her. The Rākṣasas were famous for having the ability to assume illusory forms, and they put this ability to effective use on this occasion. Rāvaṇa visited his old compatriot Mārīca and asked him to assume the form of a golden deer and lure Rāma and His brother Lakṣmaṇa away from Sītā. This would give Rāvaṇa the opportunity to carry her away. Mārīca at first refused to do this but agreed when Rāvaṇa threatened him with death if he did not comply.

> Then Mārīca, in the form of a wonderful deer with silver spots and the sheen of jewels, appeared before Sītā in the forest. His hooves were made of blue stones, and he had a little tail that shone like the rainbow. He walked this way and that, browsing on creepers and sometimes galloping. In so many ways, he drew the mind of Sītā, who asked Rāmacandra to catch him for her. Rāmacandra was, of course, cognizant that this might be the Rākṣasa magic of Mārīca, but He decided to go after the deer, and if it was actually Mārīca, He would kill him. He firmly ordered Lakṣmaṇa to stay behind with Sītā, and then pursued the deer.
>
> It became elusive, and even invisible, and Rāma resolved to kill it. He shot one deadly shaft which entered Mārīca's heart like a flaming snake. His counterfeit guise gone, Mārīca in the hideous form of a huge Rākṣasa, bathed in blood, now rolled upon the ground.[58]

The Rākṣasas were descended from the celestial sage Pulastya, who is said to live on one of the stars of the Big Dipper, a constellation known as Sapta-ṛṣi (Seven Sages) in Sanskrit. They were roughly human in form with huge stature, great muscular strength, and frightening facial features, including prominent teeth and pointed ears. With the possible exception of certain "hairy monster" cases (see pages 303–5), they do not correspond in physical form to any of the UFO entities being widely reported today. However, the powers attributed to them are typical both of UFO entities and of many Vedic humanoid races.

The Oz Factor

The British UFO researcher Jenny Randles has introduced the term "Oz factor" to refer to a peculiar, almost dreamlike, state of silence that often precedes UFO encounters.[59] This phenomenon is very commonly reported. For example, a man described the Oz factor to Budd Hopkins in the following account of the beginning of an abduction experience:

> And maybe it's my mind, but everything seemed to become very quiet at the same time. Do you know how, after an automobile accident, in the city when you hear cars crash, for a split-second after that everything stops? Well, that kind of sound happened, or lack of sound.[60]

Betty Andreasson gave the following description of the beginning of an encounter in which alien beings came into her house:

> I can see like a light, sort of pink right now. And now the light is getting brighter. It's reddish orange, and it's pulsating. I said to the children, "Be quiet, and quick, get in the living room, and whatever it is will go away." It seemed like the whole house had a vacuum over it. Like stillness all around . . . like stillness.[61]

Jenny Randles has interpreted the Oz factor as an altered state of consciousness induced by the agency behind the UFO phenomenon. In one hypothesis, she conceived of this agency as an intelligence based on another planet that can reach across the void of space by the power of consciousness and influence the brains of psychically sensitive people. The agency manipulates the consciousness of the affected person to create a UFO experience. According to Randles, this experience "is not really happening, yet it is far more than a mere hallucination."[62]

The Vedic literature also refers to experiences of abnormal stillness that are very similar to the Oz factor. These experiences are connected with the production of deliberate and vivid illusions by powerful beings. However, the beings in question are not situated far away in space. Rather they are physically present in the vicinity of the illusioned persons.

My first example of this is taken from the *Rāmāyaṇa*. After Lord Rāmacandra and his brother Lakṣmaṇa had been lured away from Sītā by the illusory deer, the Rākṣasa king, Rāvaṇa, approached Sītā in an illusory form with the aim of abducting her:

> Thereupon Rāvaṇa, in the guise of a mendicant, availing himself of the opportunity, rapidly approached the hermitage with the purpose of seeking out Vaidehī [Sītā]. With matted locks, clad in a saffron robe and carrying a triple staff and *loṣṭa,* that highly powerful one, knowing Sītā to be alone, accosted her in the wood, in the form of an ascetic, at dusk when darkness shrouds the earth in the absence of the sun and moon. . . .
>
> Beholding that monstrous apparition, the leaves of the trees ceased to move, the wind grew still, the turbulent course of the river Godaveri subsided and began to flow quietly. The ten-headed Rāvaṇa, however, profiting by Rāma's absence, drew near to Sītā in the guise of a monk of venerable appearance while she was overcome with grief on account of her lord.[63]

Here the statement that leaves and river waters became silent on "beholding" Rāvaṇa indicates that Rāvaṇa had a direct influence on these objects. In the *Rāmāyaṇa*, the unusual silence was not merely an illusion generated within Sītā's mind but was actually happening in her external environment. I also note that Rāvaṇa is described as ten-headed. His body was endowed with mystic powers, or *siddhis,* which enabled it to transcend the limitations of ordinary, three-dimensional space.

The second Vedic example is taken from the *Mahābhārata*. In this highly complex story, the hero Arjuna met Lord Śiva, who approached him in the form of a Himalayan mountain man. An intruding Daitya in the form of a boar arrived on the scene at the same time:

> When all the great-spirited ascetics had gone, the blessed Lord Hara [Śiva], who wields the Pināka, absolver of all evil, took on the guise of a mountain man. . . . The lustrous God was accompanied by the Goddess Umā, in the same guise and observing the same vow, and by excited creatures in all kinds of shapes. Garbed in his mountain man guise, the God shone surpassingly with his thousands of women, O King Bhārata.
>
> Instantly the entire wood fell silent and the sounds of streams and birds ceased. As he approached the Pārtha [Arjuna] of unsullied

> deeds, he saw the wondrous-looking Mūka, a Daitya, who had taken the form of a boar with the evil design of killing Arjuna.[64]

Arjuna and Śiva then argued over who had the right to kill the illusory boar. As we might expect, when they shot the boar, the dead body turned out to have the form of a Rākṣasa. The argument between Śiva and Arjuna resulted in a fierce battle, in which Śiva proved to be completely impervious to Arjuna's weapons. Although Arjuna was defeated, Śiva was satisfied by his fighting prowess and offered him a powerful celestial weapon.

In this story, it is hard to tell whether the "Oz factor" is due to the appearance of Lord Śiva or the Daitya, Mūka. In either case, it is connected with the projection of powerful illusions by beings who are personally present in the immediate vicinity.

7

The Story of Vimānas

In the Vedic literature of India, there are many descriptions of flying machines called *vimānas*. These fall into two categories: (1) man-made craft that resemble airplanes and fly with the aid of birdlike wings, and (2) unstreamlined structures that fly in a mysterious manner and are generally not made by human beings. The machines in category (1) are described mainly in medieval, secular Sanskrit works dealing with architecture, automata, military siege engines, and other mechanical contrivances. Those in category (2) are described in ancient works such as the *Ṛg Veda,* the *Mahābhārata,* the *Rāmāyaṇa,* and the *Purāṇas,* and they have many features reminiscent of UFOs. In addition, there is one book entitled *Vaimānika-śāstra* that was dictated in trance during this century and purports to be a transcription of an ancient work preserved in the akashic record. This book gives an elaborate description of *vimānas* of both categories.

In this chapter, I will survey some of the available literature on *vimānas,* beginning with the texts dating from late antiquity and the medieval period. The latter material is described in some detail by V. Raghavan in an article entitled "Yantras or Mechanical Contrivances in Ancient India." I will begin by discussing the Indian lore regarding machines in general and then turn to flying machines.

Machines in Ancient and Medieval India

In Sanskrit, a machine is called a *yantra*. The word *yantra* is defined in the *Samarāṅgana-sūtradhāra* of King Bhoja to be a device that "controls and directs, according to a plan, the motions of things that act each according to its own nature."[1] There are many varieties of *yantras*. A simple example would be the *taila-yantra,* a wheel that is pulled by oxen around a circular track to crush seeds and extract their oil.

Other examples are military machines of the kind described in the *Artha-śāstra* of Kauṭilya, written in the 3rd century B.C. These include the *sarvato-bhadra,* a rotating wheel that hurls stones, the *śara-yantra,* an arrow-throwing machine, the *udghāṭimā,* a machine that demolishes walls using iron bars, and many more.

These machines are all quite understandable and believable, but there are other machines that seem less plausible from the point of view of modern historical thinking. Thus Raghavan mentions a device that could create a tempest to demoralize enemy ranks.[2] Such a weapon is also mentioned by the third-century Roman writer Flavius Philostratus, who described sages in India who "do not fight an invader, but repel him with celestial artillery of thunder and lightning, for they are holy and saintly men."[3] Philostratus said that this kind of fire or wind weapon was used to repel an invasion of India by the Egyptian Hercules, and there is an apocryphal letter in which Alexander the Great tells his tutor Aristotle that he also encountered such weapons.[4]

Modern scholars tend to regard Philostratus's work as fictitious, but it does demonstrate that some people in Roman times were circulating stories about unusual fire or wind weapons in India. In ancient epics such as the *Mahābhārata,* there are many references to remarkable wind weapons such as the *vāyavya-astra* and fire weapons such as the *śataghnī* (or "100 killer"). In general, the weapons described in older works tend to be more powerful and remarkable than those described in more recent works. Some ascribe this to the fantastic imagination of ancient writers or their modern redactors. But it could also be explained by a progressive loss of knowledge as ancient Indian civilization became weakened by corruption and was repeatedly overrun by foreign invaders.

It has been argued that guns, cannons, and other firearms were known in ancient India and that the knowledge gradually declined and passed away toward the beginning of the Christian era. This is discussed extensively in a book by Gustav Oppert.[5]

Traditional Airplanes

There are many stories in medieval Indian literature about flying machines. Thus in Bāṇa's *Harṣa-carita* there is the story of a Yavana who manufactured an aerial machine that was used to kidnap a king. Likewise, Daṇḍī's *Avanti-sundarī* tells of an architect named Māndhātā who used an aerial car for such casual purposes as traveling from a distance to see if his young son was hungry. His son, by the way, was said to have created mechanical men that fought a mock duel and an artificial cloud that produced heavy showers. Both of these works date from about the 7th century A.D.[6]

In the ninth to tenth centuries, Buddhasvāmin wrote a version of the *Bṛhat-kathā,* a massive collection of popular stories. Buddhasvāmin spoke of aerial vehicles as *ākāśa-yantras,* or sky-machines, and he attributed them to the Yavanas—a name often used for barbaric foreigners. It was quite common for flying machines and *yantras* in general to be attributed to the Yavanas in Sanskrit texts.[7]

Some scholars take the Yavanas to be the Greeks, and they attribute Indian stories of machines to a Greek origin. For example, Penzer thought that the Greek philosopher Archytas (c. 428–347 B.C.) may have been the "first scientific inventor" of devices resembling the Indian *yantras,* and he pointed out that Archytas "constructed a kind of flying machine, consisting of a wooden figure balanced by a weight suspended from a pulley, and set in motion by hidden and enclosed air."[8]

No doubt there was much exchange of ideas in the ancient world, and today it is hard to know for sure where a given idea was invented and how highly developed it became. We do know, however, that fairly detailed ideas concerning airplanelike flying machines were known in medieval India.

Bhoja's *Samarāṅgana-sūtradhāra* states that the main material of a flying machine's body is light wood, or *laghu-dāru.* The craft has the shape of a large bird with a wing on each side. The motive force is provided by a fire-chamber with mercury placed over a flame. The power generated by the heated mercury, helped by the flapping of the wings by a rider inside, causes the machine to fly through the air. Since the craft was equipped with an engine, we can speculate that the flapping of the wings was intended to control the direction of flight rather than provide the motive power.

A heavier (*alaghu*) *dāru-vimāna* is also described. It contains four pitchers of mercury over iron ovens. "The boiling mercury ovens produce a terrific noise which is put to use in battle to scare away elephants. By strengthening the mercury chambers, the roar could be increased so that by it elephants are thrown completely out of control."[9]

There has been a great deal of speculation about just how power generated by heating mercury might be used to drive the *vimāna* through the air. This was discussed in an early book on UFOs by Desmond

Leslie and George Adamski.[10] Leslie proposed that the heated mercury mentioned in the *Samarāṅgana-sūtradhāra* may have something to do with the flight of UFOs.

I would suggest that the *vimānas* described by Bhoja are much more similar to conventional airplanes than to UFOs. Thus they are made of ordinary materials like wood, they have wings, and they fly like birds. Raghavan suggested that the mercury engine was intended to be a source of mechanical power for flapping the wings as in bird flight. He supported this by noting that Roger Bacon described a flying machine in which some kind of revolving engine caused wings to flap through a mechanical linkage.[11]

Ramachandra Dikshitar, however, said that according to the *Samarāṅgana-sūtradhāra*, the *vimāna* "has two resplendent wings, and is propelled by air."[12] This suggests that some kind of jet propulsion was used.

However these *vimānas* were actually powered, it seems likely that they relied on some conventional mechanical method that extracted energy from burning fuel and used it to produce a flow of air over wings. We can contrast this with the flight characteristics of UFOs, which don't have wings, jets, or propellers, and seem to fly in a manner that contradicts known physical principles.

Were the *vimānas* mentioned in *Samarāṅgana-sūtradhāra* ever actually built, or were they just products of imagination? I don't know. However, the elaborate descriptions of *yantras* found in medieval Indian texts suggest that many sophisticated machines were made in India long ago. If sophisticated mechanical technology was known in remote times, then it is quite possible that airplanes of some kind were also built.

It is interesting that the Sanskrit astronomical text entitled *Sūrya-siddhānta* mentions a mercury engine used to provide rotary motion for a *gola-yantra,* a mechanical model of the planetary system.[13] This suggests that at least one kind of mercury engine was used to produce rotary power. The text also says that the design for the mercury engine is to be kept secret. It was standard practice in ancient India for technical knowledge to be passed down only from teacher to trusted disciple. An unfortunate consequence of this is that knowledge tended to be lost whenever oral traditions depending on teachers and disciples were broken. It is thus quite possible that many arts and sciences known in ancient times have been lost to us, practically without a trace.

Additional Sanskrit works referring to flying machines are listed in a book by Dileep Kanjilal.[14] These are: the *Yukti-kalpataru* by Bhoja (twelfth century A.D.); the *Mayamatam* attributed to Maya Dānava but probably dating to the twelfth century A.D.; the *Kathāsaritsāgara* (tenth century A.D.); the *Avadāna* literature (first–third centuries A.D.); the *Raghuvaṁśam* and *Abhijñāna-śakuntalam* of Kālidāsa (first century B.C.); the *Abhimāraka* of Bhāsa (second century B.C.); and the *Jātakas* (third century B.C.). These dates are often approximate, and the material in the various works is often taken from older works and traditions.

The Vaimānika-Śāstra

The *Vaimānika-śāstra* is a highly detailed description of *vimānas,* and it is given great credence in a number of books and articles. These include the writings of Kanjilal,[15] Nathan,[16] and Childress.[17] In particular, the Indian ufologist Kanishk Nathan wrote that the *Vaimānika-śāstra* is an ancient Sanskrit text that "describes a technology that is not only far beyond the science of the times but is even way beyond the possible conceptual scientific imagination of an ancient Indian, including concepts such as solar energy and photography." [18]

It is indeed true that this book contains many interesting ideas about aerial technology. But it is important to note that it was written in the early 20th century by a psychic process similar to channeling or automatic writing.

The story behind this is briefly presented in the introduction to G. R. Josyer's translation of the *Vaimānika-śāstra*. There Josyer explained that knowledge in India used to be transmitted orally, but as this tradition died out, writing on palm leaves was used. Unfortunately, palm leaf manuscripts do not last very long in the Indian climate, and large volumes of old written material have been lost due to not being regularly recopied.

Josyer went on to say that the lost texts "remain embedded in the ether of the sky, to be revealed—like television—to gifted mediums of occult perception." The medium in this case was Subbaraya Sastry, a "walking lexicon gifted with occult perception," who began to dictate the *Vaimānika-śāstra* to Mr. Venkatachala Sarma on August 1, 1918. The complete work was taken down in 23 exercise books up to August 23, 1923. In 1923, Subbaraya Sastry also had a draftsman prepare some drawings of the *vimānas* according to his instructions.[19]

According to Subbaraya Sastry, the *Vaimānika-śāstra* is a section of a vast treatise by the sage Mahārṣi Bharadvāja entitled *Yantra-sarvasva* or the *Encyclopedia of Machines*. Mahārṣi Bharadvāja is an ancient *ṛṣi* mentioned in the *Mahābhārata* and other Vedic works, but I do not know of any reference indicating that he was concerned with machines. The *Yantra-sarvasva* is no longer extant in physical form, but it is said to be existing in the akashic record, where it was read and recited by Subbaraya Sastry. As far as I am aware, there are no references to this work in existing literature. This is discussed in Kanjilal's book on *vimānas*.[20]

Additional information about Subbaraya Sastry has been supplied by C. S. R. Prabhu, a technical director and project coordinator at the National Informics Center in Hyderabad, India. Prabhu traces Sastry's story back to 1875, when he was a young man of 20 living near the city of Bangalore in South India. Sastry had been abandoned to die of smallpox during a severe smallpox epidemic and had wandered into a forest region. He was about to commit suicide by drowning himself in a lake when he was saved by a *yogī* from the Himalayas named Bhāskarānanda. The *yogī* reportedly cured him of smallpox and kept him in his cave in the forest for about one year.

The *yogī* is said to have asked Sastry, "What do you want in life?" Sastry replied that he wanted to be renowned as a expert in *śāstras* (Sanskrit texts), and he specifically mentioned physical *śāstras,* since the standard religious *śāstras* are known by many people. The *yogī* granted his wish by transmitting to Sastry in an unknown way the texts of some 20 different *śāstras*. According to Prabhu, Sastry had been quite ordinary before meeting Bhāskarānanda.

After returning from the cave, Sastry was able to go into a trance state by closing his eyes and performing certain yogic *mudrās*. In that state he would recite elaborate Sanskrit texts on religion, science, or politics continuously, without pausing to think. One of these texts was the *Vaimānika-śāstra*.

Although the *Vaimānika-śāstra* could be a hoax, I have no reason to suppose that it was not dictated by Subbaraya Sastry in the manner described by Josyer and C. S. R. Prabhu. But is the work authentic? Even if it was existing as a vibrational pattern in the ether, during the process of psychic transmission and dictation it might have been distorted or adulterated by material from the unconscious mind of the medium.

There are good reasons for thinking this might be so, and there are also good reasons for thinking that the text might contain authentic material. I will first give the evidence suggesting that the text of the *Vaimānika-śāstra* may have been adulterated by modern material.

The text is illustrated by several of the drawings made under Subbaraya Sastry's supervision. These include cross sections of the *rukma-vimāna,* the *tripura-vimāna,* and the *śakuna-vimāna.* These cross sections show the kind of crude mechanical and electrical technology that existed in the period just following World War I. There are large electromagnets, cranks, shafts, worm gears, pistons, heating coils, and electric motors turning propellers. The *rukma-vimāna* is supposedly lifted into the air by "lifting fans" that are powered by electric motors and that are very small compared with the size of the *vimāna* as a whole. It definitely does not look as though it could fly.

These mechanical devices may well have been inspired by the technology of the early 20th century. However, C. S. R. Prabhu has reported on research showing that the text of the *Vaimānika-śāstra* contains technological information that Subbaraya Sastry is not likely to have acquired through ordinary means of communication.[21] This information consists of formulas for a number of metal alloys, ceramics, and glasses that are used in the construction of *vimānas.*

The formulas are expressed in obscure Sanskrit words, many of which cannot be found in standard Sanskrit dictionaries. Prabhu found through extensive research that some of these words can be found in rare dictionaries of Ayurveda, the ancient Indian system of medicine and chemistry. Through extensive consultation with Ayurvedic physicians and chemists, he was able to identify the actual substances referred to by some of these words. It was then possible to synthesize some of the materials mentioned in the *Vaimānika-śāstra* in the laboratory, using the instructions for mixing, heating, and cooling given in the text.

The results were remarkable. Several materials were synthesized, such as *tamogarbha loha,* a lead alloy, *arāra tamra,* a copper alloy, and *ravi śakti apakarṣana darpana,* a glass. These materials turned out to have useful properties matching the description given in the text of the *Vaimānika-śāstra.* For example, the text said that *tamogarbha loha* was a light absorbing material, and laboratory tests showed that the synthesized *tamogarbha loha* displayed a high level of absorption of laser light.[22] The synthesized materials were found to have unique properties that are new and patentable.

The formulas in the *Vaimānika-śāstra* seem to represent a science of chemistry and metallurgy that is expressed in archaic language. From what we know of the life of Subbaraya Sastry, it seems unlikely that he could have generated such formulas from readily available modern information. Perhaps these formulas do come from an ancient source.

Due to the presence of many untranslatable Sanskrit words, the *Vaimānika-śāstra* is not very intelligible. Nonetheless, this text does contain information about *vimānas* showing interesting parallels with the observed features of UFOs. To illustrate this, here are ten examples taken from a list in the *Vaimānika-śāstra* of 32 secrets that a *vimāna* pilot should know.[23] I will comment on relations between these items and common features of the UFO phenomenon.

1. Goodha: As explained in "Vaayutatva-Prakarana," by harnessing the powers, Yaasaa, Viyaasaa, Prayaasaa in the 8th atmospheric layer covering the earth, to attract the dark content of the solar ray, and use it to hide the Vimaana from the enemy.
2. Drishya: By collision of the electric power and wind power in the atmosphere, a glow is created, whose reflection is to be caught in the Vishwa-Kriyaa-darapana or mirror at the front of the Vimana, and by its manipulation produce a Maaya-Vimaana or camouflaged Vimana.
3. Adrishya: According to "Shaktitantra," by means of the Vynara thya Vikarana and other powers in the heart centre of the solar mass, attract the force of the ethereal flow in the sky, and mingle it with the balaahaa-vikarana shakti in the aerial globe, producing thereby a white cover, which will make the Vimana invisible.

Here three methods are described for hiding a *vimāna* from the enemy. They sound fanciful, but it is interesting to note that *vimānas* described in the *Purāṇas* and the *Mahābhārata* have the ability to become invisible. This is also a characteristic feature of UFOs, but this was certainly not well known in 1923.

The idea that a glow is created by the collision of electrical power and the wind is interesting. UFOs are well known for glowing in the dark, and this may be due to an electrical effect that ionizes the air surrounding the UFO. The word "shakti" (*śakti*) means power or energy.

4. Paroksha: According to "Meghotpatthi-prakarana," or the science of the birth of clouds, by entering the second of the summer cloud layers, and attracting the power therein with the shaktyaakarshana darpana or force-attraction mirror in the Vimana, and applying it to the parivesha or halo of the Vimana, a paralyzing force is generated, and opposing Vimanas are paralyzed and put out of action.
5. Aparoksha: According to "Shakti-tantra," by projection of the Rohinee beam of light, things in front of the Vimana are made visible.

Beams of paralyzing force are often mentioned in UFO accounts, as well as beams of light. The mention of a halo around the *vimāna* may be significant, since UFOs are often said to be surrounded by some kind of energy field.

6. Viroopa Karena: As stated in "Dhooma Prakarana," by producing the 32nd kind of smoke through the mechanism, and charging it with the light of the heat waves in the sky, and projecting it through the padmaka chakra tube to the bhyravee oil-smeared Vyroopya-darpana at the top of the Vimana, and whirling with 132nd type of speed, a very fierce and terrifying shape of the Vimana will emerge, causing utter fright to onlookers.
7. Roopaantara: As stated in "Tylaprakarana," by preparing griddhra jihwaa, kumbhinee, and kaakajangha oils and anointing the distorting mirror in the Vimana with them, applying to it the 19th kind of smoke and charging with the kuntinee shakti in the Vimana, shapes like lion, tiger, rhinoceros, serpent, mountain, river will appear and amaze observers and confuse them.

Although these descriptions seem completely wild, it is interesting that UFOs have been known to change shape in mysterious ways, and monstrous creatures have been known to emerge from landed UFOs and frighten people (see pages 303–5). Many of the items in this list of secrets have to do with creating illusions that bewilder enemies, and it seems that UFOs also create such illusions.

8. Saarpa-Gamana: By attracting the dandavaktra and other sevenforces of air, and joining with solar rays, passing it through the zig-zagging centre of the Vimana, and turning the switch, the Vimana will have a zig-zagging motion like a serpent.

The ability of UFOs to fly in a zig-zag fashion is well known today, but it wasn't widely known in 1923.

9. Roopaakarshana: By means of the photographic yantra in the Vimana to obtain a television view of things inside an enemy plane.
10. Kriyaagrahana: By turning the key at the bottom of the Vimana, a white cloth is made to appear. By electrifying the three acids in the north-east part of the Vimana, and subjecting them to the 7 kinds of solar rays, and passing the resultant force into the tube of the Thrisheersha mirror . . . all activities going on down below on the ground, will be projected on the screen.

The word "television" in item (9) was employed in the English translation of *Vaimānika-śāstra* that came out in 1973. The original Sanskrit text was written in 1923 before television was developed.

It turns out that there are many references to TV-like screens inside UFOs. For example, they show up in the following abduction cases described in this book: the Buff Ledge, Vermont, case (pages 118–24), the case of Filiberto Cardenas (pages 176–78), the case of William Herrmann (pages 175–76 and 180–85), and the Cimarron, New Mexico, case (pages 315–20). William Herrmann, in particular, said he was shown a screen on board a UFO that would produce close-up views of objects at ground level. With it he got a clear view of the astonished faces of onlookers who were watching the UFO from the ground.[24]

All in all, the descriptions in the *Vaimānika-śāstra* seem luridly fantastic. But there are many parallels between these descriptions and equally strange-sounding features of UFO accounts. I do not know if these parallels are significant, but it is curious that they should be there in a book written down between 1918 and 1923, before the UFO phenomenon was widely known.

It seems clear that the illustrations in the *Vaimānika-śāstra* are contaminated by twentieth century material from the medium's unconscious mind. Yet the passages I have just quoted mainly contain non-twentieth-century material, and this is expressed in terms of Vedic words and ideas. It may be largely a product of Subbaraya Sastry's imagination as applied to his extensive Vedic knowledge, or it may be a reasonably faithful rendition of an ancient Vedic text preserved as an etheric pattern.

The only way to find out about this is to obtain other obscure Sanskrit texts and see whether or not they confirm some of the material in the

Vaimānika-śāstra. Repeated confirmations would at least indicate that Subbaraya Sastry was presenting material from a genuine tradition, and further investigations would be needed to see whether or not that tradition had a basis in actual fact. The discovery of genuine metallurgical formulas in the *Vaimānika-śāstra* is certainly a step in this direction.

Vimānas in the Vedic Literature

The *Bhāgavata Purāṇa,* the *Mahābhārata,* and the *Rāmāyaṇa* are three important works in the Vedic tradition of India. I pointed out in Chapter 6 that these three texts contain a great deal of interesting material involving the aerial vehicles called *vimānas.* They also describe different races of humanlike beings who operate these vehicles, and they discuss the social and political relationships existing in ancient times between these beings and humans of this earth.

To some, this material is of no value because it seems fantastic and mythological. Thus the Indian ufologist Kanishk Nathan rejected the old Hindu religious texts because they attribute exaggerated feats to gods. He felt that they are simply poetry in which "a writer who is not reporting an actual event can let his imagination move in any direction it wishes to take him."[25] He also pointed out that these texts belong to a prescientific age, and therefore, "Given the cultural, technological and scientific knowledge of that historical period, a writer can, while enjoying generality and avoiding detail, create inventions and combinations that do not actually exist."[26]

One can reply that it has not been established that ancient writers were simply indulging in poetic imagination, with no regard for facts. There is a modern prejudice to the effect that anyone who has spiritual interests must be unscientific, and whatever he writes must be imaginary. This viewpoint makes sense as long as all observable data seem to support a mechanistic world model that excludes old religious ideas as exploded fallacies.

But if we carefully examine the UFO phenomenon, we find extensive empirical observations that completely contradict our comfortable mechanistic world view. It is noteworthy that this anomalous material—ranging from physically impossible flight patterns to beings that float through walls—fits quite naturally into the spiritually oriented cosmologies of the old Vedic texts. It is therefore worth considering that the writers of these texts may have been presenting a sound description of reality as they experienced it, rather than simply indulging in wild imagination.

General Purpose Vimānas

The preceding chapter presented the story of Śālva's *vimāna,* which is found in the *Mahābhārata* and the *Bhāgavata Purāṇa.* This was a large military vehicle that could carry troops and weapons, and it had been acquired by Śālva from a nonhuman technological expert named Maya Dānava. The *Purāṇas* and the *Mahābhārata* also contain many accounts of smaller *vimānas,* including pleasure craft that seem to be designed for a single passenger. These were generally used by Devas and Upadevas but not by human beings.

In this section, I will give a series of examples, showing how *vimānas* figure as common elements in many different stories from these texts. Each example is extracted from the midst of a larger story, and it is not feasible to present these stories fully in this book. My purpose in presenting the examples is to show that *vimānas* are frequently mentioned in the *Purāṇas* and the *Mahābhārata.* Apparently, they were as commonplace to people of the old Vedic culture as airplanes are to us today.

In the first account, Kṛṣṇa killed a pythonlike serpent who was trying to swallow his father, King Nanda. By Kṛṣṇa's arrangement, the soul of the serpent was transferred to a new body of a type possessed by the celestial beings called Vidyādharas. That soul had possessed such a celestial body before being placed in the body of the serpent, and so Kṛṣṇa asked him why he had been degraded to the serpent form:

> The serpent replied: I am the well-known Vidyādhara named Sudarśana. I was very opulent and beautiful, and I used to wander freely in all directions in my airplane. Once I saw some homely sages of the lineage of Aṅgirā Muni. Proud of my beauty, I ridiculed them, and because of my sin they made me assume this lowly form.[27]

In this passage the Sanskrit word *vimānena* is translated as "in my airplane." It seems to have been a small private vehicle.

The next story is similar. Kṛṣṇa had relieved the soul of one King Nṛga from imprisonment in the body of a lizard and had awarded him a celestial body. When the time came for the king to depart, a *vimāna* from another world came to get him:

> Having spoken thus, Mahārāja Nṛga circumambulated Lord Kṛṣṇa and touched his crown to the Lord's feet. Granted permission to depart, King Nṛga then boarded a wonderful celestial airplane as all the people present looked on.[28]

In the next case, we see the effect of a beautiful woman on the pilot of a *vimāna*. Here the sage Kardama Muni is describing the beauty of his future wife, Devahūti, to her father, Svāyambhuva Manu:

> I have heard that Viśvāvasu, the great Gandharva, his mind stupefied with infatuation, fell from his airplane after seeing your daughter playing with a ball on the roof of the palace, for she was indeed beautiful with her tinkling ankle bells and her eyes moving to and fro.[29]

It would seem that Viśvāvasu's *vimāna* was a small single-seater. Perhaps he didn't have adequate seatbelts, and he banked too steeply while trying to see Devahūti.

After Kardama Muni married Devahūti, he decided at a certain point to take her on a tour of the universe. To do this, he manifested an aerial mansion (called, as usual, a *vimāna*) that was lavishly equipped as a pleasure palace. Here the sage Maitreya relates the story of this mansion to his disciple Vidura:

> Maitreya continued: O Vidura, seeking to please his beloved wife, the sage Kardama exercised his yogic power and instantly produced an aerial mansion that could travel at his will.
>
> It was a wonderful structure, bedecked with all sorts of jewels, adorned with pillars of precious stones, and capable of yielding whatever one desired. It was equipped with every form of furniture and wealth, which tended to increase in the course of time. . . .
>
> With the choicest rubies set in its diamond walls, it appeared as though possessed of eyes. It was furnished with wonderful canopies and greatly valuable gates of gold.
>
> Here and there in that palace were multitudes of live swans and pigeons, as well as artificial swans and pigeons so lifelike that the real swans rose above them again and again, thinking them live birds like themselves. Thus the palace vibrated with the sounds of these birds.

> The castle had pleasure grounds, resting chambers, bedrooms and inner and outer yards designed with an eye to comfort. All this caused astonishment to the sage himself.[30]

The sage was astonished because he had not actually designed the aerial palace or imagined it in detail. In effect, what he did was mentally put in an order for a flying palace, and he received it from a kind of universal supply system because he had earned good karmic credit through his austerities and practice of *yoga*. To understand what was happening here, it is necessary to consider some basic features of the Vedic conception of the universe.

Over the years, many analogies have been used to describe the universe. Thus the Aristotelians compared the universe to a living organism, and the early mechanistic philosophers compared it to a gigantic clock. To understand the Vedic conception of the universe, the modern idea of a computer with a multilevel operating system is useful. On the hard disk of such a computer, there are programs that can be set into action by typing in appropriate code words. When a code word is typed, the corresponding program will execute—if the computer user has a suitable status. If he does not, then to him the code word is simply a useless name.

Typically, the user's status is indicated by the password he types when he begins to use the computer. Different users will have passwords indicating different status levels. Above all other users is a person called (in the Unix operating system) the superuser, who has full control over all programs on the system. Often this person is responsible for creating the total system by loading various pieces of software into the computer.

According to the Vedic conception, the universe has a similar organization. The superuser corresponds to the Supreme Being, who manifests the total universal system. Within that system there is a hierarchy of living beings having different statuses. A being at the ordinary human level has many remarkable powers, such as the power of speech, and a being at a higher level, such as Kardama Muni, can manifest even greater powers. When we grow up using a certain power, we tend to take it for granted, and when we completely lack access to a power, we tend to regard it as impossible or mythological. But all of the powers—including the power to call up flying palaces—are simply programs built into the universal system by the superuser.

The parallel between the Vedic conception of the universe and a computer can be made more explicit by introducing the concept of a virtual reality system. It is possible to create an artificial world by computer calculation and equip human participants with sensory interfaces that give them the impression of entering into that world. For example, a participant will have small TV screens placed in front of his eyes that enable him to see from the vantage point of the virtual eyes of a virtual body within the artificial world. Likewise, he may be equipped with touch sensors that enable him to experience the feel of virtual objects held in that body's virtual hands. Sensors that pick up his muscle contractions or his nerve impulses can be used to direct the motion of the virtual body.

Many people can simultaneously enter into a virtual world in this way, and they can interact with one another through their virtual bodies, even though their real bodies may be widely separated. Depending on their status, as recognized by the computer's superuser, the different virtual bodies may have different powers, and some of these powers might be invoked by uttering code words, or *mantras.*

An extremely powerful virtual reality system provides a metaphor for the Vedic universe of *māyā*, or illusion, in which conscious souls falsely identify themselves with material bodies. Of course, this metaphor should not be taken literally. The universe is not actually running on a digital computer. Rather, it is a system of interacting energies which, according to the Vedic conception, has features of intelligent design and organization reminiscent of certain man-made computer systems.

Returning to the story of Kardama Muni, we find that, after having acquired his marvelous flying palace, he proceeded to travel to different planets with his wife:

> Satisfied by his wife, he enjoyed in that aerial mansion not only on Mount Meru but in different gardens known as Vaiśrambhaka, Surasana, Nandana, Puṣpabhadraka, and Caitrarathya, and by the Mānasa-sarovara lake.
>
> He traveled in that way through the various planets, as the air passes uncontrolled in every direction. Coursing through the air in that great and splendid aerial mansion, which could fly at his will, he surpassed even the demigods.[31]

In the Sanskrit, the Devas are referred to here as *vaimānikān,* which means the "travelers in *vimānas.*" Thus the verse literally says that

Kardama Muni's *vimāna* excelled the *vaimānikān.* The Sanskrit word for planets is *loka,* which can refer to other physical globes and to higher-dimensional worlds not accessible to ordinary human senses.

An example of a *vimāna* used for military purposes comes up in the story of Bali, a king of the Daityas. Bali's vehicle is very similar to the one obtained by Śālva, and it was also built by Maya Dānava. It was used in a great battle between the Daityas and the Devas:

> For that battle the most celebrated commander in chief, Mahārāja Bali, son of Virocana, was seated on a wonderful airplane named Vaihāyasa. O King, this beautifully decorated airplane had been manufactured by the demon Maya and was equipped with weapons for all types of combat. It was inconceivable and indescribable. Indeed, it was sometimes visible and sometimes not. Seated in this airplane under a beautiful protective umbrella and being fanned by the best of *cāmaras,* Mahārāja Bali, surrounded by his captains and commanders, appeared just like the moon rising in the evening, illuminating all directions.[32]

My final example of a *vimāna* is taken from the story of the sacrifice of Dakṣa. It seems that Satī, the wife of Lord Śiva, wanted to attend a sacrifice arranged by her father Dakṣa, but Śiva did not want her to attend because of Dakṣa's offensive attitude toward him. Here we see Satī entreating her husband to let her go to the sacrifice after seeing her relatives traveling there in *vimānas:*

> O never-born, O blue-throated one, not only my relatives but also other women, dressed in nice clothes and decorated with ornaments, are going there with their husbands and friends. Just see how their flocks of white airplanes have made the entire sky very beautiful.[33]

All of the beings referred to here are Devas or Upadevas. We can see from this and the other examples that *vimānas* were considered to be standard means of travel for beings in these categories.

Flying Cities

Wendelle Stevens mentioned a study on the origin of UFOs carried out by a think tank in Brussels called Laboratoire de Recherche A. Kraainem. This study concluded that after reaching a certain stage of technology, a civilization will leave its home planet and "live in huge 'mother-ships,' artificial worlds, of their own creation perfectly adapted to their own needs and constantly maintained and perfected by them. . . . The artificial worlds are entirely self-sufficient and depend on no other planet or physical body for support. They are maintained and cruise [in] space indefinitely."[34]

The Vedic literature also has the idea of self-sustaining flying cities that travel indefinitely in outer space. An example is the set of three flying cities built by Maya Dānava for the sons of the Asura Tāraka. These are described as follows in the *Śiva Purāṇa:*

> Then the highly intelligent Maya built the cities by means of his penance: the golden one for Tārakākṣa, the silver one for Kamalākṣa, and the steel one for Vidyunmālī. The three fortlike excellent cities were in order in heaven, sky and on the earth. . . . Entering the three cities thus, the sons of Tāraka, of great strength and valour, experienced all enjoyments. They had many Kalpa trees there. Elephants and horses were in plenty. There were many palaces with gems. Aerial chariots shining like the solar sphere, set with Padmarāga stones, moving in all directions and looking like moonshine, illuminated the cities.[35]

It is interesting to note that shining *vimānas* were flying around the aerial cities. Here I am reminded of the many accounts in which smaller UFOs were seen in the vicinity of a large "mother-ship."

Turning to another example, the *Mahābhārata* tells the story of the flying city of Hiraṇyapura. This city was seen floating in space by Arjuna while he was traveling through the celestial regions after defeating the Nivātakavacas in a great battle. Arjuna was accompanied in his celestial journey by a Deva named Mātali, and he asked him about the city. Mātali replied:

> There once were a Daitya woman called Pulomā and a great Asurī Kālakā, who observed extreme austerities for a millennium of years of the Gods. At the end of their mortifications the self-existent God gave them a boon. They chose as their boon that their progeny should suffer little, Indra of kings, and be inviolable by Gods, Rākṣasas and Snakes. This lovely airborne city, with the splendor of good works, piled with all precious stones and impregnable even to the Immortals, the bands of Yakṣas and Gandharvas, and Snakes, Asuras, and

> Rākṣasas, filled with all desires and virtues, free from sorrow and disease, was created for the Kālakeyas by Brahmā, O best of the Bhāratas. The Immortals shun this celestial, sky-going city, O hero, which is peopled by Pauloma and Kālakeya Asuras. This great city is called Hiraṇyapura, the City-of-Gold.[36]

Here the inhabitants of the city, the Paulomas and Kālakeyas, are identified as the descendants of two rebellious relatives of the Devas named Pulomā and Kālakā. The "Snakes" are a race of mystical beings, called Nāgas, that can assume humanlike or serpentine form (see pages 289–93). The "self-existent God" is Brahmā, who is understood to be the original progenitor of all living beings within the material universe. Since Brahmā's origin is transcendental, and he has no material parents, he is said to be self-existent. The immortals are the Devas. They are referred to as immortal because they live for millions of our years. However, according to the *Vedas,* all embodied beings in the material universe have a finite life span and must die after some time.

With his superior powers, Brahmā arranged for the Paulomas and Kālakeyas to have a flying city that could not be successfully attacked by various powerful groups of beings within the universe, including the Devas. However, he left open a loophole for the Devas by declaring that the flying city could be successfully attacked by a human being.

Arjuna was half human, half Deva. His mother was an earthly woman, and his father was Indra, the king of the Devas. Indra had equipped Arjuna with celestial weapons just for the purpose of defeating enemies of the Devas who had obtained protective benedictions from Brahmā that didn't apply to humans. Thus Arjuna decided that it was part of his mission to attack Hiraṇyapura. Here is Arjuna's account of what happened after his initial attack:

> When the Daityas were being slaughtered they again took to their city and, employing their Dānava wizardry, flew up into the sky, city and all. I stopped them with a mighty volley of arrows, and blocking their road I halted the Daityas in their course. But because of the boon given them, the Daityas easily held their celestial, divinely effulgent, airborne city, which could move about at will. Now it would go underground, then hover high in the sky, go diagonally with speed,

> or submerge in the ocean. I assaulted the mobile city, which resembled Amarāvatī, with many kinds of missiles, overlord of men. Then I subdued both city and Daityas with a mass of arrows, which were sped by divine missiles. Wounded by the iron, straight-traveling arrows I shot off, the Asura city fell broken on the earth, O king. The Asuras, struck by my lightning-fast iron shafts, milled around, O king, prompted by Time. Mātali swiftly descended on earth, as in a headlong fall, on our divinely effulgent chariot.[37]

The battle between Arjuna and the Daityas began on the surface of a planet (perhaps the earth). On being strongly attacked by Arjuna, the Daityas took off in their flying city. It is noteworthy that the city could move underground and under the water, as well as through air or outer space. Many accounts describe UFOs entering and leaving bodies of water,[38] and some stories associate UFOs with underground or undersea bases. For example, Betty Andreasson's story of the Phoenix apparently took place in a subterranean realm,[39] and Filiberto Cardenas told of being taken to an undersea base.[40]

According to the *Mahābhārata,* just as the Daityas have flying cities such as Hiraṇyapura, the Devas have flying assembly houses, which are used as centers for their administrative activities. Here are two examples, beginning with the assembly hall of Indra, or Śakra, the king of the Devas. In this passage, a league is a Sanskrit *yojana,* which ranges from 5 to 8 miles:

> Śakra's celestial and splendid hall, which he won with his feats, was built by himself, Kaurava, with the resplendence of fire. It is a hundred leagues wide and a hundred and fifty long, aerial, freely moving, and five leagues high. Dispelling old age, grief, and fatigue, free from diseases, benign, beautiful, filled with chambers and seats, lovely and embellished with celestial trees is that hall where, O Pārtha, the lord of the Gods sits with Śacī. . . .[41]

It is standard for descriptions of *vimānas* to say that they are brilliantly glowing or fiery. We find the same feature in the following description of Yama's hall, which was built by Viśvakarmā, the architect of the Devas:

> This fair hall, which can move at will, is never crowded—Viśvakarmā built it after accumulating over a long time the power of austerities,

> and it is luminous as though on fire with its own radiance, Bhārata. To it go ascetics of dread austerities, of good vows and truthful words, who are tranquil, renouncing, successful, purified by their holy acts, all wearing effulgent bodies and spotless robes; . . . and so go great-spirited Gandharvas and hosts of Apsarās by the hundreds. . . . A hundred hundred of thousands of law-abiding persons of wisdom attend in bodily form on the lord of the creatures.[42]

An interesting feature of Yama's hall is that it is populated by beings of many different types. This is reminiscent of the UFO phenomenon, since it is often reported that several different types of beings will be seen on a UFO, apparently working in cooperation. In Yama's hall, in addition to Gandharvas, Apsarās, and various kinds of ascetics, there are Siddhas, those who have a yogic body, Pitās, men of evil deeds, and "those familiars of Yama who are charged with the conduction of time."

The latter are functionaries equipped with mystic powers that enable them to regulate the process of transmigration of souls. Yama is the Vedic lord of death, who supervises the process of transmigration. Strangely enough, even here we find a parallel with reported UFO phenomena. There are many reports indicating that some UFO entities can induce people to have out-of-body experiences and then exert control over their subtle bodies (see Chapter 10). This also happens to be one of the powers of the familiars of Yama.

What About Flying Horses and Chariots?

It is clear that there are extensive Vedic traditions about humanlike races of beings that can fly freely throughout the universe using vehicles called *vimānas*. But one might object that there are also Vedic stories about horse-drawn chariots that fly through the sky. Surely these stories are utterly absurd, since it makes no sense to say that an animal could run through air or outer space using its legs. Because of this absurdity, some claim, we should not take anything in the Vedic literature very seriously.

The answer to this objection is that there are indeed accounts of horse-drawn flying chariots in Vedic literatures, but these stories are not necessarily absurd. To understand them properly, it is necessary to fill in various details that will place them in context within the overall Vedic world picture. When seen in this way, both the horse-drawn chariots and the self-powered *vimānas* make sense.

I will try to fill in the needed details by referring to a number of stories from the *Mahābhārata* about the Pāṇḍava hero, Arjuna. In the first story, Arjuna is traveling through space in a literal chariot drawn by horses. This description has a number of important features, including travel through space on some kind of roadway:

> And on this sunlike, divine, wonder-working chariot the wise scion of Kuru flew joyously upward. While becoming invisible to the mortals who walk on earth, he saw wondrous airborne chariots by the thousands. No sun shone there, or moon, or fire, but they shone with a light of their own acquired by their merits. Those lights that are seen as the stars look tiny like oil flames because of the distance, but they are very large. The Pāṇḍava saw them bright and beautiful, burning on their own hearths with a fire of their own. There are the perfected royal seers, the heroes cut down in war, who, having won heaven with their austerities, gather in hundreds of groups. So do thousands of Gandharvas with a glow like the sun's or the fire's, and of Guhyakas and seers and the hosts of Apsarās.
>
> Beholding those self-luminous worlds, Phalguna, astonished, questioned Mātali in a friendly manner, and the other said to him, "Those are men of saintly deeds, ablaze on their own hearths, whom you saw there, my lord, looking like stars from earth below." Then he saw standing at the gateway the victorious white elephant, four-tusked Airāvata, towering like peaked Kailāsa. Driving on the roadway of the Siddhas, that most excellent Kuru Pāṇḍava shone forth as of old the great king Māndhātar. The lotus-eyed prince passed by the worlds of the kings, then looked upon Amarāvatī, the city of Indra.[43]

As I pointed out in Chapter 6 (pages 202–3), one important thing to notice about this passage is that Arjuna entered a region of stars where there was no light from the sun, the moon, or fire. This is what we would expect to find if we did travel among the stars. It is also stated that the stars are very large, but they seem small due to distance when seen from the earth, and this also agrees with modern ideas.

In that region, Arjuna saw that the stars were self-luminous worlds, and that they were hearths of Gandharvas, Guhyakas, and others, including "men of saintly deeds" who had been promoted to heaven. The stars themselves are spoken of as aerial chariots in this passage, and this is clearly a poetic description. They are also spoken of as persons, and this refers to the predominating persons living on them.

The next point to notice is that Arjuna was "driving on the roadway of the Siddhas," and that this roadway went past the worlds of the kings to the city of Indra. Later on, this road is spoken of as the "road of the stars" and the "path of the gods." [44] Thus it seems that Arjuna's chariot was traveling on some kind of road through outer space.

The *Viṣṇu Purāṇa* sheds some light on the actual route followed by Arjuna. It states that the Path of the Gods (*deva-yāna*) lies to the north of the orbit of the sun (the ecliptic), north of Nāgavīthī (the *nakṣatras* Aśvinī, Bharaṇī, and Kṛttikā), and south of the stars of the seven *ṛṣis*.[45] Aśvinī and Bharaṇī are constellations in Aries, north of the ecliptic, and Kṛttikā is the adjacent constellation in Taurus known as the Pleiades. Aśvinī, Bharaṇī, and Kṛttikā belong to a group of 28 constellations called *nakṣatras* in Sanskrit, and asterisms or lunar mansions in English. The seven *ṛṣis* are the stars of the Big Dipper in Ursa Major. From this information, we can form a general idea of the Path of the Gods as a roadway extending through the stars in the northern celestial hemisphere.

Another important celestial roadway is the Path of the Pitās (or *pitṛ-yāna*). According to the *Viṣṇu Purāṇa,* this roadway lies to the north of the star Agastya, and south of Ajavīthī (the three *nakṣatras* Mūla, Pūrvāṣāḍhā, and Uttarāṣāḍhā), outside of the Vaiśvānara path.[46] The region of the Pitās, or Pitṛloka, is said in Vedic literature to be the headquarters of Yama, the Deva who awards punishments to sinful human beings and whose aerial assembly house was described above. This region, along with the hellish planets, is said in the *Bhāgavata Purāṇa* to lie on the southern side of the universe, to the south of Bhū-maṇḍala, the earthly planetary system.[47]

The *nakṣatras* Mūla, Pūrvāṣāḍhā, and Uttarāṣāḍhā correspond to parts of the constellations Scorpio and Sagittarius, and it is thought that Agastya is the southern-hemisphere star called Canopus. Thus from the description in the *Viṣṇu Purāṇa* we can gain an idea of the location of Pitṛloka and the road leading to it in terms of familiar celestial landmarks.

Such celestial roadways involve large distances, and if they go through outer space, then there is the problem of the lack of a breathable atmosphere. What sort of horses could follow such roads? We can answer this question by recounting a *Mahābhārata* story in which Arjuna was offered a benediction by the Gandharva named Citraratha. Although Citraratha owned a *vimāna,* here he is concerned with horses:

> O best of men, I now wish to offer each of you five brothers a hundred horses of the type bred by the Gandharvas. The mounts of the gods and Gandharvas exude a celestial fragrance, and they move at the speed of the mind. Even when their energy is spent, they do not diminish their speed. . . .
>
> These Gandharva horses change color at will and fly at the speed they desire. And simply by your desire, they will appear before you, ready to serve. Indeed, these horses will always honor your wish.[48]

It seems that these are mystical horses that function according to laws governing subtle categories of material energy. The roadway on which they travel is presumably of a similar nature, and the fact that they can travel vast distances on this road in a short time is due to the fact that they obey the laws governing subtle energy rather than the laws governing ordinary, gross matter.

The fact that a gross human body can be carried along such a road can be understood in terms of the mystic *siddhis* called *prāpti* and *mano-java* discussed in Chapter 6. The basic idea is that the subtle laws include and supersede the gross laws. Gross matter obeying the familiar physical laws is also obeying the subtle laws. But the same subtle laws can be applied to cause gross matter to act in a way that violates the ordinary laws of physics.

Now let us consider Arjuna's chariot. Here is a description of one chariot that he used:

> The chariot had all necessary equipment. It could not be conquered by gods or demons, and it radiated light and reverberated with a deep rumbling sound. Its beauty captivated the minds of all who beheld it. Viśvakarmā, the lord of design and construction, had created it by the power of his austerities, and its form, like that of the sun, could not be precisely discerned.[49]

My tentative conclusion from this material is as follows: The technology involved in the *vimānas* and the flying horse-drawn chariots is essentially the same. It depends upon mystic powers and higher-dimensional aspects of material energy that are unknown to present-day science but are commonplace to the Devas. The *vimānas* are essentially architectural constructions that can fly, both in three dimensions and in higher dimensions, by virtue of powers that to us seem mystical. The Gandharva horses operate on the same mystical level, and the same is true of the chariots they draw.

If this is true, one might ask why the Devas and other related beings would bother with horse-drawn vehicles when *vimānas* that move by their own power are available. Judging from the *Mahābhārata* as a whole, the answer is that these beings use horses because they like them. They make use of flying architecture when that suits their purposes, but they also have a fondness for equestrian activities. Likewise, they have powerful weapons, like the *brahmāstra,* based on radiant energy, but they also have elaborate rules governing hand-to-hand fighting with maces. The general impression is that the Devas and Upadevas emphasize life and personal prowess over machines.

Are there any parallels between the celestial roadways of the Devas and information revealed in UFO accounts? There is a possible parallel in stories of people walking through space along beams of light. One example of this is in the report of Sara Shaw's abduction from a cabin in Tujunga Canyon, near Los Angeles, in March of 1953 (see page 179). After being hypnotized by Dr. William McCall, Sara told the following story of how she was taken on board a UFO:

> *McCall:* Do you stand near the ship?
> *Sara:* No, I'm starting to float. I'm starting to float toward it.
> *McCall:* What do you mean you're starting to float toward it?
> *Sara:* Well . . . they're walking with me, but my feet aren't on the ground.
> *McCall:* They were on the ground when you came out of the house. How come they're not on the ground now?
> *Sara:* Well, there's a beam of light. I'm like—it's almost like a—
> *McCall:* Now you see a beam of light?
> *Sara:* I'm *on* the beam of light. I'm standing on it, and it's angled. It's like an escala—no! It's about the same angle as an escalator would be, except it doesn't have ridges or steps. It's just a very smooth, solid beam, and you just kinda stand on it. . . .
> *McCall:* What's happening with your friends?

> *Sara:* They're all around me.
> *McCall:* They're on the beam of light also?
> *Sara:* But they're kind of—now I'm walking. All of us are walking, but in addition, the beam is conveying us. The beam is moving. In addition to that, we're kind of walking on it, too. But yet, like I don't feel anything under me. For example, it doesn't feel solid as if it were ground.[50]

If this story can be taken literally, it seems that the light beam not only nullified Sara's weight but also enabled her to balance herself in an upright stance and walk normally. Similar beam walking was apparently done by the beings in the Masse case (see page 219), who were said to "slide along bands of light."

A second example involves *swimming* up a beam of light. This phenomenon was reported by William Curtis, who experienced an abduction into a UFO in September of 1974 and recalled it in December of 1987. He had been abducted from his bedroom. At the point of being returned, he recalled being asked by his captors to jump through an opening in the bottom of the UFO, through which he could see down into his room. Here is how he described this experience:

> When I fell, it felt like . . . have you ever been on a roller-coaster? That's what it felt like. It took my breath away. But I was cushioned about two feet from the roof. I could see the distortion of the shingles. And then something picked me up, straightened me out and spun me around, and it dropped me down *right through the roof!* They laid me back on the bed, grabbed my arms and pulled me up. . . .
>
> A white light is coming through the roof and this little being is going up this light. I heard a whirring sound, like a generator coming from above. This little being as he went up was kicking his legs real fast. . . . He went up into a grey UFO. It seemed as if they had pressed a button and wanted me to see this. What I saw—the "AFM" [Alien Flying Machine] had duct work underneath it. The light went up, the ceiling went into place and that was it.[51]

In the first of these two accounts, there is an angled beam of light that a person can walk on. In the second, there is a vertical beam and a being who travels up the beam in what seems like a swimming motion. In both stories, the events, as described, seem completely bizarre from the point of view of accepted physical principles. This is especially true in the second story, where the beam of light is apparently used to transfer the man's body through the roof of his house. But in the case of Sara Shaw, the intruding beings entered her cabin by passing *through* the panes of a window (see page 224), and this is a similar phenomenon.

The parallel between these examples and the Vedic celestial roads is that the beam seems to define a pathway through space that a person can move along by using his legs. The beings that use these pathways have powers that enable them to pass through walls, and they can carry human bodies through walls also. The Vedic celestial road is also a pathway through space that one can walk on. The horses and chariots that move on it have mystical properties, and the horses can appear and disappear at will. A human being like Arjuna can also be conveyed along such a road. The point where the analogy of celestial road to light-beam path may break down is that the celestial road is cosmic in scale and seems to be relatively permanent, whereas the light beam is small and is deployed temporarily when needed.

It turns out, curiously enough, that the celestial pathways mentioned in Vedic literature are beams of light of a peculiar nature. Thus the *Bhāgavata Purāṇa* gives the following description of the travels of a mystic along the Path of the Gods:

> O King, when such a mystic passes over the Milky Way by the illuminating Suṣumṇa to reach the highest planet, Brahmaloka, he goes first to Vaiśvānara, the planet of the deity of fire, wherein he becomes completely cleansed of all contaminations, and thereafter he still goes higher, to the circle of Śiśūmāra, to relate with Lord Hari, the Personality of Godhead.[52]

The path followed by the mystic is the *deva-yāna* path, and it is referred to here as the illuminating Suṣumṇā. According to the Sanskrit dictionary, Suṣumṇā is the name of one of the principal rays of the sun. Thus the Suṣumṇā must be some kind of light beam. Clearly, however, its position in space indicates that it is not an ordinary sunbeam.

Vaikuṇṭha Vimānas

In his commentary on the *Bhāgavata Purāṇa,* A. C. Bhaktivedanta Swami Prabhupāda described three processes for moving in outer space. The first involves mechanical spaceships, and it is called *ka-pota-vāyu*. Here *ka* means ether, or space, and *pota* means ship. The phrase *ka-pota-vāyu* can also be used in a play on words, since *kapota* also means "pigeon."

The second process is called *ākāśa-patana*. "Just as the mind can fly anywhere one likes without mechanical arrangement, so the *ākāśa-patana* airplane can fly at the speed of mind."[53] Many of the *vimānas* that we have been discussing seem to use the *ākāśa-patana* process, and it may be that many UFOs also operate by mind action. Other *vimānas* and UFOs may operate by more mechanical processes that manipulate the ether, or in modern terms, the fabric of space-time.

According to the *Bhāgavata Purāṇa,* ether is the fabric of space, and all gross matter is generated by transformations of ether.[54] This is an idea reminiscent of John Wheeler's theory of geometrodynamics, which holds that all material particles are simply twists or deformations of space-time.[55] Both the *Bhāgavata Purāṇa* and Wheeler's theory imply that matter is directly connected to ether. Thus it should be possible to manipulate ether by manipulating gross matter. From this, we can see that it might be possible to build a physical machine that can manipulate space-time and provide for unusual modes of travel.

The *Bhāgavata Purāṇa* also states that the ether is the field of action of the subtle mind.[56] This suggests that it may be possible to manipulate the ether by mind action, thus allowing for the *ākāśa-patana* system of travel. Note that *ākāśa* means "ether" and *patana* means "flying."

The *ākāśa-patana* system makes use of subtle mind-energy, but it is still material. Beyond it is the Vaikuṇṭha process, which is completely spiritual. In the Vedic system, Vaikuṇṭha is the spiritual world. Whereas the material world is characterized by a duality between insentient matter and sentient spirit, in the world of Vaikuṇṭha, everything is conscious and self-effulgent. Objects in Vaikuṇṭha are made of a sentient substance called *cintāmaṇi,* which could be translated as "consciousness gem."

The Vedic literature contains many references to purely spiritual *vimānas* that originate in Vaikuṇṭha, and these should be mentioned in any account of Vedic *vimānas*. The Vaikuṇṭha *vimānas* are often compared with swans, or are said to be swanlike in shape, but they are not swans. They are flying structures that are made of *cintāmaṇi* and travel by the power of pure consciousness.

A Vaikuṇṭha *vimāna* is featured in the story of the liberation of a king named Dhruva from material bondage. Here is a description of how this vehicle appeared before Dhruva at the time of his death:

> As soon as the symptoms of his liberation were manifest, he saw a very beautiful airplane [*vimāna*] coming down from the sky, as if the brilliant full moon were coming down, illuminating all the ten directions.
>
> Dhruva Mahārāja saw two very beautiful associates of Lord Viṣṇu in the plane. They had four hands and a blackish bodily luster, they were very youthful, and their eyes were just like reddish lotus flowers. They held clubs in their hands, and they were dressed in very attractive garments with helmets and were decorated with necklaces, bracelets and earrings.[57]

Before boarding the *vimāna,* the king acquired his spiritual body, or *siddha-deha*. This is an imperishable bodily form made of spiritual energy and suitable for life in the Vaikuṇṭha atmosphere. Dhruva's travels in the *vimāna* are described as follows:

> While Dhruva Mahārāja was passing through space, he gradually saw all the planets of the solar system, and on the path he saw all the demigods in their airplanes showering flowers upon him like rain.
>
> Dhruva Mahārāja thus surpassed the seven planetary systems of the great sages who are known as *saptarṣi.* Beyond that region, he achieved the transcendental situation of permanent life in the planet where Lord Viṣṇu lives.[58]

In Vedic accounts, it is frequently mentioned that the Devas will shower down flowers on great personalities, especially at the time of great victories or other glorious events. This involves moving flower petals by the same kind of mystical transport that is typical of the Devas. A possible parallel to this is the mysterious appearance of showers of flower petals near Fatima, Portugal, at the time of visitations by an effulgent being taken by many to be the Virgin Mary. This is discussed in Chapter 8 (pages 293–301).

8

Modern Observations and Ancient Traditions

Although the main theme of this book is to explore the parallels between the UFO phenomenon and Vedic ideas, there are also close connections between this phenomenon and ideas found in other traditional systems of thought. These connections have been studied extensively in the case of old European folklore by Jacques Vallee in his books *Passport to Magonia* and *Dimensions*. Not surprisingly, there are also parallels between these traditions and the Vedic tradition. In this chapter, I will explore the triangle of mutual interrelationships connecting UFO reports, Vedic tradition, and other old traditions.

The empirical data on UFOs and related phenomena is extensive, but it tends to be incompatible with modern theoretical ideas, and therefore it is difficult to interpret. In contrast, the ancient Vedic world view is a consistent, meaningful system of philosophy and cosmology, but it does not include up-to-date observational data. My thesis is that these two things tend to complement one another: The UFO phenomena tend to corroborate the Vedic world view, and this world view, in turn, may help us understand UFO phenomena.

Vallee has made a similar point about UFO phenomena and old European folklore, but his presentation suffers from the fact that European folklore tends to lack clear-cut examples of UFO-like flying objects. His parallels are concerned mainly with the behavior and powers of the humanoid beings described in folklore and in UFO accounts. His case is therefore strengthened by the observation that there are strong parallels between the Vedic world view and the ancient world view of Europe. Since the Vedic literature contains many descriptions of flying machines, these parallels remedy the weakness in the UFO/ European-folklore comparison.

I should note that similarities between stories in two old traditions do not have the same meaning as similarities between accounts of distinct events by two contemporary witnesses. In the case of the two contemporary witnesses, one may be able to argue that they have not communicated with one another. One may then conclude that the similarities between their accounts indicate they have both had similar real experiences.

One cannot say this about similarities between ancient traditions. The very fact that the traditions are old means that there has been a great deal of opportunity for them to influence one another through ordinary means of human communication. In fact, I suspect that the ancient cultures of Europe and India had very strong ties and that much communication took place between them. Many similarities between Vedic literature and European folklore can be accounted for on this basis. At the same time, it is quite possible that some of the particular views of reality that have survived for centuries in Europe and India owe at least some of their staying power to ongoing experiences that tend to corroborate those views. According to this idea, UFO encounters would simply be contemporary examples of such experiences.

Here is one of Vallee's examples of a UFO sighting that shows a connection with old European traditions. The time was about 4:00 p.m. in the summer of 1968. A British woman was driving near Stratford with her companion when they saw a shining disc in the sky. They stopped to watch it dart and dodge, and another car also stopped to watch. After it disappeared behind the trees, she drove on, and during the drive she experienced amazing insights on the nature of reality that she said transformed her personality. After supper that evening, she then encountered a strange apparition, which she later spoke of as a "Scorpion man":

> The light from the room shone in an arc of about ten feet around the window. In that area I saw, as soon as I came to the window, a strange figure. My perception of it was heightened by the state of frozen panic it produced in me. It was for me without any doubt, a demon, or devil because of my Western oriented interpretation . . . It had dog or goat-like legs. It was covered in silky, downy fur, dark, and glinting in the light. It was unmistakably humanoid, and to my mind malevolent. It

> crouched, and stared, unblinkingly, at me with light, grape-green eyes that slanted upwards and had no pupils. The eyes shone and were by far the most frightening thing about it. It was, I think retrospectively, trying to communicate with me, but my panic interfered with any message I might have received. If it had stood to its full height it would have been about four to five feet tall. It had pointed ears and a long muzzle. It gave the impression of emaciation; its hands and fingers were as thin as sticks.[1]

Here the goatlike legs and the silky fur seem to connect the being with traditional European demon lore, while the eyes and emaciated appearance are typical of entities reported in UFO encounters. Are the traditional European demons real in some sense, and do they have some relationship with entities encountered on UFOs? Vallee extensively discussed similarities between UFO visitations and both pagan and Christian folklore regarding humanoids with paranormal powers. These include the succubi and incubi mentioned in medieval Roman Catholic writings, and the fairies and elves of old Celtic and Germanic tradition.

The Fairy Folk

The Celtic humanoids are generally called fairies in English. It is clear that this term refers to several different types of beings, although it would be difficult to arrive at clear descriptions of all of them. Some are said to have beautiful human forms, and others are said to be ugly. Some are very small, and others are as tall or taller than modern humans. The Irish word for the fairies is Sidhe (pronounced Shee), and they are also referred to as the Good People or the Little People. The *Bhāgavata Purāṇa,* uses the similar phrase Puṇya Jana or Pious People to refer to beings (such as the Yakṣas) that roughly correspond to the Celtic fairies.[2]

The Tuatha de Danann are an important type of Sidhe in Ireland. The name Tuatha de Danann means the descendants of the goddess Dana. This Dana was known as Brigit in middle Irish times, and she was apparently assimilated into Christianity as St. Brigit. According to tradition, the Tuatha de Danann were in full possession of the country when the sons of Mil, the ancestors of the Irish people, first came to Ireland. As humans invaded the island, the Tuatha de Danann remained, but they concealed themselves by their powers of invisibility. However, they continued to have intercourse with human society by communicating with seers and making themselves visible to selected humans.[3]

The ethnographer Walter Evans-Wentz made the observation that the Sidhe were sometimes thought to take birth as human kings among the Celtic people. Indeed, he pointed to literary evidence showing that the famous King Arthur was believed to be such an incarnation, and he pointed out that many of the people connected with Arthur in the Arthurian legends were either raised by the fairies or were members of their race. For example, Arthur's sword Excalibur was said to have been made in Avalon, the otherworld of the Sidhe. He was protected by a fairy woman called the Lady of the Lake, and his sister Morgan le Fay was a fairy.[4] (*Fay* is the French word for fairy.)

This shows that in early Celtic tradition human beings were thought to be in extensive and intimate contact with superhuman races that lived either on this earth or in invisible worlds directly connected with it. The same can be said of the ancient Vedic world view.

In ancient times there may have been direct cultural links between the Celtic and Vedic societies. For example, the *Bhāgavata Purāṇa* describes a race of beings called the Dānavas, or descendants of the goddess Danu. The Dānavas include several groups I have already mentioned, such as the Nivāta-kavacas, Kāleyas, and Hiraṇya-puravāsīs (inhabitants of Hiraṇya-pura). According to the linguist Roger Wescott, there is a cultural connection between the Vedic Danu and the Irish goddess Dana.[5]

Vallee showed that there are certain parallels between UFO accounts and the stories of the Sidhe in Celtic tradition. For example, UFO entities are known to mysteriously appear and disappear before people's eyes, and the same is true of the fairies. Thus Evans-Wentz heard the following from one John MacNeil of Barra, an island in the Western Hebrides of Scotland.

> The old people said they didn't know if fairies were flesh and blood, or spirits. They saw them as men of more diminutive stature than our own race. I heard my father say that fairies used to come and speak to natural people and then vanish while one was looking at them. . . . The general belief was that fairies were spirits who could make themselves seen or not seen at will. And when they *took* people they *took* body and soul together.[6]

The fairies evidently seemed physical because they could carry people off physically, and yet they seemed ethereal because they could appear and disappear at will. This apparent contradiction comes up repeatedly in connection with UFO abduction cases. It also comes up in Vedic accounts, and I will discuss this in detail in Chapter 10.

In Vedic literature, there are many accounts of humanoid beings who can appear and disappear, and who sometimes take people away to another world. They are said to do these things by means of specific powers, or *siddhis,* that involve interactions between the mind, the ether, and the gross physical elements. The stories of Duryodhana and of Arjuna and Ulūpī in Chapter 6 show that the Vedic humanoids are also said to abduct people in ways that have parallels in some UFO cases.

Abductions and Crossbreeding

Abduction is a standard theme in traditional fairy tales (which must be distinguished from the expurgated versions intended for modern children). In these stories, men and women are often abducted out of lust by fairies of the opposite sex. Children are also taken, and it is said that a fairy child, called a changeling, may be substituted for a human child. Just as we find in UFO cases, it seems that both sex desire and genetic considerations are involved in these abductions. In support of this, Vallee cited Edwin Hartland, a scholar of fairy traditions, as to the reasons people in northern European countries gave for this abduction of children:

> The motive usually assigned to fairies in northern stories is that of preserving and improving their race, on the one hand by carrying off human children to be brought up among the elves and to become united with them, and on the other hand by obtaining the milk and fostering care of human mothers for their own offspring.[7]

This interpretation was discussed by nineteenth-century mythologists on the basis of their studies of folklore. It is, of course, very similar to the explanation given by Budd Hopkins, Raymond Fowler, and others for UFO abductions, in which women are said to be impregnated—and their fetuses prematurely removed—by alien entities.

Hartland's comment about the fostering care of human mothers is especially puzzling, since one wonders why elves would need human mothering. However, a similar point comes up in studies of UFO abductions. David Jacobs, an associate professor of history at Temple University in Philadelphia, has written a book on UFO abductions that includes detailed descriptions of "presentation scenes," in which a UFO abductee is asked to come in physical contact with alien or half-alien children:

> Abductees are also required to touch, hold, or hug these offspring. . . . Apparently it is absolutely essential for the child to have human contact. Although the aliens prefer that the humans give nurturing, loving contact, any physical contact seems to suffice.[8]

In the Vedic literature, there are many accounts of sexual relations between humans and members of nonhuman races that give rise to offspring. One example is the union that took place between Bhīma, one of the human heros of the *Mahābhārata*, and Hiḍimbā, a Rākṣasa woman. Hiḍimbā had approached Bhīma out of lust for him, and to make love with him she assumed the illusory form of a beautiful human woman. The result was a child, who is described as follows:

> And while she loved Bhīma everywhere, nimble as thought, the Rākṣasī gave birth to a son by the powerful Bhīmasena. He was a terrifying sight, squint-eyed, large-mouthed, needle-eared, loathsome-bodied, dark-red-lipped, sharp-tusked, and powerful, born a great archer of great prowess, great courage, great arms, great speed, great body, great wizardry, tamer of his foes. Inhuman, though born from a human, of terrible speed and great strength, he surpassed the Piśācas and other demons as he surpassed human beings.[9]

In this case, Bhīma and Hiḍimbā remained together when the child was young, but she soon departed and took the boy with her. When he was newly born, Bhīma commented, "He is shiny as a pot!" So he was named Ghaṭotkaca, which means Shiny-as-a-Pot. Bhīma and his brothers, the Pāṇḍavas, were very fond of the boy, even though he strongly resembled his mother and looked distinctly nonhuman. His appearance is typical of Rākṣasas, and it is quite different from that of both human beings and the UFO humanoids that are commonly reported today.

No genetic manipulations were carried out by Bhīma and Hiḍimbā, but the *Mahābhārata* points out that Ghaṭotkaca had been created by Indra, the ruler of the Devas, so that he might destroy a certain warrior named Karṇa. This suggests that Indra engaged in genetic (or other) intervention at the time of Ghaṭotkaca's conception.

Indra's motive for this was to protect his own son, Arjuna, since he knew that Arjuna would eventually have to fight with Karṇa. Arjuna was one of the Pāṇḍava brothers, and as Indra's son, he was the offspring of a Deva father and a human mother named Kuntī. All of the five Pāṇḍavas were the sons of various Devas and two human mothers, Kuntī and Mādrī, who were wives of a human king named Pāṇḍu.

The Vedic accounts indicate that the various humanoid races in the universe are generally able to interbreed and produce fertile offspring. This suggests that they must all be genetically related to one another, and according to the Vedic literature, this is indeed the case. All of the humanoid races descend from male and female forms generated by Brahmā, the original created being. The Devas are among the descendants of these forms, and earthly human beings are descended from Devas along a number of different lines of ancestry.

Genetics and Human Origins

However, the meaning of the word "genetic" needs to be extended. All living organisms known to modern science contain genes made of DNA that specify the organisms' hereditary traits. The bodies of Brahmā and the Devas are made of subtle forms of energy, and thus they do not contain DNA. However, they do carry genetic information in the form of *bījas,* or seeds, that are also made of subtle energy.[10] For humans to descend from Devas, a systematic transformation is required that converts subtle energy into gross energy. This same transformation must convert the subtle *bījas* into gross genes made of DNA.

This partially confirms and partially contradicts the genetic intervention theory for human origins that I discussed in Chapter 5 (pages 186–89). The Vedic version is that earthly humans did descend from higher humanlike beings from other planets, but this was not by genetically engineered crosses between the higher beings and primitive apemen living on the earth. Rather, it involved mating between Devas that generated human offspring through preplanned genetic transformations.

In general, the descendants of Brahmā on the level of Devas and higher were able to produce offspring that were not of their own bodily type. I have not seen specific descriptions of how this was done, but I gather that it was preprogrammed by Brahmā. There is no indication that it was done by independent scientific research by the Devas. Rather, they seem to have simply made use of powers invested in them by Brahmā at an earlier stage of creation.

The Vedic conception of the origin of living species is not Darwinian. As I pointed out in Chapter 4 (pages 134–37), if humanoids of the kind reported in connection with UFOs are real, then their existence likewise poses a challenge to the Darwinian theory of evolution. According to the current understanding of natural history, such beings could not have evolved on the earth. It is also highly improbable that beings so similar to ourselves could have independently evolved on another planet.

The genetic intervention theory presented by Zecharia Sitchin proposes that humans arose from a genetically engineered cross between extraterrestrials and ape-men that had evolved on the earth. In this theory, the extraterrestrials themselves presumably evolved on another planet. However, it is difficult to explain why such beings should be close enough genetically to the ape-men to make crossbreeding a worthwhile proposition.

To illustrate this point, suppose we want to produce a new computer program by combining independently written machine language programs from two different computers. Even if the two programs did similar things, they would probably do them using completely different internal coding, and thus they would be incompatible with each other. In such a situation, even the most advanced computer expert would find it easier to create the new program from scratch than to get the two incompatible programs to work together. (Or he might prefer to produce the new program by modifying *one* of the existing programs.)

The Vedic account avoids the genetic incompatibility problem by starting with Devas and positing a transformation that alters the Deva form. The resulting human form is different from the Deva form but is apparently close enough to it that crossbreeding between humans and Devas is possible.

Note that the need to convert genetic information from a subtle form to a gross form does not constitute an insurmountable barrier. Information is abstract, and the same information can be stored using different types of energy. Converting information from subtle to gross is comparable to converting text from computer-coded electrical signals to print on paper.

Although the Deva-to-human transformation seems to have been preprogrammed by Brahmā, there are Vedic descriptions of the creation of human races by genetic manipulation. In one account, a king named Vena turned out to be a cruel tyrant, and he was killed by great sages. His mother preserved his body "by the application of certain ingredients and by chanting mantras." [11] Later on, the sages reflected that the king's hereditary qualities were valuable, and to preserve them they acted as follows:

> After making a decision, the saintly persons and sages churned the thighs of the dead body of King Vena with great force and according to a specific method. As a result of this churning, a dwarflike person was born from King Vena's body.
>
> This person born from King Vena's thighs was named Bāhuka, and his complexion was as black as a crow's. All the limbs of his body were very short, his arms and legs were short, and his jaws were large. His nose was flat, his eyes were reddish, and his hair copper-colored.[12]

I should point out that this story should not be used to support any theories of racial superiority or inferiority. According to the Vedic viewpoint, all persons are equal as spiritual beings, and it is a mistake to try to judge them on the basis of the material body, which is simply an external covering for the soul.

It would seem that the sages produced Bāhuka by carrying out an operation very similar to what is now known as cloning. According to current scientific understanding, the chemically intact DNA from any cell in a person's body contains the genetic information specifying that body. This means that it would be theoretically possible to produce a new living body using tissue from a dead body, assuming that the tissue had not begun to decay.

In this case, it would appear that a transformation of the genetic material was also carried out, and thus a dwarflike person was produced who had features quite different from those of King Vena. The sages later churned the arms of Vena's body and produced beautiful male and female persons, named Pṛthu and Arci, who were different in form and quality from both Vena and the dwarf.

Succubi and Incubi

In the UFO literature, in Western folklore, and in the Vedic literature, there are accounts of sexual relations between humans and humanoid beings that seem to be motivated by lust and that generally do not result in offspring. In this section, I will give some examples taken from these three sources of information.

Whitley Strieber wrote of meeting a strange-looking female being with huge eyes and emaciated limbs that, in his description, sounds like a cross between a human and a praying mantis. Strieber said that he had encountered this being more than once, and he felt that he had some kind of erotic relationship with her. He described her as follows to the journalist Ed Conroy:

> She is a human being, like all of her kind. But she is of the next level of man. What we call "unconscious" is among her kind full of light. I would like to take her away, for my own self. . . . She can be as innocent as a baby and as sensual as a fox. . . . When you're angry, it's like she's physically being hurt, and you have to reach into the depths of yourself to find a level of serenity that is deep enough to enable her to be calm in your presence. And when her passion comes on her, she appears out of the night. . . . I'm afraid that my succubus is quite real.[13]

The being that visited Whitley Strieber is rather different from the Celtic fairies. Typically, the fairies are described as being much more human in appearance, and many are said to be quite beautiful by our standards. However, her behavior seems to be foreshadowed by the teachings of the Catholic Church regarding succubi and incubi. Consider the following statement by St. Augustine, written in about 420 A.D.:

> There is, too, a very general rumour, which many have verified by their own experience, or which trustworthy persons who have heard the experience of others corroborate, that sylvans and fauns, who are commonly called "incubi," had often made wicked assaults upon women, and satisfied their lust upon them. That certain devils, called Duses by the Gauls, are constantly attempting and effecting this impurity is so generally affirmed, that it would be impudent to deny it.[14]

Consider the "Scorpion man," mentioned above, that was seen by a British women in connection with a UFO sighting. This being resembled a traditional faun in that it had goatlike or doglike legs. With its frightening, slanted eyes and emaciated appearance, it also resembled Strieber's visitor.

A succubus of more human form appears in a story from recent Celtic tradition related by Evans-Wentz. He said that an informant in Barra related a tale told by his grandmother Catherine MacInnis. She used to tell of a man named Laughlin, whom she knew and who was in love with a fairy woman. This woman would visit Laughlin every night, and finally he became so worn out with her that he began to fear her. To escape, he emigratēd to America, but apparently she haunted him there also.[15]

Stories of this kind are still current today in South India. A man from a brahmin family in Tamil Nadu told me that as a young man he had been dabbling in tantric occultism. In the course of this, he had a frightening encounter with a naked, not-quite-human female being who appeared before him suddenly while he waited at a midnight rendezvous. A tantric expert later explained what he had seen:

> You saw Mohinī, a demon from the underworld. Had you known how, you could have entered a pact with her for the next cycle of Jupiter (twelve years). You promise to satisfy her lust once a month, and she will do your bidding in return, protect your property, destroy your enemies, whatever.
>
> But a pact with Mohinī is very dangerous. When she comes for sexual satisfaction, she may assume eighteen forms in the course of the night, expecting you to fulfill the demands of each one. If you cannot, it will cost you your life. And if during the twelve years of your relationship with her you have an attraction to another woman, that will also cost you your life. You suddenly vomit blood—finished.

In recent years, a great deal of evidence has accumulated suggesting that some UFO abductions may have more to do with sexual exploitation than scientific research in genetics. Sexual encounters between UFO entities and humans range from Harvard psychiatrist John Mack's "blissful merging" [16] to abductee Leah Haley's encounter with a loathsome reptilian.[17] There are even reports of apparent marital relationships between humans and extraterrestrials. An example is the story of Denise and Bert Twiggs about their subconscious, four-way marriage with benevolent beings from Andromeda.[18]

There are two striking parallels between descriptions of sexual encounters in Vedic traditions and in modern UFO stories. First, as we have seen in the cases of Hiḍimbā and Mohinī, the Vedic stories typically involve the projection of attractive illusory forms. For comparison, UFO researchers David Jacobs[19] and Richard Boylan[20] have described the projection of illusory forms of human loved ones in UFO-related sexual encounters. There are also cases in which forms of beautiful women or famous personalities were apparently projected.

According to Indian tradition, beings of low consciousness such as Mohinī exploit humans with the motive of drawing energy from human emotions. This is paralleled by UFO researcher Karla Turner's comment that "a certain group of these beings in some way "feed' off our emotions, especially the strong ones that come from fear, pain, depression, and compulsive actions." [21] Ed Walters of Gulf Breeze, Florida, has also described how abducting entities somehow forced him to repeatedly recall highly emotional experiences in his life, perhaps for the purpose of enjoying them vicariously.[22] This, of course, is one way to explain the motivation behind the sexual element in UFO abductions, the display of apparent half-breed babies, and the frightening "medical" procedures.

Time Dilation and Other Worlds

In traditional Celtic stories the abduction theme is often combined with a visit to another world. In one such story, the hero Ossian was enticed into the mystical land of Tir na nog by a beautiful Sidhe princess. He married her and lived for three hundred of our years in her world. Finally, however, he felt an overpowering desire to return to Ireland and participate in the counsels of the Fenian Brotherhood. He set out on the same white horse that had taken him to the otherworld, and his fairy wife warned him not to lay his foot on the level ground.

On reaching Ireland, he searched for the Brotherhood but found that all his old companions had passed away and the country was quite changed. Only then did he realize how long he had been away. Unfortunately, at a certain point some incident caused him to dismount, and on touching the earth he immediately turned into a feeble, blind old man.[23]

In European folklore there are many stories with similar elements, including the entry into another world and the aging or death of the protagonist when he realizes how much time has passed in our world during his absence. Here is a similar story dating back to the early 19th century. In the Vale of Neath, Wales, two farm workers named Rhys and Llewellyn were walking home one night. Rhys was attracted by the sound of some mysterious music, but Llewellyn heard nothing. So Llewellyn continued home while Rhys stayed back to dance to the tune he had heard. The next day, Rhys didn't show up, and after a fruitless search, Llewellyn was jailed on the suspicion of murder.

However, a man learned in fairy lore guessed what had happened. On his advice, a party of men accompanied Llewellyn to the spot where Rhys was last seen. At this spot, Llewellyn could hear the music of harps because his foot was touching a "fairy-ring." When each of the other members of the party put his foot on Llewellyn's, he could hear the music too and could see many Little People dancing in a circle. Rhys was among them. When Llewellyn pulled him out of the circle, Rhys declared that he had only been dancing for five minutes. He could not be convinced that so much time had passed, and he became depressed, fell ill, and soon died.[24]

The otherworld of the Celts has various names, such as Avalon, Tir na nog (Land of Youth), and Plain of Delight. Examination of the stories makes it clear that this realm would have to exist in a higher dimension. To reach it, one must go to the right place in three-dimensional space, and then one must travel in a mystical fashion that we do not understand. We can speak of this as an extra dimension of travel in addition to the three we are familiar with.

Since the otherworld can be reached by mystical travel from this world, we can speak of it as a parallel reality. This idea can be understood by imagining jumping back and forth between two parallel planes that are close together. The planes represent the parallel realities, and the jumping corresponds to the higher-dimensional travel.

If we turn to Chinese folklore, we find a parallel to the story of Ossian, with its time lapse of hundreds of years. There is a book entitled *The Report Concerning the Cave Heavens and Lands of Happiness in Famous Mountains,* by Tu Kuang-t'ing, who lived from 850 to 933 A.D. This book lists ten "cave heavens" and thirty-six "small cave heavens" that were supposed to exist beneath mountains in China. Here are the reported experiences of a man who entered a passageway leading to one of these cave heavens:

> After walking ten miles, he suddenly found himself in a beautiful land "with a clear blue sky, shining pinkish clouds, fragrant flowers, densely growing willows, towers the color of cinnabar, pavilions of red jade, and far flung palaces." He was met by a group of lovely,

> seductive women, who brought him to a house of jasper and played him beautiful music while he drank "a ruby-red drink and a jade-colored juice." Just as he felt the urge to let himself be seduced, he remembered his family and returned to the passageway. Led by a strange light that danced before him, he walked back through the cave to the outer world; but when he reached his home village, he did not recognize anyone he saw, and when he arrived at his house, he met his own descendants of nine generations hence. They told him that one of their ancestors had disappeared into a cavern three hundred years before and had never been seen again.[25]

Here we find the same time dilation effect that repeatedly appears in European folklore. This effect, plus the fact that the man found himself in a land with a blue sky and clouds, indicates that the cave passageway led him to a parallel world.

In the *Bhāgavata Purāṇa,* there is a description of a parallel reality called Bila-svarga, or the subterranean heaven, which is clearly related to the Chinese story of the cave heavens. Bila-svarga is described as a very beautiful place, with brilliantly decorated cities, lakes of clear water, and extensive parks and gardens.[26] At the same time, the sun and the moon cannot be seen there, and the inhabitants have no sense of the passing of time. Bila-svarga is subdivided into seven worlds called *lokas,* and thus it is more than a mere cave within the earth fixed up with artificial lighting.

One of the *lokas* is Atala, which is said to be inhabited by three groups of women, called *svairiṇī, kāmiṇī,* and *puṁścalī*. Here is what happens to a man who manages to visit this region:

> If a man enters the planet of Atala, these women immediately capture him and induce him to drink an intoxicating beverage made with a drug known as *hāṭaka* [*cannabis indica*]. This intoxicant endows the man with great sexual prowess, of which the women take advantage for enjoyment. A woman will enchant him with attractive glances, intimate words, smiles of love and then embraces. In this way she induces him to enjoy sex with her to her full satisfaction. Because of his increased sexual power, the man thinks himself stronger than ten thousand elephants and considers himself most perfect. Indeed, illusioned and intoxicated by false pride, he thinks himself God, ignoring impending death.[27]

It is significant that Atala is referred to as a "planet" in this translation. Sometimes the word *loka* is translated as "planetary system," and the seven *lokas* of Bila-svarga are referred to as "lower planetary systems." The *Bhāgavata Purāṇa* indicates that Bila-svarga extends throughout the plane of the solar system, and for this reason it is called a *svarga,* or heaven (see pages 228–29). However, it can be reached by entering into the earth, using higher-dimensional modes of travel, and in this sense it is *bila,* or subterranean.

In Vedic literature, it is said that there is a hierarchy of planetary systems, which we can think of as parallel worlds. The highest system is Brahmaloka, the world of Brahmā, and it exhibits the most extreme degree of time dilation relative to the earth. Other, intermediate planetary systems exhibit intermediate degrees of time dilation.

The time dilation in Brahmaloka is illustrated by the following story. This story begins with mention of a submarine kingdom called Kuśasthalī that may involve a parallel reality in its own right. The people in the story are members of the Sūrya-vaṁśa, a dynasty descending from Sūrya, the presiding Deva of the sun. They are considered to be human, but they were endowed with mystic powers not possessed by ordinary humans of today. One of them, a king named Kakudmī, was able to travel to the world of Brahmā, where he experienced Brahmā's scale of time:

> O Mahārāja Parīkṣit, subduer of enemies, this Revata constructed a kingdom known as Kuśasthalī in the depths of the ocean. There he lived and ruled such tracts of land as Ānarta, etc. He had one hundred very nice sons, of whom the eldest was Kakudmī.
>
> Taking his own daughter, Revatī, Kakudmī went to Lord Brahmā in Brahmaloka, which is transcendental to the three modes of material nature, and inquired about a husband for her. When Kakudmī arrived there, Lord Brahmā was engaged in hearing musical performances by the Gandharvas and had not a moment to talk with him. Therefore Kakudmī waited, and at the end of the musical performances he offered his obeisances to Lord Brahmā and thus submitted his long-standing desire.
>
> After hearing his words, Lord Brahmā, who is most powerful, laughed loudly and said to Kakudmī, "O King, all those whom you may have decided within the core of your heart to accept as your son-in-law have passed away in the course of time. Twenty-seven *catur-yugas* have already passed. Those upon whom you may have decided

are now gone, and so are their sons, grandsons and other descendants. You cannot even hear about their names." [28]

In traditional Sanskrit texts, one *catur-yuga* is 4,320,000 years. With this information, we can estimate the rate of time dilation on Brahmaloka. If the concert given by the Gandharvas took about one hour in Brahmā's time scale, then that hour must correspond to 27 times 4,320,000 earth years. It turns out that this estimate closely matches a time dilation calculation based on another story involving Brahmā.

This is the story of the *Brahma-vimohana-līlā,* or the bewilderment of Brahmā by Kṛṣṇa. Several thousand years ago, Kṛṣṇa descended to the earth as an *avatāra* and was playing as a young cowherd boy, tending calves in the forest of Vṛndāvana (which is to the south of present-day New Delhi). To test Kṛṣṇa's potency, Brahmā used his mystic power to steal Kṛṣṇa's calves and cowherd boyfriends and hide them in suspended animation in a secluded place. He then went away for a year of earthly time to see what would happen.

Kṛṣṇa responded to Brahmā's trick by expanding himself into identical copies of the calves and boys. When Brahmā returned to see what had happened, he saw that Kṛṣṇa was playing with the boys and calves just as before, and he became completely bewildered. On checking the boys and calves that he had hidden away, he found that they were indistinguishable from the ones playing with Kṛṣṇa, and he couldn't understand how this was possible. Finally, Kṛṣṇa revealed to Brahmā that these latter boys and calves were actually identical with Himself, and He allowed Brahmā to have a direct vision of the spiritual world.

Now, it turns out that even though Brahmā was absent for one earth year, on his time scale only a moment had passed. The Sanskrit word used here for a moment of time is *truṭi.*[29] There are various definitions of a *truṭi,* but the Vedic astronomy text called the *Sūrya-siddhānta* defines a *truṭi* to be 1/33,750 seconds.[30] This tells us that one year on the earth corresponds to 1/33,750 seconds in the time of Brahmā.

As I pointed out, King Kakudmī's visit to Brahmaloka took 27 times 4,320,000 earth years. If we multiply this by 1/33,750 we find that in Brahmā's time, King Kakudmī's visit lasted 3,456 seconds, or just under an hour. This is consistent with the story that the king had to wait for a musical performance to finish before having a brief conversation with Lord Brahmā.

By the way, after Brahmā had his meeting with Kṛṣṇa, he brought the original cowherd boys back to normal consciousness. They found to their amazement that they had one year of "missing time."

Parallel Realms and UFOs

There are contemporary reports of experiences in which a person seemingly enters briefly into another world and then returns to our ordinary world to find that much time has passed. Like the story of Rhys and Llewellyn, these stories typically arise in the context of traditional belief systems involving beings with mystical powers.

For example, in June 1982 in Malaysia, a twelve-year-old girl named Maswati Pilus was going to the river at 10 a.m. to wash some clothes. Suddenly, she encountered a strange female being of her own size who invited her to see another land. "She felt no fear and found herself in a bright and beautiful place. . . . It seemed as if time had whizzed by." Two days later, she was discovered lying unconscious on the ground by relatives who had been frantically searching that very area for the whole time.[31] Malaysian tradition assigns the strange female to a group of beings called Bunians, who are known for abducting children. UFOs are not associated with them, however, and Jenny Randles reported that a search for UFO abduction cases in Malaysia turned up nothing.

The story of Ossian is typical of Celtic fairy lore in that it takes place in a parallel reality and features a time dilation effect in which time passes more slowly in the parallel world than in the ordinary world. We see the same thing in the Chinese cave story and in many Vedic accounts. The story of Rhys and Llewellyn and the story of Maswati Pilus both involve a parallel reality and a moderate time dilation effect.

As far as I am aware, in UFO accounts there are no direct parallels to the stories of Ossian, Rhys and Llewellyn, and Maswati Pilus, in which there is explicit entry into a parallel world. However, John Mack observed that "Quite a few abductees have spoken to me of their sense that at least some of their experiences are not occurring within the physical space/time dimensions of the universe as we comprehend it. . . . They experience the aliens, indeed their abductions themselves, as happening in another reality, although one that is as powerfully actual to them as—or more so than—the familiar physical world."[32] Kenneth Ring's idea of an ontologically real imaginal world also places UFO abductions in a parallel reality.[33]

The events reported in many UFO abductions are suggestive of a transition to a parallel reality. People undergoing abductions are sometimes shifted to a state in which they can pass through walls. They are sometimes taken on UFOs to very strange, unfamiliar places, such as the subterranean realm described by Betty Andreasson.[34] But it is hard to say whether these places are on this earth, on another planet, or in another dimension. Even out-of-body experiences on UFOs could be taking place in ordinary three-dimensional space, since OBEs in which a cardiac patient views his own unconscious body plainly do take place in ordinary hospital rooms. (OBEs and UFOs are discussed in Chapter 10.)

Since Betty Andreasson spoke of going through a closed door of her house and then through a normal, open door in the UFO parked outside,[35] one can argue that perhaps the UFO itself existed in another dimension. This could help explain why it is so rare for other people to witness UFO abductions from a distance. Of course, it is also possible that Betty resumed a normal physical state after passing through the door and entered a UFO made of ordinary matter in ordinary three-dimensional space.

Perhaps the strongest argument linking UFO reports with accounts of parallel realities is that both involve beings with similar mystic powers and similar modes of behavior. If certain beings can operate in a parallel world, and other, similar beings pass through walls and operate flying machines that seem to violate the laws of physics, then perhaps the flying machines can also cross into parallel worlds. Perhaps they also originate in such worlds.

Once this step is taken, an additional argument can be made as follows: The total number of reported, authenticated UFO encounters is very large, and the total number of encounters actually occurring must be much larger. It would seem that these operations must impose a great burden on the UFO entities if they have to commute regularly to the earth from another planet by ordinary three-dimensional travel limited by the speed of light. But if they live in a parallel reality, then they do not have to travel very far to reach us.

Of course, one could argue that they might travel from other stars in a quick, convenient way that avoids the limitation of the speed of light. But this would make other star systems, in effect, parallel worlds that are directly connected to our own world. The Vedic idea of the subterranean heavenly planets is perhaps similar to this. These are planets (*lokas*), and they are in the heavens. But they can also be quickly reached by entering into the earth.

The Fairy Folk and the Nāgas

In this section, I will try to establish an explicit link between the European fairy folk and some of the humanoid races mentioned in Vedic literature. One key to establishing such a link is to note that fairies were traditionally connected with the harvest of grain. Thus, Robert Rickard has observed, "Throughout the range of Indo-European cultures, the fairies were given their tithes of corn and milk at harvesting, over which they presided."[36]

In his book *Dimensions,* Jacques Vallee discussed a 9th-century French text which seems to describe a UFO abduction.[37] According to this text, three men and a woman were seen descending from aerial ships at Lyons in France. They said they had been taken on board these ships by beings called Sylphs and had been shown many wonders. Unfortunately, the local populace regarded them as evil magicians. The locals were about to cast them into the fire when Agobard, Bishop of Lyons, saved them by denying the reality of both Sylphs and magicians.

Agobard's comments about Sylphs show that they fit nicely into pre-Christian traditions of beings that preside over crops:

> We have seen and heard many men plunged in such great stupidity, sunk in such depths of folly, as to believe that there is a certain region, which they call Magonia, whence ships sail in the clouds, in order to carry back to that region those fruits of the earth which are destroyed by hail and tempests; the sailors paying rewards to the storm wizards and themselves receiving corn and other produce.[38]

Turning to India, we find that a race of beings called the Nāgas are connected with the harvesting of crops. These Nāgas should not be confused with the tribal peoples of present-day Nagaland. The Nāgas are said to be a nonhuman race descended from the celestial sage Kaśyapa and his wife Kadrū. They are sometimes described as having serpentine form, and at other times they are said to be human in form. Apparently, they have the power to assume or project various forms. The Nāgas also have standard mystical powers, such as the ability to travel through solid matter, and to appear and disappear. They live within the earth or in bodies of water, and they may be related to the dragons of Chinese tradition.

The *Rāja-taraṅgiṇī* of Kalhaṇa is a history of Kashmir that was written in about the 11th century A.D. It contains the following story about the Nāgas. The story took place in Kashmir in a beautiful city founded by a king named Nara or Kiṁnara:

> In one of the cool ponds in the main park of the city dwelt Nāga Suśravās and his two beautiful daughters. One day a poor *brāhmaṇa* named Viśākha was resting in a grove near this pond. As he was about to have some refreshments, two beautiful girls came up from the spring and, apparently ignoring him, began to hungrily eat pods of *kacchagucchа* grass which grew in abundance there. The *brāhmaṇa,* taking up his courage, asked them the reason for their poverty. He learned that they were the daughters of Nāga Suśravās, and even though they were entitled to a share of the rich crops growing around Kiṁnarapura, they could not touch it until the field-guard partook of the new harvest. But as ill luck would have it, he had taken a vow not to eat a single grain of the fresh crops; hence their miserable condition.
>
> The *brāhmaṇa* boy was moved to pity. One day he stealthily put some fresh corn into the field-guard's cooking vessel. As soon as the guard partook of this, the Nāga carried off, through thunder and storm, the rich harvest all around the city. In gratitude, the Nāga granted the *brāhmaṇa* one of his daughters in marriage.
>
> The happy couple lived peacefully in the city for some time, until King Nara learned of the beauty of the *brāhmaṇa's* wife and tried to seduce her through his emissaries. When this failed he tried to carry her off by force, and Viśākha and his wife ran for their lives and jumped into the pond occupied by Nāga Suśravās. The Nāga was furious, and he destroyed Kiṁnarapura with a severe thunderstorm. The Nāga and his daughter and son-in-law then created for their residence a lake of "dazzling whiteness resembling a sea of milk," which is known to this day as Śeṣanāga.[39]

This story has three features that link it with European fairy lore. These are (1) the Nāgas partake of human crops, (2) human men can marry Nāga women, and (3) a human being can live with the Nāgas in their world. Since the *brāhmaṇa* was able to live with his Nāga wife under a lake, we can understand that they were living in a parallel reality connected with the lake, and not simply living in the lake water. It is also interesting to note that the Nāgas could create storms to take crops, and Archbishop Agobard spoke of storm wizards who apparently did the same thing.

The Nāgas of Kashmir

There has been some controversy about whether UFO entities come from other planets or from higher-dimensional domains on the earth. According to the Vedic texts, both possibilities could be true. There are humanlike beings who live on other planets and sometimes visit the earth, either in vehicles called *vimānas* or under their own power. There are similar beings inhabiting various earthly realms that are inaccessible to most humans. Also, some groups of beings have lived both on the earth and on other planets throughout their history.

The Nāgas fall in the last category. A partial account of their history on the earth and their relations with human beings is contained in the *Nīlamat Purāṇa*.[40] This *Purāṇa* is devoted to the history of Kashmir, and it presents the ordinances given by Nila, the son of the celestial sage Kaśyapa, and the king of the Kashmir Nāgas. It gives an interesting perspective on the Vedic view of ancient human and transhuman history.

Today, Kashmir is a valley surrounded by mountains that are unbroken except for a single gorge to the south, which provides a passage for the river Jhelum. According to the *Nīlamat Purāṇa,* at one time the gorge did not exist, and the valley of Kashmir was a lake called Satīsaras. It was so named after Satī, the wife of Lord Śiva, who would sometimes enjoy boating excursions on its surface.

At one time, a demonic being called Jalodbhava (meaning "arisen from the water") took up his residence in the lake and would emerge at regular intervals to devastate the surrounding regions. The Devas asked Lord Viṣṇu to destroy the demon, and He agreed to do so. To accomplish this, His brother Balabhadra drained the lake by cleaving the mountain range bounding the valley on the south. Then, as the Devas looked on from the surrounding mountain tops, Viṣṇu attacked the demon, who was at a disadvantage due to being deprived of his natural element.

After the death of Jalodbhava, the Piśācas and the descendants of Manu (human beings) were settled in the newly drained valley by the sage Kaśyapa. The Nāgas, who were the original inhabitants, took up their abodes in lakes and springs, and other beings took up posts as the goddesses of the newly formed rivers.

The Piśācas were adapted to the conditions of extreme cold that prevailed in the valley at that time. Due to the severity of the climate, humans could initially live there only during the summer months. Finally, however, the brahmin Candradeva acquired a number of rites through the favor of Nila Nāga. These freed the country from the Piśācas and the excessive cold, and thereafter humans were able to live there permanently. Later on, the valley became known as Kashmir, a name deriving from "Kashaf Mar," meaning "the house of Kashaf."[41]

It turns out that the valley of Kashmir really was a lake in the Pleistocene period of geological history. The valley is filled with sedimentary layers called Karewas, which have been interpreted by many geologists as freshwater lake deposits. Although some geologists have interpreted these layers as river deposits, the following statement seems to sum up current opinions on the subject: "It can be stated here for certain and without hesitation that the Karewa sediments under this investigation belong to fresh water lacustrine origin of cooler climate."[42]

According to geologists, the lake continued on the Himalayan side of the valley until late Pleistocene time, after which it was drained by the formation of the river Jhelum on the valley's southern side. Radiocarbon dating indicates that this happened over 31,000 years ago.[43]

This is before the time when human beings of modern form are thought to have first arrived in the region. How then can we explain the tradition that the valley was once filled by a lake? One suggestion is that early peoples living in Kashmir deduced the previous presence of a lake on the basis of geological evidence. However, the disagreements between present-day geologists indicate that the interpretation of such evidence is far from obvious. (And some of the evidence consists of the shells of microscopic water creatures called ostracodes, which people without microscopes wouldn't be able to see.) Thus, it is just possible that the traditions have been passed down from actual historical experience of the draining of the ancient lake.

Whether this can be demonstrated or not, the main point is that the *Nīlamat Purāṇa* refers back to a time when Kashmir was an abode of Devas, Nāgas, and other nonhuman intelligent races that possess superhuman mystical powers. All of these beings were descendants of celestial sages, such as Kaśyapa, and these sages in turn were descendants of Brahmā, the first created being in the universe. The human race was similarly descended from celestial beings, and at a certain point humans were introduced into the valley by Kaśyapa, the presiding sage of that region.

This story is similar to the Celtic story of Ireland. There, too, the land was once in the possession of a race of mystically empowered beings—the descendants of the goddess Dana. Human beings entered at some point in history, but the original inhabitants remained, living in their own higher-dimensional realms and interacting in a variety of ways with their grossly embodied human cousins. The only difference between the Celtic and Vedic accounts is that the latter are more exhaustive and philosophical, with detailed descriptions of the relationship between the hierarchy of living beings within the universe and the transcendental Supreme Being. Of course, the Celtic stories are fragmentary, since the main body of Celtic teachings vanished with the advent of Christianity in Europe.

Visions and Miracles—The Fatima Case

Christianity is a source of a great deal of evidence for encounters between humans and humanlike beings endowed with mystic powers. The Roman Catholic Church, in particular, has developed a vast literature on this subject, and I have already alluded to some Catholic views regarding succubi and incubi. Here I will briefly discuss one example of an encounter with beings of a more positive spiritual nature. This is the story of meetings that took place in Fatima, Portugal, in 1917, between three children named Lucia, Francisco, and Jacinta and a brilliantly effulgent lady whom they understood to be the Virgin Mary. As we shall see, this story is relevant to the UFO question, and it is backed up by an unprecedented amount of eyewitness testimony.

The meetings occurred on the 13th of the month for six successive months in a natural amphitheater called the Cova da Iria near the town of Fatima. Revelations were made to the three children in the presence of a large throng of onlookers, which increased greatly from month to month as news spread. The actual visions of the beautiful lady could be seen only by the three children, and so our knowledge of these visions is limited to their testimony. However, during the revelations there occurred related phenomena that were witnessed by large numbers of people.

These phenomena included the appearance of a glowing, globe-shaped vehicle and the occurrence of a shower of rose petals that vanished upon touching the ground.[44] Showers of flower petals are often mentioned in Vedic accounts of celestial visitations. For example, here is an excerpt from the description of Kṛṣṇa's *rāsa* dance in the *Bhāgavata Purāṇa:*

> The demigods and their wives were overwhelmed with eagerness to witness the *rāsa* dance, and they soon crowded the sky with their hundreds of celestial airplanes. Kettledrums then resounded in the sky while flowers rained down and the chief Gandharvas and their wives sang Lord Kṛṣṇa's spotless glories.[45]

As for the aerial globe, one eyewitness, Mgr. J. Quaresma, described its appearance on September 13, 1917, as follows:

> To my surprise, I see clearly and distinctly a globe of light advancing from east to west, gliding slowly and majestically through the air. . . . Suddenly the globe with its extraordinary light vanished, but near us a little girl of about ten continues to cry joyfully: "I still see it! I still see it! Now it is going down!"[46]

In his report about what happened after these events, Quaresma said, "My friend, full of enthusiasm, went from group to group . . . asking people what they had seen. The persons asked came from the most varied social classes and all unanimously affirmed the reality of the phenomena which we ourselves had observed."[47]

The Solar Miracle

During one of the revelations, the child Lucia had requested that a miracle be shown so that people unable to see the divine lady could believe in the reality of what was happening. She was told that this would occur on the 13th of October, and she immediately communicated this to others.

On this date, it is estimated that some 70,000 people congregated in the vicinity of the Cova da Iria in anticipation of the predicted miracle. The day was overcast and rainy, and the crowd huddled under umbrellas in the midst of a sea of mud. Suddenly, the clouds parted, and an astonishing solar display began to unfold. I will describe this in the words of some of the witnesses:

> Dr. Joseph Garrett, Professor of Natural Sciences at Coimbra University: "The sun's disc did not remain immobile. This was not the

sparkling of a heavenly body, for it spun round on itself in a mad whirl, when suddenly a clamour was heard from all the people. The sun, whirling, seemed to loosen itself from the firmament and advance threateningly upon the earth as if to crush us with its huge fiery weight. The sensation during those moments was terrible." [48]

Dr. Formigao, a professor at the seminary at Santarem: "As if like a bolt from the blue, the clouds were wrenched apart, and the sun at its zenith appeared in all its splendour. It began to revolve vertiginously on its axis, like the most magnificent firewheel that could be imagined, taking on all the colors of the rainbow and sending forth multi-colored flashes of light, producing the most astounding effect. This sublime and incomparable spectacle, which was repeated three distinct times, lasted for about ten minutes. The immense multitude, overcome by the evidence of such a tremendous prodigy, threw themselves on their knees." [49]

Similar testimony was given by large numbers of people, both from the crowd at the Cova da Iria and from a surrounding area measuring about 20 by 30 miles. The presence of confirming witnesses over such a large area suggests that the phenomenon cannot be explained as the result of crowd hysteria. The absence of reports from a wider area and the complete absence of reports from scientific observatories suggest that the phenomenon was local to the region of Fatima. It seems there are two possibilities. Either remarkable atmospheric phenomena were arranged by an intelligent agency at a time announced specifically in advance or coordinated hallucinations in thousands of people were similarly arranged at this time. By either interpretation, it is hard to fit these phenomena into the framework of modern science.

Although there seems to be a remarkable amount of testimony attesting to the unusual phenomena at Fatima, it is also relatively easy to dismiss them, if that is what one wants to do. For example, consider the following statement by a skeptical newspaper reporter:

> According to what we heard, there were people who seemed to see the sun leave its supposed orbit, break through the clouds and descend to the horizon. The impression of these seers spread to others, in a common effort to explain the phenomenon, many crying out in fear that the giant orb would precipitate itself to the earth on top of them, and imploring the protection of the Holy Virgin. The "miraculous hour" passed. [50]

Here simple suggestibility and crowd hysteria are used to explain how a few people's imaginary ideas were spread and amplified by frenzied believers. However, this doesn't explain how it came about that many people in surrounding communities also witnessed the solar spectacle. For example, in 1960 the Rev. Joaquim Lourenco, a canon lawyer of the diocese of Leira, described what he saw as a boy in the town of Alburitel, some nine miles from Fatima:

> I feel incapable of describing what I saw. I looked fixedly at the sun, which seemed pale and did not hurt my eyes. Looking like a ball of snow, revolving on itself, it suddenly seemed to come down in a zig-zag, menacing the earth. Terrified, I ran and hid myself among the people, who were weeping and expecting the end of the world at any moment.[51]

It would seem unlikely that mass hysteria would give rise to the same illusions in Fatima, Alburitel, and other separated communities.

Scientists have also offered skeptical explanations of the Fatima phenomena. For example, the British meteorologist Terence Meaden has devised a theory explaining the famous English crop circles on the basis of natural ionized vortices of air (see pages 70–71). According to his theory, the ionized vortices can derange the mind and senses by electrically agitating the brain. Thus, ionized vortices can generate all kinds of hallucinations, and they can be invoked to explain any eyewitness testimony whatsoever. Here is how Meaden dealt with Fatima:

> In our theory of the luminous vortex we find an answer to the miraculous visions of the past which "faith" has supported on the part of the religious but which have gone unbelieved by the non-religious. An obvious case is the mountainside apparition of 13 May 1917 near Fatima in Portugal. The vision that the three young witnesses claimed to see could have been a plasma vortex.[52]

Meaden explained Moses's burning bush and the Star of Bethlehem in the same way. Of course, he didn't explain why ionized vortices should appear punctually at the same place for several months on the 13th of each month. Nor did he explain how they could cause people to see the spectacle of October 13, 1917, over an area of many square miles. However, the vortices may sound scientifically plausible, and thus they can be used to bring disturbing observations within a familiar framework.

The Beings Seen at Fatima

The manifestations at Fatima were strongly connected with religion, and with Roman Catholicism in particular. I should emphasize, however, that all testimony regarding communication with the effulgent lady and other paranormal beings comes solely from the three children, since only they could actually see these beings and hear them speak.

The lady would appear to the children as a dazzling, beautiful figure standing directly above a small holm oak tree that grew in the Cova da Iria. According to Lucia, "She was more brilliant than the sun, and she radiated a sparkling light from her person, clearer and more intense than that of a crystal filled with glittering water and transpierced by the rays of the most burning sun."[53] Her message to the children was couched explicitly in the terminology of the Catholic Church, and it consisted mainly of warnings that unless people give up sinful life and turn back to God, there will be terrible divine punishment, and various nations will be annihilated.

Before their meetings with the lady, the three children also had encounters with an angel. At the time of the first encounter they were tending their sheep at a rocky knoll not far from their home. They saw across the valley a dazzling globe of light like a miniature sun, gliding slowly towards them. As it approached, the ball of light gradually resolved itself into a brilliantly shining young man, who seemed to be about fourteen years old. He identified himself as the "Angel of Peace" and enjoined them to recite the following prayer: "My God, I believe, I adore, I hope, I love Thee. I ask pardon for those who do not believe, nor adore, nor hope, nor love Thee."[54] Then he disappeared by fading away.

Fatima As a UFO-Encounter Case

How should we interpret these experiences? Of course, one approach is to dismiss them as illusions or attempt physical explanations similar to Terence Meaden's. Another is to propose that they are UFO encounters. For example, Wendelle Stevens did this straightforwardly in one of his books, in a chapter entitled "The Fatima UFO Sightings."[55] Jacques Vallee also argued that the events at Fatima are UFO encounters and not divine miracles as understood by the Catholic Church.

My response to this proposal is to say "yes and no." On the one hand, the events at Fatima have many features that are also seen in

UFO accounts. There are glowing globes of light that might be vehicles and were regarded as such by some of the people who saw them. Stevens and Vallee even interpreted the spectacular spinning solar disc of October 13 as a disc-shaped UFO. This may be valid, since it moved about in a zig-zag pattern, gave off colored beams of light, and could be looked at directly without hurting people's eyes. It may well have been a flying device, which finally departed by flying in the direction of the real sun when the latter became visible through the heavy overcast.

On the other hand, there is some danger of creating a stereotyped idea of UFOs and UFO encounters, and imposing this on the evidence. Our ideas about UFOs are empirically based, and they tend to expand and transform as we learn more about the full range of relevant empirical evidence. Clearly, not all UFO encounters are the same. Both the entities involved and the technology they employ show considerable variability. Thus to force Fatima into a preconceived UFO theory might block us from getting an understanding of what was really going on.

To support his view that the Fatima events involved UFOs rather than religious revelations, Vallee stretched some of the evidence. Thus he said that the lady appearing to the children "had not said she was the Virgin Mary. She had simply stated she was "from Heaven.'"[56] She, indeed, said this during her first visitation on May 13, but she added that she would reveal her identity on October 13. On that date she identified herself as the "Lady of the Rosary"—an explicitly Catholic designation.[57]

The children's encounters with the Angel of Peace are consistent with this: "Through the white radiance of his presence the children could see that he held a chalice with a Host above it, from which drops of Blood fell into the cup."[58] If this is an honest statement of what the children saw, we must conclude that whoever the angel may have been, he was making use of explicit Catholic symbolism.

After seeing the angel with the chalice, the children experienced a sense of utter weakness and remained prostrated on the ground, praying, until evening. Vallee spoke of this as paralysis and compared it to the paralysis connected with many UFO encounters.[59] However, there are forms of weakness, including hysterical paralysis, that are caused by powerful emotional experiences and may have nothing whatsoever to do with UFOs. This seems to be the kind of weakness that is involved here. The children's experiences can be contrasted with those of UFO abductees, who describe the terror of being immobilized by weird beings who lay them out on tables and stick probes into their bodies.

There is a striking contrast between the Fatima visitations and many UFO encounters. In a large group of these encounters, the beings involved are said to have gray, pasty, or mushroomlike complexions, terrifying slanted eyes, slit mouths, and vestigial noses. They come in the night, carry people off helplessly, and subject them to sexual molestation. The victims experience psychological trauma, and in some cases develop physical diseases.

Of course, there are other cases in which people have claimed to have relatively friendly contacts with humanoids of various descriptions, including some who looked completely human. In some cases, these beings have conveyed messages to people, and some of these messages were quite complex. As in the case of Fatima, these messages have often been critical of human behavior, and they have often predicted disasters of various kinds (Chapter 5). They have also sometimes contained philosophical and theological material (Chapter 11).

Witness exclusivity is another feature that Fatima shares with many other UFO cases. The angels and the effulgent lady were seen only by the three children, although thousands of other people present in the Cova da Iria witnessed aerial light displays that tended to back up the children's story.

UFO cases can be regarded as falling along a broad continuum of types. This continuum is not linear. Rather it should be defined by several variables, which might include (1) the degree of friendliness of the contacts, (2) the degree of humanness of the contacting beings, and (3) some measure of the quality of the communicated material.

The Fatima case can certainly be added to this continuum, but I would suggest that it constitutes an outlying point that is quite far removed from the majority of other cases reported in the existing UFO literature. Thus it would rate very high in (1) and (2), especially if we measure humanness in terms of attractive qualities such as beauty. It is hard to define measure (3), but it is clear that the Fatima communications are unusual in the sense of being strongly oriented towards Roman Catholicism. Apart from this, they display an emphasis on spiritual devotion that is rarely seen in UFO communications, including those of a philosophical nature.

One interesting project would be to make a general survey of all cultural traditions in search of cases in which people have claimed to meet beautiful, effulgent beings that conveyed spiritual teachings. Many of these cases would probably display typical features connected with the UFO phenomenon, although these would often have more to do with the behavior and powers of the beings involved than with UFOs per se. Thus the UFO continuum might be seen as part of a larger mystical-humanoid continuum. Fatima would probably appear as a typical point in this larger assemblage.

Vallee summed up his viewpoint by saying, "In many UFO stories of the olden days, the witnesses thought they had seen angels from God. . . . Others thought they had seen devils. The difference may be small." [60]

On the contrary, the difference seems quite large to me. The various kinds of encounters in the UFO continuum may all be similar in that they involve similar powers and technologies lying beyond our present understanding. But the extreme points of the continuum are strikingly different with regard to the behavior and, by implication, the consciousness of the beings involved.

If the manifestations at Fatima were not "typical UFO phenomena" and did use Christian symbolism, does this mean that we should interpret them in an exclusive Christian sense? The answer is no, because the same reasoning that would induce one to take the Fatima manifestations seriously can equally well be applied to other meetings with divine beings occurring in non-Christian contexts. Some may argue that only the Christian (and specifically the Catholic) meetings are genuine, and all others are the work of Satan. However, if beings demonstrating divine qualities in all non-Christian traditions are actually clever deceptions created by a cosmic trickster, then surely the same thing can be said of the divine beings of Christianity. This brings us back to Vallee's position that the angels and devils are all the same.

The point should also be made that while the communications at Fatima were expressed in explicitly Catholic terms, they also added something new to Catholicism. In his discussion of the Fatima revelations, Francis Johnston made it clear that they have caused a great deal of controversy within the Catholic Church, and the new material (including, for example, the idea of consecrating Russia to the Heart of Mary) has not been immediately and universally acceptable. Protestants will declare that the Catholic cult of Mary is a controversial addition to Christianity, and, of course, Christianity itself arose at a certain point in history. It would seem that at Fatima a new revelation was made in the context of an older tradition, and this has also been done many times in the past.

Here is another example. In the sixth century A.D., a man in Arabia experienced a vision of the Angel Gabriel "in the likeness of a man, standing in the sky above the horizon."[61] The man was ordered to become a prophet, and over a number of years he would periodically go into trance and dictate messages that were carefully noted and memorized by his followers. The man was named Mohammed, and the messages he delivered in trance were later compiled to form the Koran.

The story of Mohammed is in one sense a typical contactee case. There is the meeting with a nonhuman being, followed by the dictation of elaborate channeled communications. As in the case of Fatima and some recent UFO cases, the being made use of existing cultural traditions, in this instance traditions of angels that were well known to the Arabs.

At the same time, Mohammed's teachings have had an enormous impact on world history, and they are clearly on a different level of quality than the teachings transmitted by many UFO contactees. I would suggest that many different communications are continuously being injected into human society using the subtle technology of the mystic *siddhis*. These revelations vary greatly in quality. They frequently make use of existing human cultural material, and they may also result in far-reaching transformations of human culture.

9

Harm's Way

Many of the UFO encounters discussed in previous chapters were traumatic and frightening to their main witnesses, and many involved physical aftereffects such as injuries and unusual infections. In most of these cases, however, it did not seem that people were being deliberately injured or terrorized with threats of violence. But there are UFO encounter cases that do have a negative, violent aspect, including some that are reminiscent of Gothic horror stories. There are also mysterious events, such as the notorious cattle mutilations, that have a threatening, sinister quality and have been linked indirectly with UFOs. This sinister aspect is an important part of the UFO phenomenon, and so I will discuss it briefly in this chapter. As it turns out, it also fits into the Vedic picture of life within the material universe.

Hairy Monster Cases

A case in the Gothic horror category occurred near Greensburg in western Pennsylvania on October 25, 1973, and was investigated by the psychiatrist Berthold Schwarz.[1] At about 9:00 p.m., a farmer named Stephen Pulaski and at least 15 other witnesses saw a bright red ball hovering over a field. Pulaski and two ten-year-old boys went toward the field to investigate, and Pulaski took along a rifle. As they approached, their auto headlights grew dim, and they saw the object descend toward the field. Continuing on foot over the crest of a hill, they saw the object glowing brilliantly with white light and either sitting on the field or hovering directly over it. Pulaski said the object seemed to be about 100 feet in diameter. It was dome-shaped and made a sound like a lawn mower.

As they were observing the object, one of the boys saw something walking along near a fence, and Pulaski fired a tracer bullet over its head. This revealed two creatures that towered one or two feet higher than the six-foot fence. Both creatures had long arms that hung down almost to the ground, and they were covered with long, dark grayish hair. They had greenish-yellow eyes, and they emanated a strong odor—like burning rubber. They seemed to be communicating by making whining sounds resembling the crying of a baby.

Pulaski fired three rounds into the larger creature, which responded by whining and lifting its right arm. At that moment, the glowing craft vanished, leaving a glowing white area in the field, and one of the boys ran away out of fear. The creatures slowly turned around and walked back toward the woods. They were not seen again, but later on, when Pulaski was joined by investigators from the Westmoreland County UFO Study Group (WCUFOSG), a dog began tracking something unseen, and several members of the group smelled a strong sulfur or chemicallike odor.

At this point, Pulaski went berserk and went running around, violently flailing his arms and growling like an animal. He had visions of a man looking like the Grim Reaper, heard his name being called from the woods, and made confused statements such as, "If man doesn't straighten up, the end will come soon." Then he collapsed.

Pulaski was examined by Dr. Schwarz, who concluded that his disoriented behavior had no precedent in his life history. Pulaski had never experienced trancelike, dissociated states, and he showed no signs of convulsive disorders such as temporal lobe epilepsy. However, he did have a history of violence on the school bully level, and he was a hunter who went for his rifle when confronted with an unfamiliar, threatening situation. Schwarz concluded that the stress of the frightening situation caused Pulaski to become temporarily unhinged and enter a dissociated psychological state known as a fugue.

Schwarz pointed out that this case was one of an epidemic of at least 79 "creature" cases occurring in a six-county area of western Pennsylvania in 1973, as documented by WCUFOSG. In all of these cases, the creatures were werewolflike entities that mysteriously appeared and disappeared, and left few traces of their existence. There were some reports that the creatures left tracks, and there were reports that they emitted foul stenches. There were also cases where they were said to have killed chickens, ripped off the hindquarters of a St. Bernard dog, and torn the throat of a pet deer, but there were no reports of injury to humans.[2]

What are these creatures? The easiest explanation is that they are all figments of overworked imaginations, but this does not explain why there should be a rash of bizarre creature incidents in a large geographical region during a particular time period. What would cause people's imaginations to become overworked in this particular way in a six-county area of western Pennsylvania?

To understand these entities, two questions must be answered. The first is, what is their ontological status? That is, what are they made of, and what causes them? The second is, why do creatures appear that seem designed to invoke terror in people who see them?

The elusive nature of both the creatures and the accompanying UFO led Schwarz to speculate that they might have been materialized and then dematerialized by some intelligent agency. As another example of such materialization, he mentioned a report by Pierre van Passen about how "his German shepherd dogs savagely fought with a poltergeist black hound, until one shepherd dropped dead."[3] As in the Pulaski case, this example involves both the paranormal and violence. It reminds one of people's old fears of ghosts, witchcraft, and black magic.

A related example of paranormal violence is provided by the account in the *Bhāgavata Purāṇa* of a person named Sudakṣiṇa, the son of a slain king of Benares, who sought revenge for his father's death. Sudakṣiṇa performed a ritual called *abhicāra,* in which the object is to summon a demonic being from a sacrificial fire and send him to attack one's enemies. This had the following result:

> Thereupon the fire rose up out of the altar pit, assuming the form of an extremely fearsome, naked person. The fiery creature's beard and tuft of hair were like molten copper, and his eyes emitted blazing hot cinders. His face looked most frightful with its fangs and terrible arched and furrowed brows. As he licked the corners of his mouth with his tongue, the demon shook his flaming trident.[4]

This description is, of course, reminiscent of ancient and medieval Western lore involving magic, and we can easily dismiss it as a fairy tale. But the creation of the fiery demon is based on a rational principle, whereby preexisting subtle forms can be manifested on a gross physical level. In Chapter 7 (pages 256–57), I compared the Vedic picture of the universe with a computer operating system, in which people with the right status can call up programs by typing appropriate key words. In a computerized virtual reality system with the right software, it would be possible to evoke fiery monsters by performing rituals.

The Vedic universe is described as a product of *māyā*, or illusion, and it can be thought of as a universal virtual reality system. One meaning of *māyā* is magic. When a magician generates an illusion, such as that of sawing a lady in half, he makes use of a suitable apparatus. Likewise, the illusory world created by a virtual reality system depends on the computer as its apparatus and the computer programmer as its magician.

In the Vedic universe, the role of the computer is played by a fundamental energy called *pradhāna*. This energy is activated by an expansion of the Supreme known as Mahā-Viṣṇu, who acts as the universal programmer. The activated *pradhāna* produces subtle forms of energy, and these in turn produce gross matter.[5] From the Vedic perspective, both types of energy are comparable to the unreal manifestations produced by a virtual reality system. But we can think of these energies as being real because they behave consistently and reliably as long as the universal system is operating.

Although subtle energy is not directly perceivable by our ordinary senses, it is just as much a product of the universal system as gross matter, and thus it is just as substantial as gross matter. In one sense it is even more substantial, since gross matter is generated from subtle energy.

In the story of Sudakṣiṇa, the fiery demon was a preexisting being with a body of subtle energy, and the fiery gross material form was generated temporarily on the basis of this subtle form. It is possible that the Pennsylvania creatures or the poltergeist black hound were similar manifestations.

This is a tentative explanation of the ontological status of these beings, but what can we say about the motives that may lie behind their sudden appearance and frightening behavior? We can shed some light on this by turning to the famous conversation between Kṛṣṇa and Arjuna called the *Bhagavad-gītā*. There it is stated that material manifestations of life and consciousness are governed by three fundamental principles: *sattva, rajas,* and *tamas*. In translation, these have been called the modes of goodness, passion, and ignorance, and Kṛṣṇa defined them as follows:

> O sinless one, the mode of goodness, being purer than the others, is illuminating, and it frees one from all sinful reactions. Those situ-

> ated in that mode become conditioned by a sense of happiness and knowledge.
>
> The mode of passion is born of unlimited desires and longings, O son of Kuntī, and because of this the embodied living entity is bound to material fruitive actions.
>
> O son of Bharata, know that the mode of darkness, born of ignorance, is the delusion of all embodied living entities. The results of this mode are madness, indolence and sleep, which bind the conditioned soul.[6]

One misconception that can arise from the word "goodness" is that the three modes have something to do with ethical distinctions between good and bad. This is not correct, and it would be better to think of the modes (called *guṇas,* or "ropes," in Sanskrit) as basic psychological programs that can be recognized by their characteristic behavioral symptoms. *Sattva* could also be translated as "pure beingness," and it refers to the introspective recognition of one's existence as a conscious self. *Rajas* could be translated as colored, reddened, and dusty, and it refers to the contamination of one's consciousness by passionate desires. *Tamas* literally means "darkness," and it refers to the tendency of conscious beings to fall into deep illusion.

The three modes are said to continuously interact with one another in the minds of individuals, with the result that different modes will become prominent at different times. According to Vedic understanding, different races of humanoid beings tend to be predominated by different combinations of the three modes. Thus human beings of this earth tend to be predominantly in the mode of passion, with some admixture of goodness and ignorance. The Devas and Ṛṣis of higher planets are predominantly in the mode of goodness and thus, in comparison with ourselves, they tend to be very peaceful and attracted to knowledge.

Among the Vedic humanoids, there are a number of groups that are predominantly in the mode of ignorance. These are broadly known as Bhūtas (a term which, appropriately enough, can be translated as "entities"). They include the Piśācas, Yakṣas, Rākṣasas, and Vināyakas, as well as the Ḍākinīs, Yātudhānīs, and Kuṣmāṇḍas. These beings are said to live in subtle form on the earth and in the region immediately above the earth's atmosphere.[7] They are known for their mystic powers, including the power of suddenly appearing in gross material form and disappearing.

It is said in the *Bhāgavata Purāṇa* that these beings are known for causing trouble to the body and the senses. They also cause loss of memory and bad dreams, and they are said to be particularly troublesome for children.[8] These problems, of course, are frequently mentioned in reports of UFO abductions. The *Bhāgavata Purāṇa* goes on to say that one can drive these beings away by chanting the name of Viṣṇu (God). In general, the chanting of the holy names of the Supreme Lord can counteract the influence of beings in *tamo-guṇa.*

The *tamo-guṇa*, or mode of ignorance, does not necessarily entail a lack of knowledge or ability. In fact, extensive knowledge of material subject matter can go hand in hand with deep illusion. The Manhattan Project in World War II is a good example of this. There the best physicists of the time used their most advanced knowledge to create a weapon that continues to threaten the security of the entire world.

One could argue that there were good reasons for building and deploying the atomic bomb. For example, it saved millions of American and Japanese lives that would have been lost in an invasion of Japan, and if we hadn't worked hard to develop it, then the Japanese or the Germans might have gotten it first. But this argument simply shows that an illusion may contain logical structure. An illusion may be so powerful and compelling that it is very difficult to see through it and perceive reality.

In the Vedic literature, Maya Dānava is the epitome of a person who is highly advanced in material knowledge and also deeply mired in illusion. He is famous for creating technological marvels, such as the *vimāna* of King Śālva described in Chapter 6, but his efforts are almost always devoted to illusory or destructive goals. In general, we find in Vedic accounts that beings in the mode of ignorance tend to be interested in acquiring advanced technology and mystic powers.

Many UFO manifestations seem to show the symptoms of the *tamo-guṇa,* and they also seem to involve advanced mastery of mystic powers and material technology. Thus the creatures seen by Stephen Pulaski were frightening monsters that appeared and disappeared, and they were accompanied by a weird, glowing UFO that also vanished suddenly.

Likewise, many UFO entities are reported to treat people in an alienated, impersonal fashion, and they are known for appearing, disappearing, and passing through walls. I argued in Chapter 5 that these beings have often presented people with communications that seem absurd or deceptive. These features are all characteristic of some of the Vedic humanoids, such as the Bhūtas, that are predominantly in the *tamo-guṇa.*

Cattle Mutilations

One puzzling development of recent years is the phenomenon of animal mutilation, in which the dead bodies of domestic animals such as cows and horses are found lying in farmers' fields with bizarre injuries. In typical cases, it is found that various organs of the victim, such as udders, genitals, or the rectum, have been removed with surgical precision. Eyes or ears may be excised, and teeth may be extracted. There are serrated "cooky-cutter" cuts, and sometimes a single joint is removed from a leg. It is almost universally noted that the body is devoid of blood, and that there are no signs of blood on the surrounding ground.

Mutilation cases first became prominent in the late 1960s, and they have occurred in great numbers since that time. Thus in Colorado, Elbert County Sheriff George Yarnell had records of 64 mutilation cases between April 6, 1975, and September 23, 1977. In the same period, over 100 reports were recorded in the Logan County Sheriff's office in northeastern Colorado. Mutilation cases were reported in all but 5 or 6 of the 48 contiguous states between 1967 and 1989, and they were also reported in 6 of Canada's southern provinces, as well as in Mexico, Panama, Puerto Rico, Brazil, Europe, the Canary Islands, and Australia.[9]

In 1975, cattle mutilations in Colorado became so common that Governor Richard Lamm spoke out against them at a meeting of his state's Cattlemen's Association. He said, "The mutilations are one of the greatest outrages in the history of the western cattle industry. It is important that we solve this mystery as soon as possible. The cattle industry is already hard hit from an economic point of view. From a human point, we cannot allow these mutilations to continue."[10] Here I cannot help but notice that there is a certain amount of irony in the governor's humanitarian concern. After all, the aim of the cattle industry is to raise animals for slaughter, a procedure that differs from the mysterious mutilations only in the matter of being carried out by known means, behind closed doors, and under economically profitable circumstances.

Many have suggested that predators are responsible for the animal mutilations, but others have pointed out that known predators do not produce long, clean cuts of the kind seen on the mutilated animals. Another theory is that satanic cults are responsible. This seems more plausible and may account for some cases. But the observed rate of animal mutilation is so high that if cultists were solely responsible, one would think that the police would have apprehended quite a few of them by now and that their stories would have become well known.

As we might expect, some have suggested a connection between the animal mutilations and UFOs. The case for this begins with the observation that the cuts on mutilated animals show features that would be difficult to duplicate out in the field using known human technology.

Testimony along these lines was provided by John Henry Altshuler, M.D., a doctor of pathology and hematology who studied at McGill University and who worked at the Rose Medical Center in Denver in 1967. In September of that year, he was shown the body of a horse named Lady on the Harry King ranch in the San Luis Valley of Colorado. The horse had been killed and mutilated on the night of September 7, 1967. On being taken to see the horse by police about ten days after it was found dead, Dr. Altshuler said, "The outer edges of the cut skin were firm, almost as if they had been cauterized with a modern-day laser. But there was no surgical laser technology like that in 1967." [11] He went on:

> I cut tissue samples from the hard, darker edge. Later, I viewed the tissue under a microscope. At the cell level, there was discoloration and destruction consistent with changes caused by burning.
>
> Most amazing was the lack of blood. I have done hundreds of autopsies. You can't cut into a body without getting some blood. But there was no blood on the skin or on the ground. No blood anywhere. That impressed me the most.
>
> Then inside the horse's chest, I remember the lack of organs. Whoever did the cutting took the horse's heart, lungs, and thyroid. The mediasternum was completely empty—and dry. How do you get the heart out without blood? It was an incredible dissection of organs without any evidence of blood. [12]

Some have reasoned that incisions requiring a technology unknown to man might be made by high-tech UFO entities. It is interesting to note that satanic cults tend to be ruled out because their members are thought to have only ordinary human powers. In an earlier age, in which people believed in the supernatural powers of witchcraft, the satanic cult theory might have seemed more persuasive.

Most of the evidence for a connection between UFOs and animal mutilations has been circumstantial, although there have been a few eyewitness accounts of direct UFO involvement. There have been many reports of strange lights in the sky at the same times and in the same places where mutilations were occurring, and UFO-like ground traces have been found near mutilated animal bodies. Strange, unmarked helicopters have also been seen—an unexplained phenomenon that has often been noted in connection with UFO encounters.

The Helicopter Connection

The helicopter/UFO connection has struck many people as rather ominous and threatening in its own right, and so I will make a few observations about it here. In the famous Cash-Landrum case, it was reported that two women and a child received apparent radiation burns in an encounter with a fiery, diamond-shaped flying object accompanied by about 23 twin-rotor helicopters. In this case, it seems that the object, and certainly the helicopters, might belong to the U.S. military, although this was denied by military officials when the matter was taken to court by the two women.[13]

In another case, UFO contactees Betty and Bob Luca reported that black unmarked helicopters repeatedly flew low over their home, followed their car, and even buzzed campgrounds where they were in residence.[14] (Betty Luca is the name of Betty Andreasson after remarriage.) This seems to be a typical experience for close-encounter witnesses. It has led to speculations that military personnel in helicopters might be pointlessly hazing UFO witnesses, or that UFOs disguised as helicopters might be harassing them. Either way, the phenomenon has the senseless, negative quality associated with the *tamo-guṇa*.

Ed Conroy, a journalist who wrote a book about Whitley Strieber, also reported being repeatedly shadowed by mysterious helicopters. On one occasion, he saw a helicopter fly behind the Tower Life Building in San Antonio, Texas, and not appear on the other side. On another, he saw two Chinook-style helicopters flying so close together that their rotors should have meshed like eggbeaters, and his neighbor Linda Winchester also saw this.[15] Curiously enough, Whitley Strieber reported a nearly identical incident: "I once saw—in the presence of two other witnesses—two such helicopters flying low over a populated area with their rotors meshed together like eggbeaters." [16]

In cases like these, the helicopters seem to be illusions. Since UFOs are reportedly capable of making themselves invisible, it hardly seems that such illusions could simply be disguises intended to make UFOs look like ordinary aircraft. As with many UFO manifestations, the mystery helicopters are difficult to explain. Perhaps the simplest interpretation is that they are intended to send people the message that there are powers in the world that they do not understand.

Ground Traces in Mutilation Cases

Ground markings similar to typical UFO traces have also been seen at mutilation sites. For example, in the case of Lady a broken bush was found about 40 feet from the horse's body, and "around the bush was a three-foot circle of six or eight holes in the ground about four inches across and three to four inches deep." [17]

Possible connections between UFOs and animal mutilations have been discussed in many newspaper articles since the late 1960s. An example is the article entitled "UFOs Zapping Cows? Why That's Out of This World," which was printed in *The Dispatch* of St. Paul, Minnesota on Dec. 27, 1974. The article related statements by UFO researcher Terrance Mitchell, who declared that "I'm convinced the taking of ears, udders and other portions of animals is part of a scientific investigation being conducted by beings using UFOs." [18]

For him, the clincher was a 400-pound heifer found dead on December 1, 1974, in a field owned by farmer Frank Schifelbien of Meeker County, Minnesota. "The heifer was found dead in a perfect circle of bare ground in a snow-covered field, with no footprints to be found anywhere in the vicinity, sheriff's deputies told Mitchell." [19] Mitchell ruled out the theory that cultists were responsible on the grounds that they should have left visible footprints. He also pointed out that photos taken from the air showed an array of perfect discolored circles, several feet in diameter, in a pasture near the heifer's body.

A similar mutilation story emerged in Cochran County, Texas, in 1975. A dead heifer was found on the farm of Darwood Marshall in the middle of a "perfectly round circle," according to a report by Sheriff Richards of Cochran county. [20] There was said to be no blood in the body of the cow or on the ground near it. While Richards was investigating this case, Darwood Marshall found a dead, mutilated steer about a quarter mile to the west. This animal was lying in the middle of a circle about 30 feet across. This circle, like the first one, was a burned patch in a field of young wheat about four inches high.

Richards speculated about a possible connection with UFO sightings:

> I have had reports of UFOs in the area, but have not seen any myself. The people that have been reporting this all tell the same story. It (UFO) is about as wide as a two-lane highway, round and looks the color of the sun when it is going down and has got a blue glow around it. When these people see this thing, in two or three days we hear about some cows that have been mutilated. I don't know what is doing this, but it sure has got every one around here uptight.[21]

Humanoid Reports in Mutilation Cases

Researcher Linda Howe has reported hearing from deputy sheriffs, ranchers, and fellow journalists many off-the-record stories connecting UFOs with cattle mutilations. She has also cited direct testimony linking UFOs and humanoid entities with cattle mutilation cases. For example, in April 1980, near Waco, Texas, a cattle rancher was walking through his land looking for a missing cow when he saw two four-foot-high creatures about 100 yards away. They were green or dressed in green, with egg-shaped heads, and were carrying a calf. He said he could see their eyes and spoke of them as "sloe-eyed, like big, dark almonds."

He had read about UFO abductions (a fact that should be considered in evaluating his testimony), and so he ran to his truck in fear of being abducted. Two days later he returned to the site with his wife and son, and found a calf's empty hide pulled inside out over its skull, along with a complete calf backbone without ribs. He reported that there was no blood associated with the remains and no sign of buzzards.[22]

In May of 1973, a close encounter reportedly took place near Houston, Texas, in which a witness observed a calf being mutilated by UFO entities. I mentioned this case in Chapter 5 as an example of UFO communications warning of adverse effects of environmental pollution and nuclear testing.

The witness, Judy Doraty, was driving along with four family members when all five people saw a very bright light in the sky that paced their car. The family members recalled Judy pulling off the road and walking to the back of the car, then returning, getting back in, and complaining of thirst and nausea. This episode apparently included a gap of an hour and 15 minutes in Judy's memory that was filled in through hypnosis administered on March 3, 1980, by psychologist Leo Sprinkle.

Under hypnosis, Judy Doraty recalled a bilocation experience, in which she seemed to be standing by her car and simultaneously standing inside a UFO where strange beings were mutilating a calf. She also saw her daughter being examined in the craft by the beings. The creatures she saw were similar to the typical "Grays" but not exactly the same. They had thin, pasty-looking skin, their noses and mouths were not noticeable, and they had large, unblinking eyes. However, the eyes, rather than being black, had vertical pupils and pale yellow irises. They spoke English with a high sing-song sound but didn't use their mouths. Apparently, this was mental communication, as usual.

The calf was taken up to the ship, squirming and struggling, by a beam of pale yellow light, which seemed to have substance to it. The light seemed as though it would be solid to the touch, and it swarmed with particles like motes of dust in a sunbeam. Once the calf was on board the craft, body parts were excised while it was still living. Fluids and other bodily materials were sucked out through tubes, and different organs were placed in different "basins" or scooped-out areas. The beings explained to Judy mentally that their purpose in cutting up cows and other animals is to monitor the spread through the environment of some kind of poison that will eventually affect humans. Finally, when they were finished, the calf's dead body was returned to the ground by the beam.[23]

This account has the surrealistic quality of many UFO close encounter stories. Whether it was real or an illusion, Judy Doraty's experience with the beings was strikingly vivid. But its content seems illogical. It hardly seems necessary to grotesquely mutilate animals in order to monitor pollution. Beings capable of producing antigravity beams would presumably be able to detect pollutants in animals without harming them. And even if it is necessary to cut up the animals, there is no need leave their bodies lying in ranchers' fields. This activity may be intended—by whoever is responsible for the mutilations—as a way of frightening people with some unknown danger.

Like the Pennsylvania creature cases, the cattle mutilations seem to combine the negative qualities of the *tamo-guṇa* with mysterious events involving paranormal powers. The Doraty testimony is particularly interesting because it involves an experience of entering into a UFO in an out-of-body state, a topic I will discuss in greater detail in Chapter 10. According to Doraty, her out-of-body experience was somehow spontaneous, and the beings told her that they had not intended to bring her on board. Nonetheless, even though they were engaged in the gross physical operation of cutting up a calf, they had no difficulty in seeing and communicating with her on a subtle level.

This power of acting on gross and subtle levels is typical of many of the humanoid beings described in Vedic literature, and the macabre acts of mutilation are reminiscent of certain groups of the beings in the mode of ignorance. If the animal mutilations are, in fact, being carried out by such beings, one might ask why they should do this at this particular time. Of course, I cannot answer this question with certainty. But one tentative possibility is that the beings are disturbed by present-day human activities involving nuclear testing and pollution of the environment. After all, this is the reason they reportedly gave to Judy Doraty. If ignorant human activities have disturbed beings in the mode of ignorance, then animal mutilations may simply be their way of showing their displeasure.

Inimical Behavior of UFO Entities Toward Humans

The mutilation of animals tends to worry people because it suggests that human beings might also be subjected to deliberate harm or death by UFO entities. There are several categories of evidence suggesting that such inimical behavior might occur. This evidence ranges in a continuum from frightening and traumatic abduction cases up to rarely reported instances of overt violence. In this section, I will give some examples of this evidence. I begin with an abduction story that is unusually traumatic, which supports the idea of a connection between UFOs and animal mutilations, and which contains dark hints of more inimical activities directed toward human beings. I should warn the reader, however, that in Chapter 10 I will give another possible interpretation of this case.

The Cimarron, New Mexico, Case

While driving home at night on a road near Cimarron, New Mexico, a twenty-eight-year-old woman and her six-year-old son reportedly saw five UFOs descend near a cow pasture. She had confused memories of a close encounter and reported a time loss of about four hours. Later, the woman was hypnotized by Dr. Leo Sprinkle in several sessions, from May 11 to June 3, 1980, in the presence of Paul Bennewitz, who was investigating for APRO (the Aerial Phenomena Research Organization).

As it turned out, a very disturbing abduction account emerged. Although this account is particularly bizarre and gruesome, I feel that it should be presented so as to give a balanced view of the experiences being recounted by UFO witnesses. The following summary is based on notes taken by Leo Sprinkle during the hypnosis sessions.[24]

Under hypnosis, the woman first reported seeing brilliant lights and witnessing the mutilation of a cow. "They're landing. Oh, God! . . . Screaming of the cattle; it's horrible, it's horrible! It's in pain. Incredible pain!" She described a silvery, tapered knife about eighteen inches long and one-half inch thick, which was plunged into the cow's chest. While the cow was still alive and struggling, the entities worked on its genitals with a circular, cutting motion.

She next reported that she and her son were captured by a number of strange-looking beings and taken into different ships. These beings wore dark brown uniforms with an orange and blue insignia having "three lines and a line across the bottom."

Initially she was unable to move, but later she regained this ability and began to put up a violent struggle. She was restrained but could kick and scream and call her captors names. She remembered being forcefully disrobed and reported a forced physical examination, including a vaginal probe. It is said that she later suffered from a life-threatening vaginal infection.

Although some of the beings treated her roughly, others showed curiosity: "They think it's funny—they love my hair. Their heads are large. But they have no hair, no eyebrows. The gentle one in the first ship was fascinated by my eyebrows and my eyelashes, too. They don't blink!"

As all this was going on, a tall, jaundiced-looking man, dressed in white and looking different from the others, entered in an angry mood. He declared that the woman wasn't supposed to have been taken: "They tell me they're sorry. They use that word, too. They apologize, 'Those things happen; it's unfortunate. The young one is all right' . . . They ask me to understand that it [the cattle mutilation] was necessary." This communication with her was telepathic, although the beings used spoken language to communicate with one another. One detail is that the tall man burned her face, perhaps unintentionally, by touching it with his hand.

He then apparently ordered the other beings to be punished: "I remember seeing them naked, waist up, thin, ribs, clavicles, more ribs than we have—I don't know. The thinness of them, their hands and yet they could pick me up. . . . Not claws, long fingernails, knotty and gnarled. Harsh looking, so small, thin, bones. . . . One has a nose that's crooked, turned up and crooked. They shuffle, drag their feet."

These beings also sported bizarre but distinctly human-looking clothing. "There was a Franciscan Monk's collar. Belt, military type boots, patch. Ugly, rowdy, rough. One was feminine. She had a collar, gathered at the neck like a pilgrim with ruffles—Victorian period? Didn't look comfortable. Patch. Squared head, holes like a small nose—or just two holes. Pea green! She honestly was green! I can't believe it still! . . . How can they be like this?"

The tall man proceeded to take her on a circuitous route through a number of strange places, perhaps on different ships. There are descriptions of a huge room with control panels and 24-inch TV screens, and some kind of elevator. At one point she saw a planet: "On the table, white light in my eyes, the takeoff, how fat my body got—lead! Heavy! . . . We get in a round cylinder—swoosh! Big room! Breathtaking! Stars everywhere. Beautiful, so beautiful. I can see a planet. It's big, white, black, white here and black. I only see the top half. They never tell me I can't go to the window, it's understood."

The sequence of events is confusing, and she also ascribed confusion to her captors. "When the man in white comes in and restores order, I feel a respect from him for me. He seems old, very old. . . . He seems agitated, but not with me. . . . I'm scared, not of him, but his confusion. It worries me that they don't know what they're doing." She was repeatedly told that it is "regrettable that they must do this."

The ship she was on seemed to land. "I'm excited, ecstatic, not scared. I think they've taken me somewhere important. . . . More people come in like the one in white, although they're not all in white. There are five of them; two look differently: narrow eyes. They're not greenish, not slitty eyes; like those on first ship, but they're not huge. They seem to be important! Maybe doctors or scientists. I don't know why they don't know what to do with me. I'm not allowed to go to the window, but I can see ships, activity, the outline of the terrain. Hilly, not big like mountains." At this point, she remarked that it was very cold, and she also noted a disturbing humming sound.

"They are very kind; the way they move is beautiful. They all shuffle their feet, long strides. Taller than me. Six feet or taller." Like the others, these beings lacked hair, but they were attractively dressed, and three looked quite human. They asked her to forgive them but said she wouldn't be allowed to say anything about her experience.

In another session, she also said that they mentioned alterations: "I don't want to talk about him again ever. Let it die, let it rest. . . . Alterations—the word 'necessary alterations' in order to bring me back." The word "alterations" apparently referred to implants that may have been placed in the body of the woman by the entities, and it has been alleged that such implants were found in the woman's body by means of CAT scans.[25]

As she was led out of the ship, she recognized the landscape: it seemed to be the Roswell area of New Mexico, to the west of Las Cruces. She was taken down an elevator to a subterranean complex swarming with the strange beings and roaring with the water of an underground river. "They like my reaction—awe! Incredible! Base city of operation." At this point, she briefly saw her son and was again separated from him.

Then, reacting in horror, she managed to break away and fled into a room with large tanks that seemed to be filled with some kind of body parts, floating in circulating fluid: "I'm looking down at pools of water. Something is horrifying me. . . . Top of a bald head. Light is dim. . . . I think I see an arm with the hand—human! Other something red and bloody looking. Oh, God! I'm so scared at seeing this. Ahh! Tongues, huge; they look real big. They're under liquid, real dark. . . . They found me, but when they found me, I was in the corner on the floor crying."

On being recaptured, she was led to a room where she experienced traumatic treatment, perhaps for the purpose of obliterating her memory. "Ow! Pain is so intense! Flashes, bright light flashing, something like two wires joined in a light bulb. Whoosh, whoosh, light! I'm screaming. My son is crying. They're doing it to him too. . . . You know something. *They don't like us.* They are something monstrous to me now. I feel like I've been in Auschwitz."

After this treatment, she and her son were taken back to one of the ships. As the ship flew through the air, she was shown her car, which was parked on board. They were placed in the car, and the car was transferred gently to the ground. She then drove home with no conscious memory of her experiences on board the UFOs.

This remarkable story contains many elements that appear frequently in UFO abduction cases, including the physical examination, telepathic communication, the strange hum heard by the woman, the display of emotions such as anger and curiosity by the humanoid entities, and the transfer of the woman's car into one of the UFOs. The possible insertion of implants into the woman's body is also a feature that this case shares with others described by Budd Hopkins and Raymond Fowler.

There are also features of this story that are unusual. Of course, one is the allusion to an underground base populated by alien beings and located somewhere in New Mexico. This is a very controversial topic, and it comes up in connection with the government-alien conspiracy theories discussed briefly in Chapter 3 (pages 110–15). There I mentioned allegations by journalist Howard Blum that Paul Bennewitz was fed disinformation about government conspiracies and underground alien bases by U.S. military agents working in league with UFO researcher William Moore.

The Cimarron case is thus wrapped up in a particularly convoluted controversy that is filled with paranoid speculations and counterspeculations. According to Blum, the hypnotic regression of the woman by Leo Sprinkle came before the period of the government disinformation campaign against Bennewitz.[26] At the same time, there seem to be discrepancies in Blum's version of the story. For example, he says that Sprinkle called in Bennewitz for consultation because he was puzzled about the woman's case,[27] whereas Linda Howe's documentation indicates that Bennewitz met the woman first and referred her to Sprinkle.[28]

In evaluating this case, we should keep this background information in mind. We should also consider that there is other evidence from UFO reports suggesting that various humanoid races may have underground or undersea bases on the earth—although these may have no connection with the U.S. Government. For example, Filiberto Cardenas described being taken to an undersea base (pages 176–78). Likewise, Betty Andreasson spoke of an abduction experience at the age of thirteen in which she was taken through water to an underground complex.[29]

One striking feature of the Cimarron case is that the alien entities seemed to be of several different physical types. The woman's statement that some of the beings have "more ribs than we have" is paralleled, strangely enough, by a description of an alien body supposedly recovered by U.S. military personnel from a flying disc that crashed in 1948 in Mexico, near the Rio Sabrinas and south of Laredo, Texas. An anonymous participant in this recovery effort maintained in letters written in 1978–80 that in this body "the entire abdomen was encased by a rib-like structure all the way to the hips."[30]

Unusual Alien Clothing

One of the most bizarre feature of the Cimarron story is the entities' clothing, which seems like something one might pick up from a party costume shop. Yet this feature is corroborated by the story of Filiberto Cardenas, whose trip to an undersea base in 1979 included an encounter with an enthroned alien personage wearing a cape and a jeweled chain.[31] It is also corroborated by another, apparently independent, encounter case that reportedly took place a few months after the Cimarron incident in the same general area of the United States.

In 1980, a week before Thanksgiving, a couple was driving north of Denver. The man, a commercial artist, reported seeing a "cerulean blue light," and the couple subsequently lost an hour of time. Hypnotic regression was carried out on July 5, 1984, by Richard Sigismund, a social scientist from Boulder, Colorado. Under hypnosis, the woman said they were picked up in their car by a light beam and transported to a nearby craft resting on struts. A hairless, tall "man" in a blue robe beckoned to them hypnotically.

In his description of this scene, the man reacted to the apparent absurdity of the being's costume:

> He's looking at us, telling us to come in. He's the leader. The leader is in a blue cape. It's stupid. It's illogical. Cape is illogical. He doesn't need a cape. Not like that. . . . He doesn't talk with his mouth. He talks with his mind.[32]

Drawn by the being's influence, they entered the craft and were examined physically by a humanoid being wearing a yellow robe and ruffled collar. The woman, who was pregnant at the time, felt violated and raped, and she had a serious illness after the abduction. Her child later turned out to have an IQ of 170, however.[33]

There are other accounts featuring entities wearing strange costumes. An example is the friendly encounter of Mrs. Cynthia Appleton with tall, fair beings wearing plasticlike garments with "Elizabethan" collars (see page 369). Other examples are the patches with serpent emblems worn by the beings in the three abduction cases involving Filiberto Cardenas, William Herrmann, and Herbert Schirmer (see page 184). Of course, in many UFO cases, beings are said to wear relatively featureless form-fitting suits, and in others spacesuits or "diving" suits are described.

The strange clothing sometimes reported in UFO encounters seems to resonate with human psychology. In particular, the Halloween-like costumes with sashes, capes, and insignia seem to fit in with the *tamo-guṇa,* which is characterized by dreams and madness. It is tempting to suppose, then, that encounters involving strange costumes are a projection of deranged human minds. But we have already discussed the evidence indicating that people evaluated as psychologically normal have reported such encounters. There is also much evidence indicating that some UFOs are physically real vehicles.

It is therefore worth considering that real beings under the influence of *tamo-guṇa* may be involved in some cases in which odd clothing is observed. It does seem unlikely that outmoded human clothing styles would just happen to be the latest fashions on distant planets. It is possible that the odd clothing is simply another feature of human culture that is being borrowed to create an impression on human witnesses.

I am not aware of any close-encounter cases in which entities from UFOs were said to wear the ordinary clothing current in the country where the cases occurred. However, there are the "Men in Black" visitations, in which contemporary suits and ties are worn in an artificial way that seems designed to create a bizarre impression (see pages 326–28).

In addition to cases where bizarre clothing is reported, there are also encounters involving beautiful clothing, or at least an overall beautiful effect. The Fatima case (pages 293–301) is an example of this, and another example is the case of the smallpox lady in Appendix 2. Vallee cited an example, which dates back to 1491 and was reported by a famous Italian mathematician named Jerome Cardan (1501–76). In his book *De Subtilitate,* Cardan presented the following account, which had been recorded by his father:

> August 13, 1491. When I had completed the customary rites, at about the twentieth hour of the day, seven men duly appeared to me clothed in silken garments, resembling Greek togas, and wearing, as it were, shining shoes. The undergarments beneath their glistening and ruddy breastplates seemed to be wrought of crimson and were of extraordinary glory and beauty.[34]

The men said that they were "composed, as it were, of air," they lived for about 300 years, and they were subject to birth and death. They conversed with the elder Cardan for over three hours on various philosophical topics and disagreed among themselves on the cause of the universe. Jerome Cardan concluded his account of his father's meeting by saying, "Be this fact or fable, so it stands."[35]

For comparison, here is a description from the *Mahābhārata* of the clothing of the Devas, presented by the sage Vyāsadeva to a king named Drupada:

> Thereupon Śrīla Vyāsa, the pure sage whose works are most magnanimous, with his ascetic strength awarded divine vision to the king, who then saw all the sons of Pāṇḍu exactly as they appeared in their former bodies. The king saw the five youths in their celestial forms as rulers of the cosmos, with golden helmets and garlands, the color of fire and sun, broad-chested, beautiful of form, with ornaments crowning their heads. There was not a particle of dust on their celestial robes, which were woven of gold, and the Indras shone exceedingly with most valuable necklaces and garlands. Endowed with all good qualities, they were like expansions of Śiva himself, or like the heavenly Vasus and Ādityas.[36]

The Devas are said to be in the *sattva-guṇa,* or mode of goodness. Their clothing tends to be spotlessly beautiful, and descriptions of them tend to stress that they are brilliantly shining. This can be contrasted with cases involving weird or frightening humanoids who are sometimes said to wear bizarre clothing. In these latter cases, features characteristic of the *tamo-guṇa* tend to cluster together.

Direct Attacks on Human Beings

Clearly, the Cimarron case has a number of features that are paralleled by other UFO cases. This, of course, does not prove that the woman's story is genuine, and one's credulity is definitely strained by her description of body parts floating in vats. These body parts might be organs from the cattle the woman saw being mutilated, but she also mentioned a floating bald head and a human arm and hand. Were these human victims? There are some accounts of human mutilations following the pattern seen in animals, but the evidence linking these cases to UFOs is not very strong.

One possible human mutilation incident occurred in India in 1958 and was reported by British UFO researcher Jenny Randles. The witness was an Indian businessman, who wished to remain anonymous and who refused to allow a taped interview of his testimony to be publicized. According to his story, a UFO was seen to land in broad daylight, and four three-foot-tall entities emerged. Two boys playing in nearby rocks were subsequently found to have disappeared. One was later discovered dead, with several organs removed, as if by "expert surgery." The other was in a catatonic trance and died five days later in a hospital without speaking.[37] This case follows the animal mutilation pattern, but the evidence behind it is weak due to the reticence of the witness.

There is an instance in Brazil of a human being who was mutilated in a way that agrees in detail with typical cattle mutilation cases.[38] However, there is no evidence in this case indicating who or what perpetrated the crime.

Cases involving attacks on human beings by UFOs armed with beam weapons have been reported from the remote forested regions of northern Brazil by Jacques Vallee. In these cases, a small, box-shaped object, known locally as a "chupa," is typically seen flying about, projecting a brilliant beam of light on the ground. The objects will sometimes attack people by "zapping" them with a narrowly focused beam of light. These beams seem to have a variety of effects. In some cases they cause illness, and in others the victim dies—either due to the direct effects of the beam or due to side effects such as heart attacks.

The case of Raimundo Souza is an example of the latter. Souza was a forty-year-old professional hunter, reputed to be in good health, who lived in Parnarama near São Luís in northern Brazil. The hunting technique used by Souza and his friends was to lurk at night in the forest in a hammock set up in the branches of a tree. When a deer came by, they would shine a flashlight in its eyes. The deer would freeze, and it could be easily shot.

While waiting for game in his hammock one August night in 1981, Souza struck a match to light a cigarette. A flying object rapidly came overhead and aimed a beam at him and his hunting companion, Anastasio Barbosa. Seeing this, Barbosa got down from his hammock and hid under some bushes, watching the object circle overhead. The next morning, he found Souza's dead body on the ground with one arm broken from his fall and with purple marks on various parts of his body, except the face. The marks were circular and smooth like a bruise, 1 to 2.5 inches in size, and there were no puncture marks. As there was no autopsy, the cause of death is not certain. However, Barbosa was not suspected of murder. Vallee obtained much of his information regarding Raimundo Souza from Police Lieutenant Magela of São Luís, Brazil, who was chief of police in Parnarama in 1981–82.[39]

These reports from Brazil are the only ones I have run across suggesting that UFO-projected light beams have caused human fatalities. However, there are corroborating stories in which people are said to have been violently attacked by such beams.

An example from America is the story of Eddie Doyle Webb, a 45-year-old truck driver from Greenville, Missouri. Early in the morning of October 3, 1973, Webb noticed that his tractor-trailer rig was being overtaken by a brightly lit, turnip-shaped object that was roughly as wide as a two-lane highway. When he leaned out of the window of his truck to get a better view of the object, he was hit in the face by a "red flash of fire" that blinded him and partially melted his glasses frames.[40] He was hospitalized with "bilateral severe reduction in vision of both eyes, undetermined etiology," but he gradually recovered his sight over a period of three weeks.[41] This incident led to an extended battle between Webb and his insurance company over a workman's compensation claim for the eye injury. Apparently the insurance company did not want to cover a UFO attack.

Here is an example of a light-beam attack in Spain. On January 28, 1976, just after midnight, a 24-year-old farmer named Miguel Carrasco was walking home from his girlfriend's house in Bencazon when he saw a powerful beam shoot from a strange craft hovering in the air. As he began to run, two tall, thin entities emerged from the craft, blinding him and paralyzing him with a beam. He awoke to consciousness at 2:30 a.m. on his front doorstep, banging and screaming, "The men from the star will come back—let me in and shut the door!" A local doctor saw strange burns on Miguel's cheek, and he was treated for them at a local hospital. However, they faded away in seven hours, and Dr. Mauricio Geara, the physician who treated them, later said, "We don't really know what they were due to."[42]

Here is another example, from Arizona. In the evening of November 5, 1975, seven woodcutters were returning home from a long day's work in the Apache-Sitgreaves National Forest. While driving along the bumpy forest road, they saw a yellow light through the trees and shortly drove within viewing distance of a hovering disc-shaped UFO. One of them, Travis Walton, got out of the truck and approached the craft, driven by curiosity. His six co-workers watched as he was knocked flat by a brilliant, blue-green beam emanating from the UFO. The men fled out of fear, and when they returned a few minutes later Walton was nowhere to be seen.

When the men told their story to law officers, they were initially suspected of murder. However, all but one of them passed a polygraph test indicating that they believed their UFO story to be true. (The man who didn't pass the test was considered to be in too agitated a state of mind for polygraph testing.) Extensive searching by police-directed search parties revealed no sign of Travis Walton, but five days later he showed up with a story of having awakened to consciousness in the presence of strange-looking aliens on board UFO. They had apparently kept him for the five days and then released him along a deserted country road. This is just the beginning of the Walton story, but here I am only interested in the fact that the account of the light-beam attack was confirmed by six witnesses.[43]

Another story involving an attack by a beam of radiation was told by Francis P. Wall to UFO investigator John Timmerman in 1989. Wall said that he was a private first class in the U.S. Army during the Korean War. In the early spring of 1951, his company was mounting an artillery attack on a village in the Iron Triangle region near Chorwon. A glowing, disc-shaped UFO approached them, and Wall requested permission from his company commander to fire on it. He shot at it with an M-1 rifle with armor-piercing bullets, and he heard the sound of metal hitting metal. The object "went wild" and began to move erratically and flash its light off and on. Then the UFO apparently prepared to attack the men by revving up some kind of generator:

> Then, a sound, which we had heard no sound previous to this, the sound of, like of, ah, you've heard diesel locomotives revving up.

> That's the way this thing sounded. And, then, we were attacked, I guess you would call it. In any event, we were swept by some form of a ray that was emitted in pulses, in waves that you could visually see only when it was aiming directly at you. . . . Now you would feel a burning, tingling sensation all over your body, as though something were penetrating you.[44]

Wall testified that at first there seemed to be no ill effects from the radiation. But he went on to say, "Three days later the entire company of men had to be evacuated by ambulance. They had to cut roads in there and haul them out, they were too weak to walk."[45] In this incident, the radiation beam was different from the one reported in the Walton case. However, the attack on Walton was also preceded by a sound reminiscent of a powerful motor. Describing what happened just before he was struck by the beam, Walton said, "Suddenly, I was startled by a powerful, thunderous swell in the volume of the vibrations from the craft. I jumped at the sound, which was similar to that of a multitude of turbine generators starting up."[46]

Men in Black

The so-called Men in Black, or MIBs, provide another category of evidence involving possible inimical behavior toward humans by alien entities. In a typical story, a UFO witness or investigator is visited by one or more strange men, who order him to suppress any information he may have about the UFO phenomenon and threaten him with violence if he doesn't comply. In many instances, these men seem to be perfectly human, and they are often thought to be government agents. This is typically the case in stories involving military personnel who disclose information about UFOs in violation of their oaths of secrecy.

There are many cases, however, in which the threatening visitors do not seem to be at all human. They are generally dressed awkwardly in black garments, they have abnormal bodily features, and they exhibit bizarre and inept modes of behavior. When this is combined with exhibitions of strange paranormal powers, the impression conveyed is of a nonhuman, ghostly sort of being who is crudely disguised as a human.

A typical MIB story of this type was related by the psychiatrist Berthold Schwarz in connection with Dr. Herbert Hopkins, a physician living in Orchard Beach, Maine. Hopkins became involved with UFO investigations when he used hypnosis to probe the memories of a close-encounter witness named David Stephens. This apparently brought him to the attention of a rather strange personality.

On Saturday night, September 11, 1976, while his wife and son were out to see a drive-in movie, Hopkins's phone rang. The caller identified himself as the vice president of a New Jersey UFO organization that later turned out to be nonexistent. He wanted to come and discuss the abduction case Hopkins was investigating, and Hopkins invited him over.

The man arrived almost immediately at Hopkins's door—apparently without time for travel from wherever he made his phone call. He was dressed like an undertaker in an impeccable black suit, was bald, and lacked eyebrows and eyelashes. He sat motionlessly like a clothing store dummy and proceeded to ask Hopkins a number of questions in flawless English, speaking evenly spaced words in an expressionless monotone. When he rubbed his straight, lipless mouth with his gloved hand, it turned out that he was wearing lipstick.

After the conversation had gone on for some time, the man said Hopkins had two coins in his left pocket, which was true, and he asked him to remove one. Hopkins took out a penny, and the man asked him to lay it on his palm. According to Hopkins, "The shiny new penny was now a bright silver color . . . the coin slowly became light blue in color, and then it began to become blurred to my vision. . . . It became more blurred, and then became vaporous and gradually faded away."[47] Hopkins declared that this was a "neat trick" and asked the man to make the coin return. He replied, "Neither you nor anyone else on this *plane* (not planet) will ever see that coin again."[48]

The man then asked Hopkins if he knew why Barney Hill had died. Hopkins said that he thought this was due to a long illness. But the man replied that, no, Barney died because he had no heart, just as Hopkins no longer had his coin. With this, he ordered Hopkins to destroy all tapes and other materials in his possession relating to the Stephens abduction case, and out of fear Hopkins later complied.

At this point, the man seemed to run down and said slowly, "My energy is running low—must go now—goodbye."[49] He walked out and went very unsteadily down the steps. As the man went around the corner of his house, Hopkins saw a strange bluish-white light shining up his driveway. The man walked up the driveway away from the road, even though it was a dead end with no means of egress. Then he was seen no more.

According to Betty Hill, Barney Hill died of a stroke, not a heart attack.[50] Thus the visitor's statements about the removal of Barney Hill's heart were incorrect. Nonetheless, they had their intended effect, for Hopkins immediately destroyed his tapes and other records relating to the Stephens abduction case.

What would constitute proof that Herbert Hopkins's story was true? His wife testified that when she returned home after the experience, she found that he had turned on all the lights in the house and was sitting at the kitchen table with a gun.[51] This unusual behavior plus his sincerity in telling his story was enough to convince her. But even if there were other eyewitnesses to Hopkins's story, this would not constitute proof. One can argue that eyewitnesses might be lying or hallucinating. And if someone claimed to have photographed the Man in Black, one can argue that the photograph might be a hoax.

Proof is not possible, but the story might be true. If so, it is an example of behavior that is certainly inimical and manipulative. Since the strange man appeared abnormal and stunted, and used occult powers to create fear, he fits nicely into the Vedic category of beings in the *tamo-guṇa,* or the mode of darkness. This is additional evidence tying in at least some UFO entities with beings in this category.

Inimical Behavior of Vedic Humanoids Toward Humans

Thus far, I have briefly touched on the following types of evidence for inimical activities by unknown intelligent beings: (1) cases involving frightening monsters, (2) macabre animal mutilations, (3) horrific abduction experiences, (4) attacks with beam weapons, and (5) visitations by ghoulish "Men in Black." There are also some stories suggesting that human beings have been captured and killed by alien beings, and there is a large literature on mysterious disappearances that I have not tried to review here.

All of these events seem to take place on the margins of human social consciousness. Although some of them create a stir in the news media for some time, none have ever become prominent enough to be openly recognized as real by official civic and academic bodies. This may be partly because people have a strong tendency to deny things that seem incomprehensible or threatening. This denial, of course, begins on the level of individuals, and it can be formalized by policies set within governmental and academic institutions.

It is also evident that the threatening events I have been discussing are truly marginal in the sense that they do not seriously interfere with human activities at the present time. Things would be different, for example, if an alien invasion force took over London, or even if people were regularly zapped by aerial beam weapons in the streets of New York City.

The natural question is, "If unknown beings actually exist and are doing all these things, then what is their plan, and what are the prospects for the future? Are they going to invade and take over, as we might do in their position, and if not, then why not?" It is difficult to answer these questions by examining UFO reports, for these reports show that the mysterious UFO entities are not inclined to clearly explain their plans to the people they contact.

However, the Vedic literature contains a great deal of information on the relationships between various human and humanlike races, both on the earth and in interplanetary space. Since there are strong parallels between reported UFO phenomena and Vedic accounts of humanoid races, this information may convey some insight into the threatening actions that people have associated with UFOs. In this section, I will therefore discuss some examples from the *Mahābhārata* and the *Rāmāyaṇa* of inimical treatment of human beings by nonhuman races.

Star Wars and Their Consequences

One common theme of the Vedic literature is that there are wars in the heavens between the Devas and the Asuras. There is a cosmic hierarchy that rules the universe according to divine law, and there are also rebellious elements that oppose this hierarchy. As I mentioned in Chapter 6 (pages 207–9), the upper level of the universal hierarchy is predominated by sages (called Ṛṣis and Prajāpatis) who are mainly interested in meditation and spiritual development and who do not concern themselves with political struggles. The lower level, however, is controlled by the Devas, who do engage in politics.

Generally, the Devas act as universal administrators under the authority of the sages, who in turn act under the authority of Brahmā, the first created being in the universe. However, certain relatives of the Devas rebelled against this system, and their descendants have engaged in repeated and extended wars with the Devas. These beings are known as Asuras, and they include various subgroups such as the Daityas, the descendants of Diti, and the Dānavas, the descendants of Danu.

As we might expect, the wars between the Devas and the Asuras involved various reverses caused by political and technical developments on either side. This is illustrated by the three flying cities of Maya Dānava that I mentioned in Chapter 7 as Vedic equivalents of "mother-ships." The following quote from the *Bhāgavata Purāṇa* shows how these cities upset the balance of power between the Devas and the Asuras:

> Maya Dānava, the great leader of the Asuras, prepared three invisible residences [*pura*] and gave them to the Asuras. These dwellings resembled airplanes made of gold, silver, and iron, and they contained uncommon paraphernalia. My dear King Yudhiṣṭhira, because of these three dwellings the commanders of the Asuras remained invisible to the Devas. Taking advantage of this opportunity, the Asuras, remembering their former enmity, began to vanquish the three worlds—the upper, middle and lower planetary systems.[52]

Here the word *pura* can mean residence or city. In this case, the Devas were saved by Lord Śiva, who destroyed the three flying cities and thereby obtained the name Tripurāri ("Enemy of the three cities").

Here is another reference to interplanetary warfare in the *Bhāgavata Purāṇa*:

> When the atheists, after being well versed in the Vedic scientific knowledge, annihilate inhabitants of different planets, flying unseen in the sky on well-built rockets prepared by the great scientist Maya, the Lord will bewilder their minds by dressing himself attractively as Buddha and will preach on subreligious principles.[53]

The commentator, Śrīla Jīva Gosvāmī, pointed out that the Buddha referred to here is not the historical Buddha that we know but one who lived in a different age. Here the word "atheists" is used to translate *deva-dviṣām,* which literally means those who are inimical toward the

Devas. In this case, the enemies of the Devas again obtained remarkable flying machines from Maya Dānava. They were thwarted, however, by an incarnation of Buddha, who captivated them through external material features such as beautiful clothing and who then persuaded them to adopt the philosophy of nonviolence.

An important feature of the wars between the Devas and the Asuras is that they were never allowed to get too far out of hand. Higher authorities would periodically intervene to restore the divine order, and this often provided the occasion for *avatāras* of the Supreme Being to present lofty philosophical teachings and engage in remarkable pastimes.

At times, however, these wars would have repercussions involving the earth and its human population. For example, Indra, the king of the Devas, once slew Vṛtrāsura, the ruler of a group of Asuras. The followers of Vṛtrāsura were thoroughly defeated, and one contingent of them, called the Kāleya Dānavas, decided to seek revenge by terrorizing humans on the earth. They made a plan to do this by setting up a base of operations within the oceans of the earth and coming out at night to attack the sages and ascetics who at that time provided guidance to human society:

> In the Hermitage of Vasiṣṭha the miscreant band devoured a hundred and eighty-eight *brāhmaṇas* and nine other ascetics. They went to the holy hermitage of Cyavana, which is visited by the twice-born, and ate one hundred of the hermits, who lived on fruit and roots. This they did in the nighttime; by day they vanished into the ocean. At the Hermitage of Bharadvāja they destroyed twenty restrained celibates who lived on wind and water. In this fashion the Kāleya Dānavas gradually invaded all the hermitages, maddened by their confidence in the strength of their arms, killing many hosts of the twice-born, until Time crawled in upon them. The people did not know about the Daityas, best of men, even as they were oppressing the suffering ascetics. In the morning they would find the hermits, who were lean from their fasts, lying on the ground in lifeless bodies. The land was filled with unfleshed, bloodless, marrowless, disemboweled, and disjointed corpses like piles of conch shells. . . .
>
> While men were wasting away in this manner, O lord of men, they ran from fear into all directions to save themselves. Some hid in caves, others behind waterfalls, some were so fearful of death that fear killed them. There were also proud and heroic bowmen who did

> their utmost to hunt down the Dānavas; but they could not find them, for they were hidden in the ocean; and the bowmen succumbed to exhaustion and death.[54]

There is at least a superficial resemblance between this story and the modern accounts of cattle mutilations and of attacks on humans by UFOs. In both cases, death is inflicted by unknown beings that operate at night, using powers that are remarkable from an ordinary human point of view. In both cases, there are corpses drained of blood. It is also interesting to note that UFOs are often seen to enter and exit from oceans and lakes, as though perhaps they were maintaining some bases of operation hidden within the waters. (See Sanderson, 1970.)

The attack of the Dānavas on the ascetics clearly had a much stronger impact on the human society of that time than cattle mutilations and UFO activities do today. However, it was still on the level of terrorism. Even though the Dānavas had been fighting the Devas for outright supremacy, they did not try to openly invade the earth and take over but simply tried to frighten people with gruesome scare tactics. This is similar to what happens with cattle mutilations and the more frightening UFO manifestations, and one might ask why things should be done in this way. Some reasons will emerge as we consider additional examples.

The Plot of the Rāmāyaṇa

The basic plot of the *Rāmāyaṇa* is that a powerful being named Rāvaṇa had taken over a region called Laṅkā on the surface of this earth, and from that base of operations he was causing considerable trouble to many different groups of beings. For this reason, a group of Devas, Gandharvas, and sages, who were concerned with affairs on the earth, gathered together and made the following appeal to Lord Brahmā:

> O blessed Lord, having been favored by thee, the Rākṣasa Rāvaṇa perpetually troubles us since thou hast granted a boon to him, and we are helpless and forced to endure his fearful oppression! The Lord of the Rākṣasas has inspired terror in the Three Worlds, and, having overthrown the Guardians of the Earth, he has even humbled Indra himself. Provoking the Sages, the Yakṣas, Gandharvas, *brāhmaṇas*,

> and other beings, he tramples them under foot, he who has become insufferable through pride, being under thy protection.[55]

This statement makes the important point that Rāvaṇa had conquered the earth. At least, he had overthrown the Guardians of the earth. In those times, there were some beings on the earth and in its general vicinity who were nearly as powerful as Rāvaṇa himself, and he would fight with them to establish his hegemony. He would also send Rākṣasa marauders to make nocturnal attacks on *brāhmaṇas* and ascetics who were living in the forest, away from major centers of population. This, of course, is reminiscent of the stories about people in remote regions of Brazil being zapped by UFOs wielding beam weapons.

However, Rāvaṇa did not try to take over human lands and herd people onto reservations, as European settlers did with the American Indians. Instead, he would simply enjoy luxuries in his aerial mansion, while sending out henchmen to engage in acts of terrorism. I would suggest that two observations can be made about this: The first is that life on the earth in the human environment was not attractive to Rāvaṇa. By conquering the Guardians the earth was his, but he and his people had no interest in invading the human ecological niche.

The second is that the pattern of terrorizing people at night says something about Rāvaṇa's psychology. He, along with the Rākṣasas and Dānavas in general, had a strong streak of the *tamo-guṇa,* or mode of ignorance. On the ordinary human level, this is the same sort of psychology that one sees in psychotic killers or mad dictators. Rāvaṇa was particularly concerned with tormenting *brāhmaṇas* and ascetics, because these persons were worshipers of the Devas, who were Rāvaṇa's old enemies.

We can get some more insight into the viewpoint of Rāvaṇa from the answer given by Brahmā to the Devas, Gandharvas, and Sages:

> Here is a way of bringing about the end of that perverse being! "May I not be destroyed by Gandharvas, Yakṣas, Gods or Rākṣasas" was Rāvaṇa's request, but thinking man to be of no account, he did not ask to be made invulnerable in regard to him; therefore, none but man can destroy him.[56]

Rāvaṇa thought that human beings were completely insignificant, and this gives us another clue as to why he didn't particularly concern himself with them. But this turned out to be his downfall. Following the advice of Brahmā, the assembled celestial beings asked Lord Viṣṇu to incarnate on earth as an apparent human being to slay Rāvaṇa. Lord Viṣṇu agreed, and He took birth as Rāma, the son of King Daśaratha of Ayodhyā.

In due course of time, Rāvaṇa learned about the beauty of Sītā, the wife of Rāma, and he devised a plan to abduct her (see pages 238–39). This created a conflict between Rāvaṇa and Rāma, and eventually Rāma slew him with celestial weapons in a great battle.

This brings up another point concerning human beings. From the point of view of celestial beings such as Rāvaṇa, humans are utterly inferior and unimportant. Why then did Lord Viṣṇu, the original source of Brahmā and all the Devas, agree to live among them as one of them?

The answer is that according to Vedic literature the human form of life is uniquely advantageous for making spiritual advancement. Subhuman forms of life lack the intelligence required for spiritual contemplation, and superhuman beings tend to become wrapped up in the enjoyment of great power, beauty, and longevity. But the human form, with all its trials and tribulations, provides a gateway through which the soul can readily ascend to higher spiritual stages. Since the primary concern of Lord Viṣṇu is with the destiny of the soul, it was natural for Him to be concerned with the human race.

Curiously enough, this very idea comes up in one of the channeled UFO communications—whatever their real source may be. Here is a quotation from a communicator named Hatonn, who said that he represents the "Confederation of Planets in Service of the Infinite Creator:"

> Many of us who are now circling your planet would desire to have the opportunity that you have, the opportunity to be within the illusion and then, through the generation of understanding, use the potentials of the illusion. This is a way of gaining progress spiritually and has been sought out by many of our brothers.[57]

Here is a quotation from the *Bhāgavata Purāṇa* that makes a very similar point:

> Since the human form of life is the sublime position for spiritual realization, all the demigods in heaven speak in this way: How wonderful it is for these human beings to have been born in the land of Bhārata-varṣa. . . . We demigods can only aspire to achieve human births in

> Bhārata-varṣa to execute devotional service, but these human beings are already engaged there.[58]

Bhārata-varṣa is the domain of the short-lived human form of life, and thus it refers to this earth planet.[59] Since the human race is important from a spiritual point of view, it tends to be protected by higher authorities within the universe, and this is one reason why it is not easily taken over by more powerful beings. This idea also comes up in another form in the following description of the Gentry, or fairy folk, recorded in Ireland by the ethnologist Evans-Wentz:

> The folk are the grandest I have ever seen. They are far superior to us, and that is why they are called the *gentry*. They are not a working class, but a military-aristocratic class, tall and noble-appearing. They are a distinct race between our own and that of spirits, as they have told me. Their qualifications are tremendous. "We could cut off half the human race, but would not," they said, "for we are expecting salvation."[60]

In summary, the Vedic literature, many UFO communications, and Celtic folklore all suggest that human society may sometimes be affected by the activities of more powerful beings who are primarily concerned with their own affairs. While pursuing their own agendas, these beings may occasionally intervene in human society in ways that seem mysterious from a limited human perspective, but make sense within their own complex framework of activity. These interventions may be harmful or beneficial, depending on the underlying motives of the beings involved. They fall short of displaying the full powers of these beings for a variety of reasons, ranging from spiritually-based laws of noninterference to contempt for the weakness of insignificant humans.

The Plot of the Mahābhārata

Thus far, I have discussed two Vedic examples of alien invasions of the earth. In each case, most human beings experienced these invasions in the form of sporadic nocturnal attacks by terrifying beings who seemed to come out of nowhere. The attacks were highly disturbing to people who heard about them and devastating to those who experienced them, but they didn't have much effect on human society as a whole. There is one example, however, of an attempt by Daityas and

Dānavas to take over and rule human society, and this forms the main plot of the *Mahābhārata*.

The story begins at a time long ago, when human society was prospering. People were dedicated to principles of virtue, and they did not decline into decadence as they began to experience material success. However, this auspicious situation did not last. Just as in the story of the Kāleya Dānavas, human society began to be affected by events occurring in celestial planetary systems. Here is what happened, as narrated to King Janamejaya by the sage Vaiśampāyana:

> But then, O best of monarchs, just as humankind was flourishing, powerful and demonic creatures began to take birth from the wives of earthly kings.
>
> Once the godly Ādityas, who administer the universe, fought their wicked cousins the Daityas and vanquished them. Bereft of their power and positions, the Daityas began to take birth on this planet, having carefully calculated that they could easily become the gods of the earth, bringing it under their demonic rule. And thus it happened, O mighty one, that the Asuras began to appear among different creatures and communities.[61]

As in the case of the Kāleya Dānavas, this attempt involved covert activities rather than an out-and-out invasion of the earth by alien armies. The technique adopted by the invading forces was to enter in their subtle bodies into the wombs of the wives of kings and thereby take birth in royal families. In this way they seized control of earthly governments and were able to exploit the earth as they liked.

> As these demonic creatures continued to take birth on the earth, the earth herself could not bear the weight of their presence. Having fallen from their positions in the higher planets, the sons of Diti and Danu thus appeared in this world as monarchs, endowed with great strength, and in many other forms. They were bold and haughty, and they virtually surrounded the water-bounded earth, ready to crush those who would oppose them.
>
> They harassed the teachers, rulers, merchants, and workers of the earth, and all other creatures. Moving about by the hundreds and thousands, they began to slay the earth's creatures, and they brought terror to the world. Unconcerned with the godly culture of the *brāhmaṇas,* they threatened the sages who sat peacefully in their forest

> *āśramas*, for the so-called kings were maddened by the strength of their bodies.[62]

In response to this invasion, Bhūmi, the earth-goddess, approached Lord Brahmā and asked him to save the earth. Brahmā responded by ordering the Devas to incarnate on earth just as the Asuras had done: "In order to remove the burden of the earth, each of you is to take birth on the earth through your empowered expansions to stop the spread of the demonic forces."[63] Brahmā also requested Lord Viṣṇu to appear on the earth as an *avatāra* to oppose the demonic forces, and He agreed to do so.

In due course of time, various Devas appeared on the earth, either by entering personally into the wombs of earthly mothers or by impregnating earthly women and producing offspring that partook of their own nature. Then Lord Viṣṇu appeared as Kṛṣṇa, the son of Vasudeva and Devakī.

With the aid of the incarnate Devas, Kṛṣṇa gradually annihilated the forces of the Dānavas. This involved many complex developments, and one of them, the fratricidal struggle between the Pāṇḍavas and the sons of Dhṛtarāṣṭra, is the main subject of the *Mahābhārata*. In this struggle, the celestial war between the Devas and Asuras was reenacted on the earth, and by Kṛṣṇa's arrangement the forces of the Asuras were eventually defeated.

Several points can be made about this complex story. The first point is that at the present time much has been written about beings from other planets who reincarnate in human bodies as "Wanderers" with the aim of carrying out some higher purpose. There is also talk of "Walk-ins," or souls that take over existing bodies and displace their original souls. These concepts are similar to the idea presented in the *Mahābhārata* that the Devas and Asuras could take birth on earth with specific missions to perform.

To understand this idea, it is necessary to have a preliminary understanding of the soul, the subtle body, and the process of reincarnation. Curiously enough, these topics come up repeatedly in UFO close-encounter cases, and I will discuss this in the next chapter.

Another point is that invasions of the earth by inimical forces often provide an occasion for the introduction of profound ethical and spiritual teachings into human society. Thus Rāvaṇa's invasion resulted in the descent of Lord Rāmacandra, who taught the life of an ideal king. Likewise, the *Mahābhārata* invasion culminated in the speaking of the *Bhagavad-gītā* by Kṛṣṇa. An interesting question is: Will something similar happen as a result of today's situation?

10

Gross and Suble Energies

In August, 1975, a 48-year-old man was undergoing open-heart surgery. A half hour after being taken off cardiopulmonary bypass, he suffered from cardiac arrest and had to be revived by an injection of epinephrine into the heart and two electric shock treatments. On awakening, he recalled the following experience:

> I was walking across this wooden bridge over this running beautiful stream of water and on the opposite side I was looking and there was Christ and he was standing with a very white robe. He had jet-black hair and a very black short beard. His teeth were extremely white and his eyes were blue, very blue. . . . He looked different from any pictures I had seen before. . . . My real focus was on the white robe and if I could prove it to myself that this was really Christ. . . . And during all this time the knowledge, the universal knowledge, opened up to me and I wanted to capture all of this so that when I was able, I could let people know what was really around them. Only trouble was that I was unable to bring any of it back.[1]

This is a typical example of an out-of-body experience, or OBE. In such an experience, a person has the impression of leaving his physical body while continuing to see, hear, and think as a conscious being. Out-of-body experiences often occur when a person is in a near-fatal physical condition, and so they are also called near-death experiences, or NDEs. With the recent development of techniques for reviving a person who is near death, there has been a great increase in reports of such experiences, and a number of books have been written about them by doctors, psychologists, and psychical researchers.[2]

Many OBEs are difficult for an external observer to distinguish from dreams, although persons experiencing them may regard them as real because of their great vividness and profound psychological impact. This is true of the OBE mentioned above, which took place entirely in some dreamlike otherworld. There are many instances, however, in which a person having an OBE sees his own unconscious body from a distance. In some of these cases, persons who were supposedly unconscious due to cardiac arrest were able to give accurate descriptions of medical procedures being used to revive them.

Here is how cardiologist Michael Sabom summed up one man's description of his own resuscitation from a heart attack, as seen during an OBE:

> His description is extremely accurate in portraying the appearance of both the technique of CPR [cardiopulmonary resuscitation] and the proper sequence in which this technique is performed—i.e., chest thump, external cardiac massage, airway insertion, administration of medications and defibrillation.[3]

Sabom said that he came to know this man quite well and that the man gave no indication that he possessed more than a layman's knowledge of medicine. The man also testified that he had not watched cardiac resuscitations on TV before his experience. In his OBE, the man had seen the resuscitation procedures in vivid detail, even though his heart was not functioning at the time and his brain was deprived of oxygen. If the experience was a dream, then how did the man acquire the accurate knowledge of detailed medical procedures that this dream contained?

OBEs and UFOs

There has been a great deal of controversy over how to interpret out-of-body experiences, with some favoring theories based on dreams or hallucinations and others advocating paranormal explanations. One topic that is generally not introduced into these discussions is the subject of UFOs. However, it has recently emerged that out-of-body experiences may take place in connection with UFO encounters. Many witnesses have reported experiencing out-of-body travel during UFO abductions, and some have also reported spontaneously experiencing OBEs in the aftermath of UFO encounters. This leads to a completely new controversy over out-of-body experiences—one that inevitably brings in the body of observation and theory that has built up around the UFO phenomenon.

One theory is that UFO-related out-of-body experiences are actually misperceptions of physically real abduction experiences. Another is that UFO abductions are essentially illusory. According to this theory, OBEs are also hallucinatory experiences that are generated by the mind, perhaps due to the influence of some external agency. UFO abductions and OBEs go together because both are of a similar illusory nature.

A third theory is that UFO abductions are real events that can take place on a gross or subtle level of material energy, and OBEs are real events involving the temporary separation of the subtle mind from the physical body. In a UFO abduction, the physical body may be taken onto a UFO, and during this experience an OBE may or may not take place. Also, some UFO abductions may take place within OBEs. In these cases, the subtle mind is taken on board a UFO and the gross body is left behind. I will discuss these theories after giving a few examples of out-of-body experiences associated with UFOs.

First of all, there are reports indicating that OBEs are sometimes induced by humanoid entities of the kind associated with UFOs. One example of this is an experience reported by Betty Andreasson. She said that in July of 1986 she was lying on the couch of her trailer-home reading a Bible when she heard a whirring sound and saw a strange being appear next to the couch.

At this point, she had the experience of seeing her own body from an external vantage point:

> I see myself standing and I see myself laying on the couch! The being had put a small box or something on the couch first and then I saw myself appear there. I see myself standing up. . . . And I see myself moving toward the being. And then I turn toward the couch and I reach down to touch myself and—Ahhhh!—when I do, my hand goes right through me! [4]

In this case, the being was of the standard "Gray" type.[5] After entering the out-of-body state, Betty underwent a strange experience involving visions of crystal spheres, the passing shadow of a gigantic bird, and a hovering spherical craft. This is in some ways reminiscent of the otherworldly experiences that often occur in OBEs. However, the "Gray" aliens were present throughout the experience.

Whitley Strieber recounted a very similar experience, in which he awoke at about 4:30 in the morning in his country cabin and tried to achieve an OBE using methods recommended by Robert Monroe, a well-known investigator of out-of-body states. He said that he saw an image of a long, bony, four-fingered hand of a "Gray" being pointing towards a two-foot-square box on a gray floor. He then experienced an inappropriate wave of sexual feeling, followed by an OBE. He found himself floating above his body. He saw his cat, which should have been in New York City, and he saw the face of a "Gray" visitor outside one window. He found that he could move about in his out-of-body state, and he described his adventures of passing out through a closed window and back. During all this he experienced himself as a "roughly spherical field."[6]

There are also cases in which a person has an out-of-body experience with no obvious cause, enters a UFO in the out-of-body state, and has an encounter with UFO entities. For example, Betty Andreasson, after her remarriage to Bob Luca, reported a joint OBE in which they both entered a UFO occupied by typical "Gray" beings. There she encountered featureless human shapes glowing with light and found that she was also in this condition. She also saw the featureless light-forms changing into balls of light and then back into human light-forms.[7]

In another out-of-body UFO experience, a person named Emily Cronin had the experience of standing by her car and seeing her dormant body within the car. Here she recounts this experience under hypnosis:

> *Emily:* Not in the car. But I *am* in the car. It's silly.
> *McCall:* Don't worry about it. Just tell me what's going on.
> *Emily:* That's silly! You can't do that!
> *McCall:* You can't do what?
> *Emily:* You can't be in the car and out of the car, too. That's silly! *But I am.*[8]

After this, she saw a large, glowing "bubble" hovering over some trees by the roadside. She communicated telepathically with unseen intelligences associated with this object, and she had a realization that all life is one and that the intelligences were not actually alien.[9]

In the case of Judy Doraty discussed in Chapter 9 (pages 313–14), the witness, Judy, experienced standing by her car and simultaneously entering a UFO and observing what was happening inside it. In this case, the UFO had been previously seen by Judy and other witnesses from a normal, physically embodied point of view. Within the UFO, Judy reportedly observed humanoid entities cutting up a living calf and then dropping its body to the ground using a shaft of light. This would suggest that the scene within the UFO was grossly physical and that Judy was viewing it just as persons having OBEs sometimes observe their own bodies.

People in their normal bodily condition usually cannot perceive someone who is present near them in an out-of-body state. However, Judy reported that the entities she saw in the UFO were aware of her presence, and they communicated with her telepathically.

The White-Robed Beings

Betty Andreasson recalled under hypnosis that as a teenager she was taken into an alien craft, which entered a body of water and emerged in an underground complex. Up to this point, she seemed to be traveling in her physical body, since the trip involved what appeared to be great g-forces of the kind produced by ordinary acceleration.[10] In the underground complex, she was told by "Gray" beings that she was going to be taken home to see the One. Here is the experience that unfolded next, as relived through hypnotic regression:

> *Betty:* We're coming up to this wall of glass and a big, big, big, big, big, *door.* It's made of glass.
> *Fred Max:* Does it have hinges?
> *Betty:* No. It is so big and there is—I can't explain it. It is door after door after door after door. He is stopping there and telling me to stop. I'm just stopping there. He says: "Now you shall enter the *door* to see the *One.*"
> And I'm standing there and *I'm coming out of myself!* There's two of me! There's two of me there! . . . It's like a twin.[11]

After thus entering an out-of-body state, she went through the door:

> Betty: I went in *the door* and it's very *bright. I can't take you any further.*
> *Fred Max:* Why?
> *Betty:* Because . . . I can't take you past this *door.*
> *Fred Max:* Why are you so happy?
> *Betty:* It's just, ah, I just can't tell you about it. . . .Words cannot explain it. It's wonderful. It's for *everybody.* I just can't explain this. I understand that *everything is one.* Everything fits together. It's beautiful! [12]

This sounds like a typical description of the experience of Brahman realization, a state of consciousness that has been sought by *yogīs* and mystics the world over. In the Vedic tradition, there are several schools of philosophical thought regarding the nature of Brahman realization. I will discuss this topic in some detail in Chapter 11. For now, I am interested in what happened after this experience was over.

After leaving the door of the One, Betty reported encountering mysterious white-robed beings: "Okay, I'm outside the door and there's a tall person there. He's got white hair and he's got a white nightgown on and he's motioning me to come there with him. His nightgown is, is glowing and his hair is white and he's got bluish eyes." [13] This person looked like a normal human, in contrast to the small "Gray" beings she had encountered thus far.

In his discussion of this case, Raymond Fowler pointed out that possibly similar beings were reported by Italian Navy personnel during a UFO sighting on the slopes of Mount Etna on July 4, 1978. Here a red, pulsating, domed disc landed, and the witnesses encountered "two tall golden-haired, white-robed beings accompanied by three or four shorter beings wearing helmets and spacesuits." [14] In this case the tall white-robed beings were seen by military personnel who were presumably in their physical bodies, in a more or less normal state of consciousness.

In OBEs not connected with UFOs, there are frequent references to beings in white robes. An example would be the cardiac patient's OBE mentioned at the beginning of this chapter. It is significant that although this witness thought the white-robed being he saw was Christ, he remarked, "He looked different from any pictures I had seen before." Evidently, he felt some doubts about this identification. It is also interesting that the man's encounter with this being was accompanied, as with Betty Andreasson, by a mystical experience involving intimations of universal knowledge.

So here we have three cases in which white-robed beings are described. In one, a person had apparently been physically abducted by ufonauts, then had an OBE involving a mystical experience, and finally met a being of this type. In another, these beings were seen along with a UFO by military personnel who were walking about in an apparently normal state. In yet another, a being of this type was encountered in an OBE that occurred during a medical emergency and was not connected with UFOs.

Jenny Randles discussed a possibly related case, in which a medically instigated OBE led to a meeting with a tall white-haired being from a UFO. This experience was reported to her by Robert Harland, a professional magician and, alas, a self-confessed phony medium. Harland told Randles that he went to a dentist in 1964 for major oral surgery. An anesthetic gas was administered, and he had an OBE. From an out-of-body perspective, he saw that the dentist banged his knee, and the dentist later verified this.

Thus far, this was a typical OBE of the kind associated with physical trauma. But then Harland saw a tall being with long white hair drift through the ceiling and explain telepathically that they must go together. They floated through the roof into a UFO. He was shown around. The UFO's operation was explained, and he was given a message to convey about a terrible holocaust in which the Earth's crust would split apart. He was then told he would have to fight his way back to his body. Indeed, small, ugly creatures tried to prevent his return, but he made it and awoke to see the dentist thumping him and looking very worried. He had nearly died in the chair.[15]

One can always hypothesize that the beings seen in these four cases were simply dreams or hallucinations, although this raises the question of why people would independently have such similar dreams. If we leave the dream hypothesis in the background and consider that the beings might actually exist, then the question is: Are these beings operating in gross physical bodies or bodies made of some kind of subtle energy? The observations of the Italian naval personnel would suggest the former, while the stories of the cardiac patient and Mr. Harland would suggest the latter.

Physical Form or Subtle Form?

One interpretation of this bewildering data is to interpret all UFO abductions as strictly physical and reject OBEs as a mistaken idea. This approach has been taken by David Jacobs, an associate professor of history at Temple University in Philadelphia and an active investigator of UFO abductions. Jacobs wrote the following about abductees' perceptions:

> Part of these anomalous memories and dreams might be the unaware abductees' knowledge that they have had Out of Body Experiences. It is common for abductees to feel that they in some way left their body, usually during the night in bed. . . . A few unaware abductees

> claim that they have not only had Out of Body Experiences but that they have experienced Astral Travel as well. They know that they have in some mysterious way experienced a strange displacement in location. . . . The only way that they can reconcile what has happened to them is through the only available explanation—astral travel, no matter how ill-defined that might be.[16]

Jacobs's idea was that abductions by UFO entities really happen but that out-of-body experiences are a "new age" misconception adopted by "unaware" abductees. He maintained that abductees will generally abandon their false ideas about OBEs when they become aware of what really happened to them. Thus, "knowledge of the abductions finally gives them the answers they were seeking and the majority of them let go of previously held belief structures that were never fully satisfactory."[17]

This interpretation seems unsatisfactory because it blurs the distinction between (1) abduction experiences in the physical body during which an OBE takes place, and (2) abduction experiences occurring entirely in an out-of-body state and accompanied by memories of seeing the gross body as it is left behind.

The same can be said of the interpretation of all abduction experiences as being entirely psychical or mental. For example, Jenny Randles has used accounts such as Robert Harland's to argue that UFO abductions are entirely mental experiences induced in psychically susceptible and visually creative persons by alien beings that "have harnessed the power of consciousness to cross the gulfs of space and seek out new life forms."[18]

This interpretation also blurs the distinction between points (1) and (2). If all abduction experiences occur entirely in the mind, then why do some seem to the witnesses to occur on the bodily platform of experience, while others, such as Harland's or Emily Cronin's, occur in an out-of-body state?

Physical Aftereffects of UFO Abductions

Of course, a further objection to the all-mental theory is that scars and infectious diseases have been reported in connection with UFO abductions. Budd Hopkins is well known for his claim that abductees sometimes bear scars, which they associate directly or indirectly with UFO encounters.[19] One example is Virginia Horton, whose illusory deer encounter is mentioned above (page 236). She also told of a deep, profusely bleeding but painless cut that she received as a six-year-old child. In her conscious recollections, the cut was memorable because at the time she was unable to explain to her elders how she had gotten it. Under hypnosis, she related an elaborate abduction scenario in which aliens of the typical "Gray" variety took her into a circular room illuminated by diffuse, pearly gray light and made the cut with some kind of machine. They explained to her that "we need a little, bitty piece of you for understanding."[20]

The noted UFO researcher Raymond Fowler has also described under hypnosis a nightmarish, dreamlike experience in which he seemed to be manipulated by beings that he could not see.[21] This would seem to be a good candidate for a purely mental experience but for the fact that it occurred on the night before a mysterious, unexplained scar appeared on his leg. This scar was said by a dermatologist to resemble the mark made by a punch biopsy.[22]

Fowler cited research into scars and other medical sequelae of UFO encounters that was carried out by Dr. Richard N. Neal, a specialist in obstetrics and gynecology at the Beach Medical Center in Lawndale, California. Neal maintained that scars tend to show up on the bodies of abductees in a consistent manner. Thus, "scars have been observed on the calf (including just over tibia or shin bone), thigh, hip, shoulder, knee, spinal column and on the right sides of the back and forehead."[23] These scars tend to be either thin, straight hairline cuts about 2–3 inches long or circular depressions about 1/8-inch to ¾-inch in diameter and as much as ¼-inch deep.

Other kinds of bodily marks have also been noted, such as rashes on the upper chest or the legs that are often geometrical in shape. First or second degree burns have been noted, as well as infections and unusual growths. For example, in Chapter 9 (pages 316, 320), there are two examples of women reporting severe vaginal infections after UFO abductions that involved gynecological examinations.

The very fact that abduction witnesses report invasive physical examinations suggests that their experiences are not simply mental. Dr. Neal pointed out, "Aliens have taken blood, oocytes (ova) from females and spermatozoa from males, and tissue scrapings from their subjects' ears, eyes, noses, calfs, thighs and hips."[24] Sometimes tubes are inserted into women's navels—an operation that was described to

Betty Hill as a pregnancy test by her captors. It has been pointed out that this operation is similar to a gynecological testing procedure called laparoscopy that was developed years after Betty and Barney Hill's abduction experience in September of 1961.[25]

Finally, we shouldn't overlook the controversial topic of probes that are inserted into the nose by alien entities. As Dr. Neal put it, "Many abductees have described a thin probe with a tiny ball on its end being inserted into the nostril—usually on the right side. They are able to hear a "crushing' type sound as the bone in this area is apparently being penetrated. Many will have nosebleeds following these examinations."[26] Fowler and Hopkins give examples of this, and it seems to come up repeatedly in UFO accounts.

From time to time, investigators have claimed that such probes have been recovered from people's bodies for examination. However, I have yet to see any reliable publication describing a systematic study of a recovered probe.

One could postulate that people may imagine physical experiences, such as having probes inserted into their bodies. However, it is not clear what their inner motive would be for imagining such things. Many UFO abductees reporting these experiences have been tested psychologically and found to be quite normal. So their testimony cannot be attributed to abnormal mental processes.

One could also postulate that beings acting on a subtle level might be able to invoke in people's minds traumatic experiences that would result in physical symptoms. There are cases in which people have developed bleeding wounds called stigmata, apparently under the influence of intense religious emotions. There are also reports that a particular pattern of reddened skin, such as a cross, can be produced by hypnotic suggestion.[27] Could it be that the physical symptoms of UFO abductions are similarly produced by some form of psychical influence?

One reply to this is that some abduction cases involve physical traces on objects or on the ground that suggest the presence of some physically real agent. Examples would be the ground traces reported by Budd Hopkins in the Kathie Davis case,[28] or the strange shiny spots appearing on the car of Betty and Barney Hill after their UFO experience.[29] Also, there are abduction cases, such as those of Travis Walton, William Herrmann, and Filiberto Cardenas, in which the abductee was dropped off by the UFO miles from his pickup point.

Near-Death Experiences with Administrative Bungling

There is certainly a great deal of evidence indicating that UFOs can become manifest as physically real vehicles, and there is also much evidence suggesting that people are sometimes physically taken on board these vehicles. However, since some UFO abductions do seem to involve out-of-body experiences, the idea that trauma on a subtle, mental level can bring about gross physical effects should be carefully considered. To illustrate what might happen, consider the following account of a near-death experience occurring in India:

In the late 1940s, an Indian man named Durga Jatav suffered for several weeks from a disease diagnosed as typhoid. At a certain point his body became cold for a couple of hours, and his family thought he had died. But he revived and told his family that he had been taken to another place by ten people. After attempting to escape from them, they cut off his legs at the knees to prevent further attempts. Then they took him to a place where about forty or fifty people were sitting. They looked up his "papers," declared that the wrong man had been fetched, and ordered his captors to take him back. When he pointed out that his legs had been cut off, he was shown several pairs of legs and recognized his own. These were somehow reattached, and he was warned not to "stretch" his knees until they had a chance to heal.

After his revival, his sister and a neighbor both noticed that he had deep folds or fissures in the skin on the fronts of his knees, even though such marks had not been there previously. The marks were still visible in 1979, but X-rays taken in 1981 showed no abnormality beneath the surface of the skin.[30] Could it be that the experience of having his legs cut off in a subtle realm caused these marks on his physical legs?

Ian Stevenson has assembled a large amount of evidence indicating that young children who spontaneously remember previous lives sometimes bear birthmarks on their bodies corresponding to injuries received during those lives. He has about 200 cases of this type, and he says that in fifteen he has been able to match up birthmarks with postmortem reports describing the previous body. Regarding these birthmarks, he made the following observation: "Some marks are simply areas of increased pigmentation; in other cases, the birthmark is three-dimensional, the area being partly or wholly elevated, depressed, or puckered. I have examined at least two hundred of this kind, and many of them cannot be distinguished, at least by me, from the scars of healed wounds."[31] The point about scars is especially significant in connection with UFO abductions.

In the case of Durga Jatav, one can imagine that some psychical influence injected into his brain the idea that his legs had been cut off, and this in turn resulted in the fissures in his knees. However, if a wound in one life can affect a body in another, then more must be involved than just the brain.

An explanation can be devised if we introduce the idea that the soul, encased in a body made of subtle energy, is able to transmigrate from one gross physical body to another. In that case, one can suppose that the fatal injury in one life traumatized the subtle body, and this resulted in birthmarks in the developing embryo in the next life. One could likewise suppose that Durga Jatav's subtle body was traumatized in a subtle domain, and this resulted in the knee fissures when his subtle body was returned to his gross body.

A wide variety of physical effects can apparently be produced by subtle action. Here is an example involving a man named Mangal Singh who experienced an NDE in about 1977 while in his early 70s. He described his experience as follows:

> I was lying down on a cot when two people came, lifted me up, and took me along. I heard a hissing sound, but I couldn't see anything. Then I came to a gate. There was grass, and the ground seemed to be sloping. A man was there, and he reprimanded the men who had brought me: "Why have you brought the wrong person? Why have you not brought the man you had been sent for?" The two men ran away, and the senior man said, "You go back." Suddenly I saw two big pots of boiling water, although there was no fire, no firewood, and no fireplace. Then the man pushed me with his hand and said, "You had better hurry up and go back." When he touched me, I suddenly became aware of how hot his hand was. Then I realized why the pots were boiling. The heat was coming from his hands.[32]

On returning to consciousness, Mangal felt a severe burning sensation in his left arm. This area developed the appearance of a boil and left a residual mark after healing. He was apparently unable to describe the appearance of the "men" he had met.

The stories of Durga Jatav and Mangal Singh are part of a group of sixteen Indian near-death accounts collected by Satwant Pasricha and Ian Stevenson. They observed that in these cases messengers typically come to take the witness, in contrast to Western NDEs, in which the witness generally meets other beings only *after* being translated to "another world." Pasricha and Stevenson noted that their Indian subjects naturally identify these messengers with the Yamadūtas, the agents of Yamarāja, the lord of the dead in traditional Hinduism.

They also pointed out that the evident cultural differences between Indian and Western NDEs do not necessarily demonstrate that these experiences are simply unreal mental concoctions. It is possible that persons near death are treated differently in different cultures by personalities on the subtle level. There could be different policies for groups of people with different karmic situations.

According to Vedic literature, the transmigration of souls is regulated by the Yamadūtas, or servants of Yamarāja. The Yamadūtas serve as functionaries in the celestial hierarchy, and they are equipped with mystic powers, or *siddhis,* that enable them to carry out their duties. They are described as having a very negative, fearful disposition. Nonetheless, they are employed by higher authorities for the positive purpose of reforming the consciousness of souls entangled in material illusion.

Generally, when the Yamadūtas take a person, he doesn't return to tell the tale. But Vedic accounts do mention some cases where someone returns. There is the story from the *Bhāgavata Purāṇa* of Ajāmila, a sinful man who uttered "Nārāyaṇa," a name of God, when seeing the Yamadūtas at the time of death. As a result of this action, several effulgent servants of Nārāyaṇa intervened and told the Yamadūtas not to touch Ajāmila. There followed a debate between the Yamadūtas and the servants of Nārāyaṇa on the laws regarding the treatment of departed souls. Finally, the Yamadūtas accepted defeat in this debate and departed from the scene, and Ajāmila was revived from apparent death.[33]

There are UFO encounter cases involving the capture-by-mistake theme of the Indian NDEs. In Chapter 9 (pages 315–20), I presented the story of a woman and her son who were abducted by strange beings and taken on board a UFO while driving near Cimarron, New Mexico. In this case, the woman and boy were physically dragged away by strange "men." The woman was subjected to a harrowing physical examination, after which a tall, authoritative "man" appeared on the scene and angrily declared that the woman should not have been taken and should be sent back. Not only that, but the tall man placed his hand on the woman's forehead, and she was burned by it. This is reminiscent of the Indian NDE cases, and of the case of Mangal Singh, in particular.

However, the woman came down with a severe vaginal infection after the experience, apparently as a result of the examination she received on the UFO. Was this due to a subtle examination, or was it caused by a botched physical examination?

Another example illustrating the theme of capture by mistake is an encounter story related by Emily Cronin. (This is a different encounter than the one mentioned previously, on page 342.) On this occasion, Emily, her young son, and her friend Jan were resting by the side of a road called Ridge Route near Los Angeles. She consciously remembered seeing a bright yellow light, hearing a high-pitched whine that seemed to have a paralyzing effect, and feeling the car shaking. Under hypnosis, she said that a strange, tall figure in black was looking in the back window of the car and was shaking it. Two other similar beings were standing to one side, telepathically telling the first being that this was a mistake and they shouldn't be there. When Emily managed to move one finger by strongly focusing her will, the noise stopped, the light and figures vanished, and everything was back to normal.[34] Here, the way the experience ended suggests that it occurred on a subtle level.

UFOs and the Recycling of Souls

Western OBEs occurring during medical emergencies are naturally related with death, and the persons experiencing them often connect them with the fate of the soul in the next life. In India, of course, these experiences are associated with the process of transmigration, whereby the soul, riding in the subtle body, is transferred to a new situation at the time of death. Given all the parallels that exist between OBEs and UFO abductions, could it be that some UFO entities are involved with the transmigration of the soul? It turns out that ideas along these lines have been discussed in the UFO literature.

For example, Whitley Strieber has said that his visitors told him, "We recycle souls."[35] Strieber's visitor experiences inspired him with the following general idea: "Could it be that the soul is not only real, but the flux of souls between life and death is a process directed by consciousness and supported by artistry and technology?"[36] This idea is completely Vedic, and so is the corollary that our actions are watched and appraised by beings who control our destination after death. Appraising modern attitudes, Strieber noted, "Because we have deluded ourselves into ignoring the reality of the soul, we imagine everything we do to be some kind of secret," and he asked, "Who watches us?"[37]

The following story gives some indication of how Strieber arrived at these ideas. He related that his visitors invisibly spoke to him, repeatedly warning him not to eat sweets. After several weeks of these warnings, he asked why he shouldn't eat sweets, and they said, "We will show you."

Six days later, he learned through an acquaintance about a woman in Australia who was dying from diabetes. During the previous evening, the woman had seen seven little men "like Chinese mushrooms" who appeared and descended from the ceiling. They lifted the sick lady to the ceiling, and as she protested they put her on the floor. Then she had a vision of sitting in a park, putting on a flowing blue robe, and watching the sun set as a desolate wind blew—all symbols of death. After this experience, the woman declined quickly.[38] Strieber was told that the woman was very conservative and probably had given no thought at all to such topics as UFOs and humanoid visitors.

Strieber took this unexpected story from Australia as a graphic answer to his question as to why he shouldn't eat sweets. The story involved beings similar to his visitors; it involved diabetes, a disorder of the body's sugar metabolism; and it came from a bare acquaintance on the other side of the world shortly after he asked his question. Since the woman's encounter with the beings involved symbolic intimations of her death, it struck him that his visitors might have some connection with what happens to people after death.

The relationship between Strieber's visitors and the Vedic Yamadūtas is difficult to ascertain. There are differences between these two groups indicating that they play different roles, and there are also similarities suggesting that they may be closely related. For example, one difference is that the Yamadūtas normally act only on the subtle level, whereas Strieber maintained that when he was abducted on one occasion, he was able to physically take his cat with him—an indication that his trip took place on the physical platform. Nonetheless, there are also similarities. For example, the Yamadūtas look strange and frightening, they emanate a strongly negative mood, they can travel invisibly and pass through walls, and they can induce OBEs in human subjects.

Similar remarks can be made about the beings who repeatedly abducted Betty Andreasson, but in her case there are additional complications. For example, during one UFO abduction she had a classical mystical experience, and then she saw white-robed beings similar to those connected with mystical insights in Western NDEs. To understand fully what is going on here, we will need much more information. I suspect that we are seeing a few traces of a complex universal control system involving many different types of intelligent beings.

Soul Recycling and the Government

It may come as no surprise that references to the soul, OBEs, and reincarnation come up in the lore on UFOs and the U.S. Government. In addition, some of this material shows connections with Whitley Strieber's testimony. Here is the story:

Strieber described dreams or visions in which his visitors were found to live in a strange desert setting where ancient buildings were built into cliffs under a tan sky.[39] Now according to Linda Howe, an Air Force intelligence officer named Richard Doty informed her in 1983 about EBEs—Extraterrestrial Biological Entities—that were allegedly in contact with the U.S. Government. Supposedly, these EBEs come from a desert planet where they live in buildings like those of the Pueblo Indians. One of them is said to have informed an Air Force Colonel that "our souls recycle, that reincarnation is real. It's the machinery of the universe."[40]

This provides a link between the Strieber visitors, the highly physical aliens spoken of in connection with the U.S. Government, and reincarnation. The similarities are so close that we seem to be faced with two alternatives. Either Strieber wrote material from Government/EBE stories into his book, or he was independently reporting experiences that tend to corroborate some of those stories.

There is another story connecting UFOs, OBEs, and the U.S. Government. This involves the thoroughly physical case occurring in October of 1973 in which a UFO was said to approach an Army Reserve helicopter flying from Columbus, Ohio, to Cleveland. At about 11:02 p.m. the crew members saw a red light on the eastern horizon that seemed to be on a collision course with the helicopter. The pilot, Capt. Lawrence J. Coyne, tried to radio a nearby airport, but after an initial response he couldn't get through. To avoid collision, he sent the helicopter into a dive. A cigar-shaped, metallic object took up a position directly over the helicopter and flooded the cockpit with green light. After a short interval, the object continued to the west, but Coyne found that the helicopter was at 3,500 feet and climbing at 1,000 feet per minute, even though they had initiated a dive from 2,500 feet to 1,700 feet. Once the object had departed the radio worked.

There were ground witnesses. A family consisting of a mother and four adolescent children were driving on a rural road below. They saw the encounter between the object and the helicopter and noted the green light. Also, Jeanne Elias, who was in bed at home watching the TV news, heard the diving helicopter and hid her head under her pillow. Her 14-year-old son woke up and saw the green light, which lit up his whole bedroom. The object was explained as a meteor by the famous UFO debunker Philip Klass.[41]

In the aftermath of this case, Capt. Coyne reported receiving a call from the "Department of the Army, Surgeon General's office," asking whether he had had any unusual dreams after the UFO incident. As it happened, he reported a vivid dream of an OBE.

Sgt. John Healey, one of the helicopter crewmen, reported, "As time would go by, the Pentagon would call us up and ask us, well, has this incident happened to you since the occurrence? And in two of the instances that I recall, what they questioned me, was, number one, have I ever dreamed of body separation, and I have—I dreamed that I was dead in bed and that my spirit or whatever was floating, looking down at me lying dead in bed, . . . and the other thing was if I had ever dreamed of anything in spherical shape. Which definitely had not occurred to me."[42] He went on to say that the Pentagon would often call Coyne with such questions, asking about all the crew members, and the Pentagon people seemed to believe what they were told. One wonders who in the Pentagon might be interested in the UFO/OBE connection.

The Physical, the Subtle, and Beyond

In summary, the available evidence suggests that UFO abductions and close encounters may occur both in an ordinary bodily state and in an out-of-body state. In the former, the subtle senses of the witness operate through the medium of the gross sense organs (such as eyes and ears), and in the latter, perception occurs directly through the senses of the subtle body. Experiences involving a combination of in-body and out-of-body phases may also occur, and the Doraty case (pages 313–14) suggests that it is possible to perceive through gross bodily senses and through subtle senses at the same time. This has been called bilocation.

The evidence also suggests that the UFO occupants themselves can operate both on a physical and on a subtle level. They can perceive the subtle form of a human being, and they can arrange things so that a human being can see them in the out-of-body state. They can make themselves physically manifest and visible to ordinary eyes, or they can become unmanifest and invisible. They can also make their vehicles and other paraphernalia visible on either a gross or subtle level.

There is also evidence indicating that UFO entities can enter into a person's mind and control it in a manner reminiscent of traditional spirit possession. In her survey of UFO abductees, Karla Turner noted that "in some cases there seems to be a merging and the abductee then begins to feel or think what the ET is feeling or thinking." [43] She also observed that "We have ET takeover of a human's body. . . . The person is still there but they're not in control. Sometimes they're not even aware until somebody tells them afterwards . . . that they were doing or saying things that are not characteristic of the person." [44]

In the *Bhāgavata Purāṇa* a mystic *siddhi* is described which enables a grossly embodied being to leave his gross body behind and enter in subtle form into another person's body.[45] This is illustrated by the following story in the *Mahābhārata:*

A king named Kalmāṣapāda once arrogantly insulted and struck the sage Śakti because the latter would not give way to the king on a narrow forest path. Śakti, a son of the famous sage Vasiṣṭha, then cursed the king to become a man-eater.

While the king and Śakti were quarreling, Viśvāmitra, an enemy of Vasiṣṭha and a powerful *yogī,* approached invisibly with the aim of gaining something for himself. After seeing what happened and evaluating the condition of the king's mind, Viśvāmitra waited until the king returned to his capital city and then ordered a Rākṣasa to approach him. By the sage's curse and the order of Viśvāmitra, the Rākṣasa was able to enter the king and possess him.

The king was severely harassed by the Rākṣasa within him, but he was able to protect himself with his own willpower. Later the king was asked by a *brāhmaṇa* for a meal with meat. The request slipped the king's mind, but late that night he remembered it and asked a cook to prepare the meal for the *brāhmaṇa,* who was waiting at a certain place. Unable to find any meat, the cook asked the king what to do.

The Rākṣasa then exerted his influence, and the king ordered the cook repeatedly to get human meat. The cook did this, using flesh from an executed prisoner. The *brāhmaṇa,* on seeing the resulting meal, realized that it was unfit to eat, and he also cursed the king to become a man-eater. As a result of this second curse, the Rākṣasa was able to completely take over the king, and driven by madness and a desire for vengeance, the king began to kill and devour first Śakti and then the other sons of Vasiṣṭha.[46]

The Rākṣasas were mentioned in Chapter 6 (page 240) in connection with the illusory deer that Rāvaṇa used to abduct Sītā, and in Chapter 8 (pages 276–77) in connection with Bhīma and his Rākṣasī wife, Hiḍimbā. They were beings with powerfully structured gross bodies, and they were also known for their mastery of mystic powers.

Before meeting Hiḍimbā, Bhīma engaged in an intense hand-to-hand struggle with her brother Hiḍimba and killed him by strangulation after exhausting him in the fight.[47] This battle was thoroughly physical. But in the story of king Kalmāṣapāda, the Rākṣasa ordered by Viśvāmitra was able to act on a subtle level and possess the king in the manner of a traditional evil spirit.

This story illustrates the idea that beings of essentially inimical motivation may have the power to act both on the subtle and gross platforms of existence. The Vedic literature also describes a completely transcendental level of existence, and it is similarly possible for suitably qualified beings to function on both the transcendental and the physical planes. I will present three accounts illustrating this that date back roughly 500 years. As with the UFO stories that we have been considering, these stories display a bewildering combination of what appear to be physical phenomena and phenomena occurring on another plane of existence.

All three accounts are religious in nature, which means that they have to do with spiritual worship and meditation. Although some would categorically reject such material as admissible evidence, I disagree. If so many strange phenomena mentioned in this book could be true, it doesn't make sense to think that phenomena reported in religious contexts must all necessarily be false. In fact, I think that an imbalanced picture will be created if events of a positive spiritual nature are excluded, while those of a negative or at best neutral character are extensively presented.

The first example involves the Vaiṣṇava saint Narottama Dāsa Ṭhākura, who lived in India in the 16th century. Narottama would regularly meditate on living in the spiritual world in his *siddha-deha,* or perfected spiritual form. There he would perform the service of boiling milk for Kṛṣṇa, and he would actually experience this as real in all respects. In Vaiṣṇava philosophy, Kṛṣṇa is the Supreme Lord, and He lives in the transcendental realm in an eternal personal form. In that realm, many simple acts of service serve as media for the exchange of intense love between Kṛṣṇa and His devotees.

On occasion, the milk would boil over, and in his meditation Narottama would burn his hands while trying to stop it. It turned out, however, that upon awakening from his reverie, he would find that his hands were actually burned.[48]

This story can be compared with the two near-death experiences mentioned above, in which physical effects resulted from subtle experiences. One might argue that in all these cases, the physical effects were somehow impressed on the body by the power of the mind, as a consequence of intense mental experiences. From the Vedic point of view, this idea is acceptable as long as we understand that the mind of the individual involved had actually been functioning in another realm of existence. But more is involved than some kind of psychosomatic influence of the mind on the body. To illustrate this point, consider the next story.

The Vaiṣṇava saint Śrīnivāsa Ācārya was a contemporary of Narottama Dāsa Ṭhākura's. On one occasion, he was meditating on the pastimes of Lord Caitanya, who is an incarnation of Kṛṣṇa. Śrīnivāsa was meditating on Kṛṣṇa's form as Lord Caitanya by placing a garland of aromatic flowers around His neck and fanning Him with a *cāmara* whisk:

> As Śrīnivāsa served the Lord in this way, he could not keep his composure and, looking at the Lord's magnificent form, he began to exhibit ecstatic symptoms. This pleased Lord Caitanya, who then took the same garland of flowers that Śrīnivāsa had given Him and placed it around Śrīnivāsa's neck. After the Lord made this loving gesture, Śrīnivāsa's meditation broke; but the garland was still adorning his own chest. Its fragrance was unlike anything he had ever experienced.[49]

In this case, an object that was observed in trance in another world appeared in physical form in this world. This is certainly not a psychosomatic effect, but one might imagine that the mind of Śrīnivāsa, charged with intense spiritual emotion, might have paranormally manifested the garland as a physical object. Now, however, I turn to an example in which a human being in this world first meets someone from a higher realm and later visits that realm through meditative trance and again meets the same person.

In this account, a Vaiṣṇava saint named Duḥkhī Kṛṣṇadāsa was performing the daily service of sweeping a certain sacred area in the town of Vṛndāvana, a famous pilgrimage place in India. While doing this one day, he found a golden anklet that seemed to emanate a remarkable aura. Impressed by the influence that it had on his consciousness, he considered it to be very important, and he buried it in a secret place.

Shortly thereafter an old lady came to him, asking for the anklet and saying that it belonged to her daughter-in-law. Because of its spiritual influence, Duḥkhī Kṛṣṇadāsa was convinced that the anklet must really belong to Rādhārāṇī, the eternal consort of Kṛṣṇa. After a long discussion, the old lady finally admitted that this was so, and revealed that her true identity was Lalitā-sundarī, one of Rādhārāṇī's servants.

At this point, Duḥkhī Kṛṣṇadas wanted to see his visitor in her true form, but she said he would be unable to bear such a revelation. After being convinced of his sincere desire, however, she finally acquiesced to his request and revealed her true, incomparable beauty. After giving him several benedictions and receiving the anklet from him, she disappeared, and he was unable to find where she had gone.

One of the benedictions given to Duḥkhī Kṛṣṇadāsa was a special *tilaka* mark on his forehead, and a new name, Śyāmānanda. Since Lalitā had sworn him to secrecy about their meeting, it was difficult for Śyāmānanda to explain the *tilaka* and new name to his *guru,* who thought that he had simply concocted-them. In the course of dealing with this difficult situation, Śyāmānanda again met Lalitā-sundarī. This time, however, he met her by entering into her transcendental realm in a state of meditation.[50]

In this case, Duḥkhī Kṛṣṇadāsa met Lalitā-sundarī in this world, in his physical body, and he also met her in another world that he entered in his spiritual form by meditation. Thus both Duḥkhī Kṛṣṇadāsa and Lalitā-sundarī were able to operate on different planes of existence. It is significant also that Lalitā-sundarī was able to assume a disguised form.

Thus in both ancient and recent Vedic traditions there are accounts of beings who can operate on different planes of existence. These beings may be materialistic in orientation, like Viśvāmitra Muni and the Rākṣasa, or they may be spiritually advanced. The UFO literature likewise seems to contain examples of activity on both subtle and gross physical planes.

11

UFOs and Religion

In previous chapters the topic of religion and UFOs has come up on a number of occasions, but I have skirted around it while discussing other matters. In this chapter I will try to confront this topic directly and arrive at a coherent picture of the relationship between religion and the revelations connected with UFOs. One way to begin is to observe that there is one prominent view of reality that seems to be conspicuously absent in UFO-related communications. This is the world view of modern science.

According to the modern scientific perspective, the physical universe is the total observable reality. It is composed of matter and energy that transform according to laws that can be expressed in mathematical equations. In the theories of modern physics, all phenomena in the universe reduce to shifting vibrational states of a universal quantum field. We can visualize this crudely by imagining waves colliding with one another on a choppy sea. In quantum field theory, all phenomena can be thought of as wave patterns endowed with a certain fuzzy, partially defined quality known as quantum uncertainty.

Some scientists have advocated the idea that the quantum field is conscious, and they have even tried to identify it with the universal, unified consciousness. For example, this has been done in the works of Fritjof Capra, John Hagelin, and David Bohm. However, these are all attempts to modify the scientific world view by superficially grafting onto it some ideas taken from the Vedic philosophical school of Advaita Vedānta. In practice, the calculations of the physicists make no reference at all to consciousness. These calculations deal strictly with material causation—with interactions between various kinds of waves.

Other scientists hold that God is the basis of reality, but they insist that God acts solely as the sustainer of physical causality. This is sometimes proposed in an attempt to add Judeo-Christian theological categories to the scientific world view, but here again the additions are simply cosmetic. All phenomena occur in accordance with the laws of physics, and therefore they can be understood solely on the basis of those laws with no real need to bring God into the picture. To mainstream scientists, all objectively observable phenomena can, in principle, be explained on the basis of blind physical causation.

In this view, life is understood to be a byproduct of physical processes that occur under very special circumstances, on planets of the right size and composition, situated at the right distance from suitable stars. On such a planet, a soup of organic chemicals accumulates in a primordial ocean. Molecules collide, form bonds, and somehow gradually develop into living cells.

Then a process of Darwinian evolution takes place. Over hundreds of millions of years, cells gradually evolve into multicellular organisms. Some of these develop senses and nervous systems, and only then does the first glimmer of consciousness arise. On some planets, evolution may eventually produce creatures, such as human beings, that are capable of conscious, introspective thought.

However, this consciousness is simply a byproduct of physical interactions of matter in the brain. As soon as the brain is destroyed, or begins to seriously malfunction, consciousness is snuffed out. Nothing survives the death of the body but the body itself, and this is simply a collection of molecules that are eventually broken down and perhaps incorporated into other bodies.

According to this philosophy, humankind is the only technically advanced life form to have evolved in this solar system. However, intelligent life may have arisen on other planets in the universe, and the program called SETI (Search for Extraterrestrial Intelligence) has been developed to listen for radio signals from extraterrestrial civilizations.

Many scientists doubt that other intelligent beings have been able to overcome the obstacles to interstellar flight. But if nonhuman beings really are operating flying machines in the earth's atmosphere, then, according to conventional scientific ideas, these beings must have come from distant stars by means of advanced technology. There is no other possibility.

Ātmā, Brahman, and the Evolution of Consciousness

This philosophy of scientific materialism is certainly accepted by many people today. If it is true, the ufonauts must be cosmic superscientists, and we might expect them to do and say things that are incomprehensible. But we would not expect them to make understandable statements that clearly contradict fundamental scientific principles. Yet this is exactly what they are reported to be doing.

In many UFO encounters, the ufonauts are said to be uncommunicative. However, there are other encounters in which they are said to deliver elaborate philosophical discourses. This frequently happens in encounters involving friendly contact. It also occurs as a phase of some UFO abductions, including those that are otherwise frightening and traumatic. The philosophy presented by the entities tends to follow a consistent pattern, and it radically contradicts modern science. This philosophy can be summed up as follows:

There is life throughout the universe, and this includes vast numbers of beings that are very similar to ourselves in form and behavior. We can call these beings humanoids. They are conscious, and they have humanly recognizable emotions. They also generally have highly developed psychical abilities.

These beings, like ourselves, are souls inhabiting material bodies. As souls, they transmigrate from one physical body to another. There is a process of cosmic evolution of consciousness, whereby souls gradually progress in spiritual development by undergoing experiences in a succession of material bodies.

Spiritual advancement involves developing love and compassion for all beings, and it also involves the development of knowledge, intelligence, and psychical powers. Beings at high levels of spiritual advancement work together cooperatively in an organized system of universal government. In contrast, most humans of this earth are regarded as crude barbarians who are retarded in spiritual development.

In addition to the gross body made of familiar material elements, there is a subtle body made of finer energies unknown to modern science. There are also different planes of existence, which can be thought of as parallel or higher-dimensional realities. These planes are inhabited by humanoid beings, and some of these beings are able to travel from one plane to another. Some of these beings can also exert control over the gross and subtle bodies of human beings and cause them to move and transform in remarkable ways. (For example, they can move a human body through a solid wall.)

The life forms in the universe have all come into being through a process of creation. This process is not clearly explained, but the basic idea is that there is a universal Creator from whom living beings are naturally generated. This explains how humanlike forms can arise throughout the universe, even though this seems highly implausible from the viewpoint of Darwinian evolutionary theory.

This philosophy is pantheistic. The Creator is present everywhere, and acts everywhere through nature. The Creator is often regarded as impersonal and is said to be nearly incomprehensible and inaccessible.

At the highest level, the Creator is regarded as the One—as eternal, nondual being, full of consciousness, love, and light. It is said that the evolution of consciousness will eventually bring one to the point of experiencing the One or of entering into It.

This, in brief, is the philosophy that emerges fully or partially from many UFO-related communications, including those obtained by channeling and those received in direct encounters with UFO entities. This philosophy sharply contradicts scientific materialism in many important ways. It is also far from being alien. It is expounded in a vast human literature, and it is well known to many people.

In India this philosophy of merging into the impersonal Absolute is prominent in Buddhism and in the philosophical system of Advaita Vedānta. The latter has been identified with Hinduism by many people in the West, but in India both Advaita Vedānta and Buddhism stand in contrast to the philosophy of personal monotheism, called Vaiṣṇava Vedānta, that is presented in Vedic texts such as the *Bhāgavata Purāṇa.* This philosophy maintains that the Absolute Truth is personal in nature, and it regards the One of the Advaita Vedāntists as an incomplete conception of the Supreme Being. I will say more about Advaita Vedānta and Vaiṣṇava Vedānta later on in this chapter.

To me, at least, the idea that ostensibly nonhuman beings are promoting a philosophy similar to Advaita Vedānta was unexpected and astonishing. Nonetheless, there is a great deal of evidence suggesting that this is happening, and I have presented some of it in previous chapters. Here I will review some of this evidence and introduce some additional material. Then I will make a few remarks about what this all implies regarding philosophy, science, and religion.

Transmigration and Higher Planes

I mentioned in Chapter 5 that the UFO abductee Betty Andreasson spoke of alien beings as living in other "planes" or dimensions. Thus, when asked if these beings could travel to other stars, she said they could travel to some near our earth and to others beyond. She clarified this by saying, "Beyond ours there are others, but they are in a different plane. They're in a heavier space." She also pointed out that they can see the future, and that "time to them is not like our time, but they know about our time."[1]

She reported being conducted through a UFO by a being of the "Gray" type who identified himself as Quazgaa. This being told her telepathically about the intentions of his group, saying, "Because of great love, they cannot let man continue in the footsteps that he is going . . . They have technology that man could use . . . It is through the spirit, but man will not search out that portion. . . . If man will just study nature itself, he will find many of the answers that he seeks . . . Man will find them through the spirit. Man is not made of just flesh and blood."[2]

Betty Andreasson is a fundamentalist Christian, and thus we might expect her to make remarks about "love" and "the spirit." However, the idea of a "different plane," or a "heavier space," does not play any role in traditional Christian thinking. It turns out that there are also other indications of a non-Christian source for her alien communications.

For example, on one occasion it appears that Betty's speech was taken over by her alien visitors during a hypnotic session. At this point she said, with mechanical intonation, "You try to seek in wrong directions. Simplicity 'round about you. Air you breath, water you drink, fire that warms, earth that heals. Simplicity, ashes, things that are necessary taken for granted. Powers within them overlooked. Why think you are able to live? Simplicity."[3]

This statement refers to the elements air, water, fire, and earth. These elements are an important part of the Sāṅkhya philosophy of India, of ancient Greek philosophy, and of the medieval Hermetic traditions. But today they are regarded as outmoded categories by scientists, and it seems doubtful that Betty Andreasson was ever taught otherwise in school or in her church. Betty's experience of the burning of the Phoenix (page 189) also involved a theme that is certainly not prominent in Christianity today but which was a part of old Egyptian tradition.

The idea that there are other planes or dimensions came up in a rather harrowing fashion in the experience of a commercial artist and his wife who reported being mentally lured into a UFO by a tall, hairless man in a bizarre blue cape (page 320). The artist reported being subjected to a typical examination. In the course of this, his mind was forcefully invaded, and he received insights into higher dimensions of reality:

> It's like they're picking my mind . . . like I don't have any control. My brain, it's like there's a tunnel that goes through my mind to theirs. . . . Our minds are connected. It's like a tube, maybe it's light? It's like a grey light, grey-brown light, brownish-grey. It's like everything's pulled out of my head. . . . There's a terrible sound, but I can't tell what it is— only it's piercing, high pitched. . . . It's coming from my head! My head is gone . . . it's like I can see all my thoughts, like goo. Everything in my mind is stripped. I've got it, but they've got it, too.[4]

Then they put it back with additions:

> There's more to it than anybody knows. There's more to life, more to the world. There's more to everything than anybody knows. More dimensions, things co-existing. There are other dimensions . . . more than three dimensions. Everywhere, it all works together. Everything co-exists. There's different dimensions we can't go into.[5]

The UFO researcher Don Elkins has compiled a great deal of channeled material that purportedly comes from entities from other worlds.[6] He observed that this testimony displays highly consistent patterns, even though it comes from many, widely separated individuals from all walks of life. References to the soul, reincarnation, and higher planes or dimensions are commonplace. For example, one channeled entity known as Sut-ko is said to have communicated the following information:

> From time immemorial, teachers of Light came to planet Earth, incarnating from other planets, from other systems, even from other galaxies, and from the realms known to you as the nonphysical or supernal realms of existence; and great companies of Light came into incarnation, carrying with them the banner of Truth and Love and Light.[7]

This statement is a typical reference to higher beings incarnating on earth to help suffering humanity. Elkins cited another contactee's statement describing reincarnation as it applies to people on the earth:

> As we progress on to higher planes of life, we shall incarnate in bodies far more ethereal than those now used by us, just as in the past we used bodies almost incredibly grosser and coarser than those we call our own today.[8]

Many of the communications cited by Elkins contain wild and dubious statements, and it is not possible to prove that any of them really do originate from otherworldly beings. But the thematic consistency of these communications is striking, and I do not think this can easily be explained in ordinary terms. Elkins said, "Since I have observed over 100 people go through this [channeling] process and have read millions of words of contactee literature, both published and unpublished, I believe that I am now in a position to select highly correlative material from the masses of communications."[9] It appears that either the typical American subconscious harbors ideas of reincarnation and ethereal worlds, or "someone" is trying to send paranormal messages about these topics.

Turning to a case of face-to-face contact, a Southern Baptist minister in Puerto Rico claimed to have had many meetings with humanoid beings from the planet Koshnak, in the direction of the constellation of Orion. These beings were very similar in appearance to the familiar "Gray" entities. They had melon-shaped, expressionless faces with thin lips, undeveloped noses and ears, and large wraparound eyes without pupils. The eyes were green with scintillating flashes, and they were said to be intense and arresting.[10]

However, unlike the typical encounters with "Grays," this was a classical contactee case. The beings treated the man in a very friendly way. One of them, who was named Ohneshto, took him on rides in one of their vehicles, showed him undersea bases on the earth, and telepathically presented him with long philosophical discourses about time, space, and the reasons for human existence.[11] This included references to higher dimensions:

> He said they travel in the seventh and eighth dimensions, unknown to Earth humans, and that they are aware of 13 dimensions of being. Ohneshto pointed out references in our Bible pertaining to UFOs. He

> said that their normal span of life is about 800 to 1,000 of our years. . . . He explained that they could continue life forever with only one cell of the body. Ohneshto said that the axis of the Earth has changed four times as far as they have checked this out, that it tilts about every 20 to 25 centuries.[12]

This case was similar in many ways to the case of Filiberto Cardenas discussed in Chapter 5 (pages 176–78). There, also, the contacting entities spoke of other dimensions. Cardenas testified that during a voluntary visit on one of their ships, they told him that "they are beings of other dimensions, of other worlds, but that they are not gods, and they do not want to be considered such." [13]

Pantheism and Impersonalism

There are a number of reports of UFO-related communications that directly mention some idea about God. To my knowledge, these practically all present a pantheistic or an impersonal conception of the Supreme. Impersonal conceptions describe the Supreme as an ultimate force, energy, or state of being that is the source of all phenomena but is devoid of all personal attributes. They include the pantheistic conception of the Supreme, which identifies God with the universe. These impersonal ideas can be contrasted with the idea that God possesses absolute personal features, as well as various impersonal energies and aspects.

The mention of some conception of God in UFO communications is consistent with the stress many of these communications place on spirituality. Theological ideas naturally tie in with the idea that human beings have a spiritual dimension, and they are incompatible with strictly mechanistic views of life.

At the same time, strictly impersonal conceptions of God are incompatible with devotion to a personal Supreme Being. Quite a few UFO-related communications tend to denigrate the personal conception of the Supreme, and some do this by claiming that earth humans used to mistakenly worship visiting extraterrestrials as God or as gods. The implication—which is sometimes explicitly spelled out—is that anthropomorphic ideas of the Deity began with extraterrestrial contacts with humanlike beings. Of course, another possibility is that all humanlike beings within the universe have derived their form from an original, humanlike Creator, and thus the human form is actually "deomorphic."

In some cases, humanoid entities are reported to make brief theological comments in the course of apparently accidental close encounters. For example, a 25-year-old man told of an encounter with strange beings in July 1968 at the Grodner Pass in the Italian Dolomites. He said that he met tall, thin beings with domed heads and beautiful Oriental eyes. The beings, who were accompanied by a small robot, telepathically told him, "We come from a planet in a far galaxy," and, "Everything is God." They also warned that a pole shift is coming, the earth's crust will crack, and life will be in great danger. [14]

Another theological revelation was given to Mrs. Cynthia Appleton, a 27-year-old mother of two children living at Aston in Birmingham, England. At 3 p.m. on November 18, 1957, she was about to check on her baby daughter. Suddenly she sensed an oppressiveness, like that preceding a thunderstorm, and saw a "man" materialize with a whistling noise near the fireplace. This apparition was initially blurry and then clear. He was tall and fair, with a tight-fitting plasticlike garment featuring an "Elizabethan" collar. He answered her questions telepathically, revealing that he had come from a world of peace and harmony in a saucer-type craft. He was able to convey a picture of this in a mysterious fashion.

On a second occasion, two similar figures spoke to her in a strange style of English, informing her that they were projections and should not be touched. One point they made was that "the Deity itself dwells at the heart and core of the atom." It is said that there were no books in Mrs. Appleton's house, only newspapers. Those who interviewed her described her as a pleasant and sincere young woman.[15]

Although this statement about the Deity may seem pantheistic, it can also be given a broader interpretation. In the *Brahma-saṁhitā* it is said that God dwells within each atom (in Sanskrit, *paramāṇu,* or "smallest particle") and that innumerable universes simultaneously exist within God.[16] Here the idea is that God is a Supreme Person who is distinct from the universal manifestation and at the same time is fully present within every particle of matter.

The psychic Robert Monroe, who is known for his investigations of out-of-body travel, reported receiving a communication that strongly negated his existing conceptions of God. This involved a mysterious beam of radiation that seemed to emanate from a point in the sky:

> I suddenly felt bathed in and transfixed by a very powerful beam that seemed to come from the North, about 30 degrees above the horizon. I was completely powerless, with no will of my own, and I felt as if I were in the presence of a very strong force, in personal contact with it.
>
> It had intelligence of a form beyond my comprehension, and it came directly (down the beam?) into my head, and seemed to be searching every memory in my mind. I was truly frightened because I was powerless to do anything about this intrusion.[17]

Jacques Vallee compared Monroe's beam with the beams of light shown in religious art carrying revelations from God. It is interesting that during one of its appearances, the beam conveyed a very cold, impersonal concept of God to Monroe. This was so overpowering that it caused Monroe to weep bitterly. He said, "Then I knew without any qualification or future hope of change that the God of my childhood, of the churches, of religion throughout the world was not as we worshiped him to be."[18]

Monroe had apparently been thinking of God as a person who might show concern for an individual worshiper. But whoever was responsible for the beam went to the trouble of disabusing him of this concept.

There are many reports of mind-probing beams from the sky, and to show the possible relationship between these beams and UFOs, I will mention another example. This involved a woman living in Westchester County, New York. The woman reported to J. Allen Hynek's team of investigators that in April of 1983 she was awakened by a beam of light that came through the window of her bedroom. The beam seemed to penetrate through her body, and she felt paralyzed. She said:

> I lay there about ten minutes, and all the time I felt as if my insides were being probed, like a doctor was probing my insides. I was terrified, but there was nothing I could do.
>
> Then, as I lay there, these images started to flash in my mind . . . images of lights flashing all different colors. Then it seemed as if someone was trying to speak to me. I saw this image of a being with claylike skin and a large head with large eyes. He had no hair and no mouth. He assured me that I would come to no harm and said that I was being tested.[19]

This, of course, is similar to many descriptions of humanoid beings seen in UFO close encounters. Could it be that such beings transmitted impersonal theological doctrines to Monroe? There are other accounts that are consistent with this idea.

I noted in Chapter 10 that the U.S. Government has allegedly hosted an "Extraterrestrial Biological Entity," or EBE, who stated that reincarnation is real and that "it's the machinery of the universe." On October 14, 1988, a television documentary entitled "UFO Cover-up? Live" was broadcast across the United States.[20] This show presented testimony by a supposed U.S. intelligence agent named Falcon, who made numerous statements about this EBE and his race of beings. When asked if these aliens believe in a Supreme Being, Falcon answered, "They have a religion, but it's a universal religion. They believe in the universe as a Supreme Being."

The EBE stories, of course, are tied in with a complex mass of allegations regarding UFO cover-ups, government conspiracies, and disinformation (pages 110–15). They also include the story that the EBEs created Jesus Christ.[21] Since Christ is worshiped by Christians as a personal God, it would seem that whoever is behind the EBE stories has some interest in undermining personal theism and replacing it with pantheism.

We can get some further insight into the matter of UFOs and pantheism by considering another contact story that is somewhat different from the ones I have considered thus far. In all of these stories, the contacting entity has ostensibly been a nonhuman being. Now I will turn to a story in which this entity is said to be human.

In November of 1919, Alice Anne Bailey was sitting under a tree on a hillside in California, resting after sending her three children off to school. Suddenly she sat up, startled: "I heard what I thought was a clear note of music which sounded from the sky, through the hill and in me. Then I heard a voice which said, "There are some books which it is desired should be written for the public. You can write them. Will you do so?'" [22] At first she refused, but later on she allowed herself to be persuaded. In due course, she wrote a number of thick volumes on metaphysics and occultism that were dictated to her telepathically by a personage known as "the Tibetan." These include *A Treatise on Cosmic Fire,* which runs to 1,282 pages.[23]

Bailey's experience on the hillside is reminiscent of many UFO encounters, in which a person receives a telepathic message that is preceded by a high-pitched sound. However, in her case, the telepathic message supposedly came from a living human being in Tibet. (She also reported receiving communications from a Tibetan adept through a beam of light that struck her room.[24])

Tibetan mystics? Even a person accustomed to UFO humanoids might be tempted to reject this story and denounce Bailey as a fool or a charlatan. However, if telepathic communications from strange humanoids are possible, why should we rule out telepathic communications from human *yogīs*?

As often happens, it is easier to dismiss the story the less one knows about it. I have read Bailey's autobiography, and it seems to portray her as a rational and honest person.[25] She seems as reliable as many of the witnesses mentioned in this book, and she may well have written her books in the way she described. I should emphasize, of course, that in saying this I am not granting these books any particular authority. It is one thing to receive a message by telepathy (or by any other means), and it is another thing for the message to be true.

Bailey's books consist of ideas from Christianity, Buddhism, Vedic texts, and Western occult traditions that have been woven together by a sophisticated intellectual. In my opinion, they represent a clever synthesis of ideas, but they tend to strongly distort some of their source materials. They can be seen as an attempt to assimilate Christianity into Buddhism.

To see their possible connection with the UFO phenomenon, consider the following points from the teachings of the Tibetan:

1. He made predictions of great disasters, including serious disturbances in Alaska and California.[26] Volcanic action, in particular, was mentioned.
2. He made predictions about various future political and technical developments. For example, he said the energy of the atom will be harnessed.[27]
3. He said that humans of this earth have deplorably bad qualities: "The selfishness, the sordid motives, the prompt response to evil impulses for which the human race has been distinguished has brought about a condition of affairs unparalleled in the system."[28]
4. He presented Love as the "impelling motive for manifestation" on individual and universal levels.[29]
5. He said that the Masters are occult adepts who are part of the planetary control system and who live for fabulous lengths of time.

Although they are highly advanced, they are still evolving. He deplored personality cults that grew up around the Masters.[30]

6. He said that devotional religion is to be eliminated: "The Master Jesus is . . . working in collaboration with certain adepts of the scientific line, who—through the desired union of science and religion—seek to shatter the materialism of the west on the one hand and on the other the sentimental devotion of many devotees of all faiths." [31]

Statements along these lines are frequently made in reported UFO communications. Predictions of disasters are standard, and political predictions may also be made. UFO humanoids almost always point out the deplorable qualities of human beings, and in some cases they also stress the importance of universal love. They often say that they have been visiting the earth for thousands of years and that they have extremely long life spans. They point out that they have been worshiped in the past as gods but that they are not gods and are still evolving towards perfection. For example, the Tibetan's points can be compared with the following six alien communications reported by Filiberto Cardenas:

1. The aliens who abducted Cardenas predicted great disasters and said California will sink into the sea. [32]
2. They made various predictions about public figures and international politics. [33]
3. They criticized the vanity of humans and spoke of the difficulty in trying to deal with them.[34]
4. They spoke at length about universal Love.[35]
5. The aliens said they have been visiting human society for 4,000 years. People used to worship them as gods, but they are not gods.[36]
6. They criticized earth religions.[37]

It is significant that the avowed objective of Bailey's communications is to "shatter the materialism of the west on the one hand and on the other the sentimental devotion of many devotees of all faiths." In particular, the personal conception of God is to be replaced by the remote, abstract conception of the Supreme as "He About Whom Naught May Be Said." [38]

One can argue that one of the primary effects of the UFO phenomenon, for those who take it seriously, is to fracture their Western, scientific view of reality. In addition, reported UFO communications containing theological material often promote an impersonal or pantheistic conception of God, and some specifically attack the foundation of particular devotional faiths—notably Christianity.

One hypothesis to explain all this is that there exist beings endowed with mystic powers who are trying to indoctrinate human society with a spiritual philosophy based on cosmic evolution of consciousness and an impersonal conception of the absolute. Some of these beings are humanoids that visit people in UFOs, and others may be humans like ourselves who have acquired mystic powers through the practice of *yoga.* For the latter, their impersonal philosophy may have its roots in historical traditions such as Buddhism and the Indian philosophy of Advaita Vedānta. The purpose of the indoctrination program may be to save humanity and the earth from the perils caused by modern materialism.

According to the Vedic perspective, a joint indoctrination program carried out by human sages and UFO humanoids is definitely not out of the question. The Vedic literature describes a world in which ordinary people and powerful *yogīs* coexist with a variety of humanlike races. These races are endowed with mystic *siddhis,* and many are accustomed to traveling in *vimānas*. Spiritual philosophies based on impersonal conceptions of God are said to be quite prominent among both the *yogīs* and these mystically endowed beings.

If some of these persons are disturbed by the present state of affairs on the earth, it is natural that they would use their own philosophies and technologies in an effort to deal with this situation. This might involve loosely coordinated activities on the part of beings from a number of different groups, with personal qualities ranging from *tamo-guṇa* to *sattva-guṇa* (pages 306–7).

Brahman Realization

In Chapter 10 (pages 343–44), I recounted a UFO abduction in which Betty Andreasson was brought to a huge door in an underground complex. At that point she went out of her body, passed through the door, and had an experience of meeting the One. This experience created great happiness, but she was unable to explain it:

> *Betty:* It's—words cannot explain it. It's wonderful. It's for everybody. I just can't tell you this.

> *Fred Max:* You can't? Okay, why can't you?
> *Betty:* For one thing, it's too overwhelming and it is . . . it is undescribable. I just can't tell you. Besides, it's just impossible for me to tell you.
> *Fred Max:* Were you *told* not to share it with me? [39]

It seems doubtful that this testimony was evoked by leading questions from the hypnotist, Fred Max. He seemed to think that Betty could not describe her experience because the aliens were controlling her mind—a standard idea among investigators of UFO abductions. However, it seems clear that she could not describe the experience because it was literally beyond words.

A standard method of trying to get around mental blocks inhibiting a person's memory is to ask the person, while under hypnosis, to visualize the blocked experience as though seeing it on TV. When this was tried with her experience of seeing the One, Betty responded by saying:

> Ohhhh! There's a bright light coming out of the television! This is weird! There's rays of light, bright white light, just [pause] like they've got a spotlight coming out of the television! It's hurting my eyes! [40]

In Vedic literature, Brahman is spoken of as an indescribable white light that is characterized by oneness, eternity, and unlimited happiness. The UFO investigators who were interviewing Betty Andreasson did not seem to know about this, and it seems probable that she also did not know about it. Realization of ultimate oneness is described by Catholic mystics such as Meister Eckhart, but it does not generally figure in fundamentalist Protestant traditions. It seems quite likely that her experience of the One actually took place.

The three children who received the revelations at Fatima also had what appears to be an experience of Brahman. After the initial conversation between the children and the effulgent lady, "she opened her hands and streams of intense light flowed from them which overwhelmed the children's souls, causing them to feel 'lost in God' Whom they recognized in that light." [41] This description makes sense from the Vedic standpoint, and it also illustrates the idea that a higher being can cause a more or less ordinary human to have a temporary experience of Brahman. Something similar seems to have happened in the Andreasson case.

The contactee Orfeo Angelucci also reported an experience of Brahman realization that occurred while he was on board a UFO. Angelucci's UFO experiences were of a positive, spiritual nature, and they tend to be rejected as bogus by many ufologists. Generally, reports of enjoyable or self-fulfilling experiences are thought to be less credible than reports of unpleasant or embarrassing experiences. However, positive experiences are reported, and they should be included in any comprehensive account of the UFO phenomenon. As always, the strategy is to study the patterns that show up in large numbers of reports.

Angelucci said that on July 23, 1952, he felt ill and stayed home from his job as a mechanic at the Lockheed Aircraft Corporation in Burbank, California. In the evening he took a walk in a lonely place near the concrete embankments of the Los Angeles River, and he was troubled by strange prickling sensations and a dulling of consciousness. Suddenly he saw before him a luminous, misty, igloo-shaped object that gradually increased in solidity. There was a door leading into the brightly lit interior of the object. Upon entering, he found himself alone in a vaulted room about eighteen feet in diameter, with shimmering mother-of-pearl walls. He saw before him a reclining chair made of the same translucent substance, and he felt impelled to sit down in it. The door then shut, leaving no sign that there was a door at all, and the object apparently traveled into outer space.

A short while later, a window opened in the wall of the room, and he saw the earth from a distance of about a thousand miles. A voice began to speak to him, describing the unfortunate position of the materialistic people of the earth and enjoining him to tell them about their real spiritual nature. The voice said, "Each person upon Earth has a spiritual, or unknown, self which transcends the material world and consciousness and dwells eternally out of the Time dimension in spiritual perfection within the unity of the oversoul."[42]

After listening to these teachings for some time, Angelucci underwent the following experience:

> A blinding white beam flashed from the dome of the craft. Momentarily I seemed to partially lose consciousness. Everything expanded into a great shimmering white light. I seemed to be projected beyond Time and Space and was conscious only of light, Light,

> LIGHT! Every event of my life upon Earth was crystal clear to me—*and then the memory of all my previous lives upon Earth returned . . .*
>
> I am dying, I thought. I have been through this death before in other earthly lives. This is death! Only now I am in ETERNITY, WITHOUT BEGINNING AND WITHOUT END. Then slowly everything resolved into radiant light, peace and indescribable beauty. Free of all falsity of mortality I drifted in a timeless sea of bliss.[43]

When Angelucci awoke to normal consciousness, he realized that the object was returning to the earth. On returning to his home, he remembered a burning sensation that he had felt beneath his heart while on board the craft. He found that he had been marked by a stigmata consisting of a reddish dot surrounded by a reddish circle about the size of a quarter. This was the only tangible evidence which remained to show that his experience had actually occurred.

A more subdued, philosophical account of the One was presented by the "Ra" entity in channeled communications received by Carla Rueckert (page 187). Ra claimed to be a telepathically linked complex of beings that had once lived on a higher-dimensional level on Venus and that had communicated monotheistic ideas to the Pharaoh Ikhnaton in ancient Egypt.[44] Now, however, the concern of Ra is to merge with the One and teach others about this possibility. Thus Ra said, "We cannot say what is beyond this dissolution of the unified self with all that there is, for we still seek to become all that there is, and still are we Ra. Thus our paths go onward."[45]

Don Elkin's compilation of contactee material contains many examples expressing this idea. Here are three examples taken from three different contactees:

> Separation is an illusion. All things are one thing: the creation. . . . you and those whom you serve are the same: you are one.[46]

> It is impossible to separate yourself from the creation, it is impossible to isolate yourself from the creation. You are it, and it is you.[47]

> And then, my friends, you and the Creator are one, and you and the Creator have equal powers. For this is truth. Each of us is the Creator.[48]

These statements bring to mind the famous Indian philosophy of Advaita Vedānta, which teaches that the ultimate goal is to merge the individual ego into the one Brahman. This school follows the Vedic teachings, and thus it maintains that there is a celestial hierarchy of inhabited realms and that souls transmigrate through gross and subtle forms in these realms. But it also holds that at the ultimate level of understanding, all of these realms are illusory and nothing exists but the One Consciousness, or Brahman. Thus ultimate understanding means to become identical with Brahman, which is all that is.

However, there is more to Indian philosophy than just the school of Advaita Vedānta. According to Vaiṣṇava philosophy, Brahman is the effulgence of the transcendental body of the Supreme Lord, and it forms the atmosphere of the spiritual world. Brahman realization is simply the starting point of higher spiritual experience, and the idea that it is the ultimate goal is an impediment to spiritual progress.

The *Bhāgavata Purāṇa* is one of the principal Vedic texts presenting the Vaiṣṇava philosophy of personal monotheism. Here is a description from this text of a journey by Arjuna and Kṛṣṇa into Brahman and beyond:

> Following the Sudarśana disc, the chariot went beyond the darkness and reached the endless spiritual light of the all-pervasive *brahma-jyoti*. As Arjuna beheld this glaring effulgence, his eyes hurt, and so he shut them.
>
> From that region they entered a body of water resplendent with huge waves being churned by a mighty wind. Within that ocean Arjuna saw an amazing palace more radiant than anything he had ever seen before. Its beauty was enhanced by thousands of ornamental pillars bedecked with brilliant gems.
>
> In that palace was the huge, awe-inspiring serpent Ananta Śeṣa. He shone brilliantly with the radiance emanating from the gems on His thousands of hoods and reflecting from twice as many fearsome eyes. He resembled white Mount Kailāsa, and His necks and tongues were dark blue.
>
> Arjuna then saw the omnipresent and omnipotent Supreme Personality of Godhead, Mahā-Viṣṇu, sitting at ease on the serpent bed. His bluish complexion was the color of a dense rain-cloud, He wore a beautiful yellow garment, His face looked charming, His broad eyes were most attractive, and He had eight long, handsome arms. His profuse locks of hair were bathed on all sides in the brilliance reflected

> from clusters of precious jewels decorating his crown and earrings. He wore the Kaustubha gem, the mark of Śrīvatsa and a garland of forest flowers.
>
> Serving that topmost of all Lords were His personal attendants headed by Sunanda and Nanda; His *cakra* and other weapons in their personified forms; His consort potencies Puṣṭi, Śrī, Kīrti, and Ajā; and all His various mystic powers.[49]

Here the word *brahma-jyoti* means Brahman effulgence. The potency called Ajā is the energy of material creation. The understanding is that this scene lies completely beyond the material realm.

If the *brahma-jyoti* is simply the atmosphere of a higher spiritual region, then, to paraphrase Ra, there is something beyond the dissolution of the unified self with all that is. According to the Vaiṣṇava philosophy, once the bondage of material ego is dissolved, the soul becomes free to act on a purely spiritual platform.

Since the soul emanates from the Supreme Being, there is a natural relationship of love between the soul and the Supreme. This natural love is obscured when the soul is in a state of material consciousness. When the soul attains to Brahman, it reaches a neutral state, and its natural loving tendency is manifest without an object. But going beyond this neutral state, this love becomes expressed in the form of service to the transcendental Supreme Lord. It is also expressed in the form of compassion towards souls in material bondage, who are all parts and parcels of the Lord but are lost in forgetfulness.[50]

This idea of love in relation to the Supreme Person can be compared with the "universal love" mentioned in many UFO communications. These communications often define the ultimate One as an impersonal energy or force. Yet love is something having to do with persons. If the One is impersonal in nature, then how can love have a truly universal role? Note that this problem does not come up in modern scientific theories. According to modern science, love is simply a recent outgrowth of hominid evolution in Africa, and it has nothing to do with ultimate causes. But if the personal quality of love is a fundamental feature of nature, then it is natural to ask how this could be. If there is a Transcendental Person behind the universe, then the answer is that the universe was crafted according to the loving intentions of that person.

The Role of Māyā

Many actions and communications connected with UFOs are consistent with the ancient Vedic world view. Some UFO entities present philosophies representing particular subsets of Vedic thought. There are discussions of the soul and its evolution in consciousness. There are practical demonstrations of different kinds of travel of the gross and subtle bodies, and of spiritual states of consciousness up to Brahman realization. There are also theoretical descriptions of these states, especially in channeled communications. I am not aware, however, of any discussions of direct, personal love of God.

This material conveys a positive overall message. But at the same time, many UFO encounters have a less auspicious aspect. It is not uncommon for reported communications from UFO entities to contain doubtful or absurd information, such as the technical gibberish reported by William Herrmann, or the statement that the aliens come from "a small galaxy near Neptune." In addition, there are the disturbing UFO abductions and the indications of overtly harmful UFO activity that I discussed in Chapter 9.

How are we to understand all of this? One intriguing idea that appears in some communications is that there is a cosmic law of confusion that regulates the dissemination of information to human beings. This law may help explain why UFO communications seem to contain a bewildering mixture of nonsense and possibly valid information.

The Ra entities mentioned this law in connection with their story of how they built the Great Pyramid of Egypt. When they said that they built it out of thought-forms, they were asked why it was created in separate blocks, as though it had been assembled from quarried stones. Ra replied:

> There is a law which we believe to be one of the more significant primal distortions of the Law of One. That is the Law of Confusion. You have called this the Law of Free Will. . . . We did not desire to allow the mystery [of the Great Pyramid] to be penetrated by the peoples in such a way that we became worshiped as builders of a miraculous pyramid. Thus it appears to be made, not thought.[51]

Regardless of how the Great Pyramid was really built, this Law of Confusion is worth thinking about. The basic idea is that in order to preserve the free will of human beings, it is necessary to withhold information from them and even bewilder them with false information.

This concept may help explain not only the bewildering character of UFO communications but also the elusive nature of UFO evidence in general. Often this evidence is strong enough to be impressive, but it is never so overwhelming that a skeptic would be denied his own free will in deciding whether or not to accept it. One can conceive of scenarios, such as a mass landing of UFOs in Washington, D.C., that would be so convincing as to rule out this exercise of free will. Could it be that the Law of Confusion is being applied to the UFO phenomenon so as to preserve people's freedom to reject or disregard it, while at the same time providing useful information for people who are prepared to accept it?

Vedic ideas can throw a great deal of light on the nature of the Law of Confusion. According to the *Vedas,* the material world is fashioned out of an energy called *māyā. Māyā* means illusion, magic, and the power that creates illusion. The basic Vedic idea is that the universe is created as a playground for souls who seek to enjoy life separately from the Supreme Being. If these souls were in full knowledge of reality, then they would know the position of the Supreme, and they would know that such separate enjoyment is impossible. The universe is therefore created as a place of illusion, or *māyā,* in which these souls can pursue their separate interests.

Another aspect of the Vedic world view is that the Supreme Being wants the materially illusioned souls within the universe to return to Him. But for this to be meaningful, it must be voluntary. The real essence of the soul is to act freely out of love. Thus if the soul is forced to act by superior power, then this essence cannot be realized. For this reason, the Supreme Being tries to give the soul the knowledge of how to return to the Supreme in a delicate way that does not overpower the soul's free will.

Here is the perspective of the *Bhāgavata Purāṇa* on the relationship between the Supreme and the world of illusion:

> I offer my obeisances to Vāsudeva, the Supreme, All-pervading Personality of Godhead. I meditate upon Him, the transcendent reality, who is the primeval cause of all causes, from whom all manifested universes arise, in whom they dwell, and by whom they are destroyed. He is directly and indirectly conscious of all manifestations, and

> He is independent because there is no other cause beyond Him.
>
> It is He only who first imparted Vedic knowledge into the heart of Brahmā, the original living being. By Him even the great sages and demigods are placed into illusion, as one is bewildered by the illusory representations of water seen in fire, or land seen on water. Only because of Him do the material universes, temporarily manifested by the reactions of the three modes of nature, appear factual, although they are unreal.[52]

In Chapter 7 (pages 256–57), I compared the universe to a virtual reality manifested within a computer by a master programmer. The inhabitants in a virtual reality actually exist outside of the false, computer-generated world, but they experience the illusion that they are within that world. If they were to forget their actual existence, then the illusion would become complete, and they would identify themselves fully with their computer-generated virtual bodies. According to the *Vedas,* this is the position of conditioned souls within the material universe.

Within the overall illusion of *māyā,* there are many subillusions. The overall illusion causes one to forget the omnipotence of the Supreme, and the subillusions cause one to forget the cosmic managerial hierarchy set up by the Supreme within the material universe. All of these illusions allow the individual soul to act by free will, even though he is actually under higher control.

At the same time, the illusions are not so strong that an individual who wants to seek out the truth is unable to do so. If *māyā* were so strong as to stop any effort to find the truth, then this too would block people's free will. According to the Vedic system, the Supreme Being arranges for teachers to descend into the material world to give transcendental knowledge to the conditioned souls. By the arrangement of *māyā,* people will always have plentiful excuses for rejecting these teachers if they so desire. But if they desire higher knowledge, they will also be provided with adequate evidence to distinguish that knowledge from illusion.

Within the last few centuries on this earth, the view has been developed that life is simply a physiochemical process that evolved gradually over millions of years. According to this view, we are the topmost products of evolution on this planet. If there is life elsewhere in the universe, it also had to slowly evolve on planets with suitable conditions. Therefore, intelligent life forms that might be superior to us are likely to be safely far away, and we don't have to worry about them.

This view is highly conducive to a program of free enjoyment and exploitation for people of this earth. But unfortunately, such a program causes damage to the earth's biosphere, and it blocks the path of advancement for those who might want to learn about their spiritual nature. This means that even though the modern materialistic world view expedites the free will of some persons, it blocks the free will of others.

Perhaps the UFO phenomenon is one way in which the modern materialistic outlook is being gently revised by higher arrangement. Scientists are given their comeuppance by being confronted with impossible flying machines that break the laws of physics. Beings with magical powers appear to show us that we are not the topmost living species. Yet at the same time, the UFO phenomena are elusive, the communications are contradictory, and there is always room for doubt.

If this is what is happening, I suspect that it involves complex arrangements involving many different forms of life. Some UFO phenomena may be directly caused by mode-of-darkness beings that frighten people but at the same time expand their understanding of life and its powers. Some of these phenomena may involve a genuine protest by beings that live in our own world and are disturbed by our technological misadventures.

Other phenomena may involve preaching programs carried out by beings who have a message to convey. After all, religious proselytizing does not have to be limited to ordinary humans. These messages may vary in quality and in depth, and ultimately individuals will have to use their own discrimination to decide what to accept and what to reject. I suggested above that some beings who produce lights, high-pitched sounds, and telepathic communications may even be human *yogīs* with highly developed mystic powers. The events at Fatima (pages 293–301) suggest that persons from higher planets may also be appearing on the earth, moved by compassion for human suffering.

All of these possibilities are consistent with the Vedic tradition. According to ancient Vedic texts, there was a time when people of this earth were in regular contact with many different kinds of beings, from negative entities in the mode of darkness to great sages in advanced states of spiritual consciousness. The modern phenomena tend to confirm the Vedic picture, and this may also be part of the plan behind these phenomena. The teachings of the ancient sages are still available, but they have become eclipsed by the modern developments of materially oriented science and technology. Perhaps the time is coming when they will again be taken seriously.

APPENDICES

Appendix 1

On the Interpretation of Vedic Literature

In this book I have presented Vedic literature in a straightforward way according to the direct meaning of the texts. It is natural for many people to question the justification for this and ask whether or not there might be other, better ways of interpreting Vedic literature. In this appendix I will briefly address this issue.

People who take an interest in Vedic literature can be roughly divided into several groups, which include the following:

1. Strict followers of the major *sampradāyas.* A *sampradāya* is a school of thought based on the Vedic literature as presented by a great *ācārya,* or teacher. The major *sampradāyas* all take for granted the existence of the Supreme Controller (God), the soul, the subtle body, transmigration, subtle and spiritual worlds, and superhuman beings who inhabit these worlds. They also accept the validity of Vedic historical accounts. Some of the *sampradāyas* disagree with the others about the nature of God, but they all accept the Vedic literature as the authority on which conclusions should be based.

2. Those who adopt the accommodational approach to Vedic literature. In Christian apologetics, "accommodation is an interpretive device or principle which allows an interpreter to preserve valid meaning found in a text without a sterile literalism."[1] The method is to regard the text as genuine knowledge expressed in figurative language and perhaps overlain with fanciful poetic embellishments. With the introduction of Western science into India, the Vedic world view encountered a strong challenge. Many modern Hindus have responded by using the accommodational approach to bring the Vedic literature into line with modern science.

3. Those who regard the Vedic literature as mythology. People in this group tend to see this literature as a body of fantasy that was built up gradually by prescientific poets. This view is held by many scholars who study Vedic literature in fields such as Indology, anthropology, and comparative religion.

I will now briefly discuss the approaches to textual interpretation used by myself and by the people in these three groups. First of all, I would like to distinguish between a direct presentation of a text, a literal interpretation, and a figurative interpretation. In a direct presentation, one considers the text as it is, without insisting that one fully understands what it means. In a literal interpretation, one assumes that one can fully understand the text according to the dictionary definitions of its words. In a figurative interpretation, one interprets words according to indirect meanings.

When studying a text from an unfamiliar cultural tradition, I would recommend making a direct presentation. I would not recommend trying to insist too strongly on a literal or figurative interpretation—at least not at first. The reason is that it is not easy to understand the meaning of written material coming from another culture.

The true meaning of a text can be defined as the meaning intended by the author. The true meaning may be literal from the viewpoint of the author, or it may be figurative. But if we try prematurely to arrive at our own literal or figurative interpretation of the text, we may completely miss the author's intended meaning.

What was the author's understanding of his own words? If the author was working within an established cultural tradition, then he presumably used words according to the meanings accepted in that tradition. But a person who approaches the tradition from a foreign vantage point may not find it so easy to understand those meanings. To understand them, the outsider may have to immerse himself in that cultural tradition for a long time and gradually grasp meanings through usage and context.

Some meanings may differ so strongly from what a person is accustomed to thinking that he will fail to grasp them for a long time. He will tend to take words out of the context understood by the author and force them into a context dictated by his own cultural conditioning. This may cause him to reject the text as absurd, and such rejection can create an impediment to true understanding.

For example, consider the Vedic statement that beings called Devas live in Svargaloka. What does Svargaloka mean? According to Apte's Sanskrit-English dictionary, it means "1. the celestial region, 2. paradise." [2] Based on modern ideas of astronomy, we may think that this refers to some region of outer space. We may therefore think that it's absurd to say that Devas live in Svargaloka, since that would mean they must be floating out in space along with comets and cosmic rays.

In earlier chapters I have interpreted Svargaloka as referring to a higher-dimensional domain that cannot be included in ordinary three-dimensional space. But what is the justification for introducing this idea?

The dictionary doesn't help us here, because the English words "celestial region" and "paradise" do not tell us whether the region in question is three-dimensional or higher-dimensional. There is no point in delving into the deeper meanings of "celestial" or "paradise," since these meanings relate to Western culture rather than Vedic culture.

If we turn to the Vedic texts, we find that they don't directly resolve this issue, since, to my knowledge, there are no Sanskrit words that directly correspond to the term "higher-dimensional." So what is the justification for introducing this term? My reason for bringing in the mathematical idea of higher-dimensional space is that it provides a consistent explanation of many detailed points in Vedic texts involving modes of travel and relations between places in the universe.[3] I suspect that this idea brings us closer to the intended meaning of the Vedic texts than the idea of Svargaloka as an ordinary region of outer space above our heads. However, I am sure that it is only an approximation. To appreciate the true meaning, one would have to become deeply acquainted with the Vedic world view.

Is this idea of Svargaloka as higher-dimensional a literal interpretation, or is it figurative? The answer is that it is a figurative use of the English term "higher-dimensional," but it is an attempt to approximate the true meaning of Svargaloka intended by the Vedic authors. In general, my recommendation is to directly present the original texts while trying to appreciate their intended meaning on the basis of overall context. This is a slow process, and one's understanding at any given time should be regarded as tentative.

Someone might object that the Vedic texts must have been composed by primitive people. Therefore Svargaloka could not refer to something as sophisticated as a higher-dimensional realm. It must simply refer to the sky immediately over people's heads.

The problem with this is that even people who are widely regarded as primitive, such as the Australian aborigines, have sophisticated ideas that a Westerner may find hard to grasp. What is an aborigine's understanding of the sky? It may turn out to be something quite difficult for us to understand. This is even more true of the Vedic literature, which is undoubtedly sophisticated in many ways.

The question might be raised that if it is difficult to ascertain the true meanings of Vedic texts, isn't it wise to consult authorities who have deeply studied these matters and stick to the interpretation they recommend? No doubt this is a good idea, but what authorities shall we select? The three options listed above introduce three sets of possible authorities: (1) the traditional *sampradāyas,* (2) persons who try to accommodate traditional ideas to modern ideas, and (3) Indological scholars.

If we at all want to understand the original intended meaning of the Vedic literature, we cannot neglect group (1). In most of the traditional *sampradāyas,* the emphasis is on direct presentation. I have particularly studied the teachings of the Gauḍīya Vaiṣṇava *sampradāya,* which was founded by Caitanya Mahāprabhu in the 16th century and descends from the earlier school of Madhvācārya.

I have seen on many occasions that if a commentator in this school comes across two points in a text that seem contradictory, he will simply present them as they are, contradiction and all. Someone might say that this is not an intelligent thing to do. But the intention of the commentator is to simply preserve the tradition as it is. Understanding may or may not come, but that understanding should be based on the original texts forming the basis for the tradition.

In contrast, the approach of people in group (2) is to try to make the Vedic literature acceptable by indirectly interpreting Vedic statements that seem to disagree with modern ideas. This involves interpreting some statements figuratively and dismissing others as embellishments made by overimaginative poets.

Consider how a person with a modern education might react to the story of Arjuna's abduction by Ulūpī (see Chapter 6). If this person is a Hindu, he might want to accept that Arjuna existed as a historical personality. But he might object that the story of Arjuna being pulled into Nāgaloka by Ulūpī must be a fanciful poetic concoction. According to the story, Arjuna was pulled down into the Ganges, but instead of hitting the river bottom, he entered the world of the Nāgas. An educated person tends to reject this, because he knows such things are impossible.

However, the UFO data reviewed in this book contains many modern accounts in which a person seems to be taken through solid matter by a mysterious being. Arjuna's abduction by Ulūpī may seem to be within the realm of possibility for a person acquainted with this data. At the same time, such a person may continue to regard other features of Arjuna's story as impossible.

Since our ideas about what seems possible may change as our knowledge changes, we cannot use these ideas as the basis for a fixed indirect interpretation of the Vedic literature. Therefore, instead of presenting such an interpretation, it is better to present the Vedic texts as they are and allow understanding of their meaning to gradually emerge. This is particularly true in a book like this one, in which the introduction of empirical UFO evidence may tend to change one's idea of what is possible or impossible.

But the objection can be raised that many scholars in universities regard the Vedic literature as unscientific mythology having no basis in reality. Therefore, it cannot help us understand the UFO phenomenon. Extensive study of this literature in comparison with UFO data probably cannot be justified in the expectation that science will be advanced thereby.

It would not be difficult to find scholars in the field of Indology who would support this position. But I should point out that the established views of scholars in this field are perhaps not as objective and impartial as one might hope. In fact, this field has a history of religious and ethnic bias. To see this, it is helpful to consider the early history of Indology.

When the British began to colonize India in the 18th century, they came into contact with the Vedic teachings. This immediately gave rise to a conflict between their own Christian faith and the religion of the Hindus. This conflict involved both a perceived threat to Christianity from Hinduism and an opportunity to spread Christianity by converting the Hindus. The threatening aspect of Hinduism was pointed out by the early Indologist John Bentley in his criticism of an Englishman (probably John Playfair) who had written in praise of the Vedic writings. Bentley wrote,

> By his attempt to uphold the antiquity of Hindu books against absolute facts, he thereby supports all those horrid abuses and impositions found in them, under the pretended sanction of antiquity. . . . Nay, his

> aim goes still deeper; for by the same means he endeavors to overturn the Mosaic account, and sap the very foundation of our religion: for if we are to believe in the antiquity of Hindu books, as he would wish us, then the Mosaic account is all a fable, or a fiction.[4]

A strategy was adopted of translating the Vedic books into English so that they could be used to convince the Hindus of the inferiority and falsity of their religion. For this purpose, Colonel Boden willed a large sum of money to Oxford University on August 15, 1811, to endow a chair in Oriental Studies. Monier Williams, who held this chair until his death in 1899, wrote,

> The special object of his [Boden's] munificent bequest was to promote the translation of the scriptures into English . . . to enable his countrymen to proceed in the conversion of the natives of India to Christian Religion.[5]

The German scholar Friedrich Max Müller came to England to take up this work, and he published many translations of Vedic texts which are still considered standard today. Müller wrote to his wife in 1886,

> I hope I shall finish the work, and I feel convinced, though I shall not live to see it, yet the edition of mine and the translation of the Veda will hereafter tell to a great extent on the fate of India and on the growth of millions of souls in the country. It is the root of their religion, and to show them what the root is, I feel sure, is the only way of uprooting all that has sprung from it during the last three thousand years.[6]

The technique used by Müller and his colleagues was one which is familiar to anyone who has studied the UFO field: ridicule what you don't understand and explain it away in ordinary terms. For example, the "gods" are personified natural forces converted into vain idols by wily priests who perpetrated pious frauds.

Today, Indologists are not generally concerned with converting the Hindus to Christianity. However, they have inherited a legacy of ridicule and misunderstanding from the founders of their field which continues to exert its influence. I will close by quoting a remark from the standard English translation of the *Viṣṇu Purāṇa,* by Horace H. Wilson (originally published in 1865). Regarding the *Purāṇas,* Wilson said,

> They may be acquitted of subservience to any but sectarial imposture. They were pious frauds for temporary purposes: they never emanated from any impossible combination of the Brahmans to fabricate for the antiquity of the entire Hindu system any claims which it cannot fully support. A very great portion of the contents of many, some portion of the contents of all, is genuine and old. The sectarial interpolation or embellishment is always sufficiently palpable to be set aside, without injury to the more authentic and primitive material. [7]

Here Wilson uses the kind of negative language that is typical of the founders of Indology. But he admits that the *Purāṇas* contain genuine and old material. This so-called "primitive" material may provide us with a novel perspective on reality that will help elucidate the nature of the UFO phenomenon.

Appendix 2

Contemporary Indian Cases

In this appendix I will present four stories from modern-day India that are related to the themes discussed in this book. The first two of these stories are about the personal experiences of Kannan (pseudonym), a South Indian man in his forties. In these two stories the only witness was Kannan himself. The second two stories were also told by him, but they involve multiple witnesses, and the first one did not directly involve Kannan himself. My purpose in presenting these stories is to show that phenomena are being reported in India today that show parallels both to American and European UFO phenomena and to traditional Vedic themes.

Kannan had a traditional Hindu upbringing, but he rebelled against this as a young man by adopting popular ideas of atheism and rational skepticism. In the late 1960s and early 1970s, he worked for TVS and Sons, a major South Indian automobile company. During this period he resumed his earlier interest in spiritual questions, and he began exploring various popular Indian religious movements. He spent some time as an associate of Satya Sai Baba, and later on he became involved with the Kṛṣṇa Consciousness Movement (ISKCON). He spent several years as a teacher at a boy's school (*gurukula*) run by ISKCON in the village of Māyāpura, West Bengal, the birthplace of the sixteenth-century religious teacher Caitanya Mahāprabhu.

The Smallpox Lady

The first story has to do with encounters Kannan had in his childhood with a mysterious woman who cured him and some of his friends of smallpox. People in South India have traditionally worshiped a goddess, sometimes called Mariamma, who is said to have control over this disease. The story suggests that Kannan had encounters with this goddess or with a similar being. Whoever she may have been, I will call her the smallpox lady.

This lady has a number of features reminiscent of commonly reported UFO entities. At the same time, Kannan's description of her closely matches traditional Vedic accounts of female Devas (or goddesses). The following six points sum up the salient features of the smallpox lady:

1. She appeared at times of smallpox epidemics, and she would mystically cure people of smallpox.
2. She looked like a classical celestial woman, as portrayed in South Indian temple sculptures. She had a big forehead, a very thin waist, and very prominent breasts. She was dressed very elegantly in traditional Vedic fashion. She had an air of authority, like a very aristocratic person.
3. Kannan could see that she was breathing. But at the same time her impact on him was more like that of a beautiful painting or sculpture than that of a human being of flesh and blood.
4. She floated through the air and passed through objects. She seemed to be "on a different track," and she seemed to use human doors and stairways only as a matter of convention.
5. She communicated telepathically. On one occasion she seemed to be speaking normally, but the movement of her lips did not match the perceived sound. Kannan compared this to dubbing in a movie.
6. She was able to block a person's thinking.

Items 4, 5, and 6 show up repeatedly in accounts of UFO entities, and I pointed out in Chapter 6 that these items are paralleled by Vedic *siddhis,* or mystic powers. One interpretation of these parallels is that UFO entities, the smallpox lady, and the humanlike beings described in Vedic literature may all have something in common. They may all be real beings of the same nature.

The physical appearance of most reported UFO entities is only roughly similar to that of classical Vedic humanoids, and UFO entities often exhibit weird clothing ranging from silvery jumpsuits to something one might find in a party costume shop. In contrast, the smallpox lady fits perfectly into classical Vedic iconography. This inevitably raises the question of cultural influence. Was the experience of Kannan influenced by his Indian cultural conditioning? It is interesting to note that Kannan himself knows practically nothing about Western UFO encounters. But when I told him about UFO abductions, he suggested that the beings people were reporting were a product of Western cultural conditioning.

Here are three possible relationships between reported encounters with entities and the cultural conditioning of the witnesses:

1. People report imaginary beings with features determined by the people's culture.
2. Real beings appear to people in forms that the people expect to see on the basis of their culture.
3. Real beings appear to people according to the beings' own cultural norms, and this influences the development of human culture over the centuries.

I have discussed the evidence indicating that many UFO reports do involve real beings, and I would suggest that a detailed survey of current Indian encounter cases might indicate that many of these also involve real beings. This suggests that there must be many encounter cases where option 1 does not apply, although some people may indeed experience fantasies in which it does apply.

I suspect that many close encounter cases may involve a combination of options 2 and 3. Option 2 seems to apply in cases where beings adopt human clothing styles that are strictly limited to a particular moment in history. An example would be UFO encounters in which entities are reported to wear spacesuits or modern Western clothing.

Option 3 may apply to cases where beings appearing in traditional Indian society display ancient Vedic clothing styles. There may also be other traditional cultures in which particular groups of beings have influenced the culture by their own cultural norms over long periods of time. (Celtic and Native American cultures come to mind.) In such cases, the human society and the visiting groups of beings may form an extended cultural unit. Certainly the ancient society described in Vedic literature seems to be an example of this.

In Chapter 10, I pointed out that Indian near-death experiences follow a different pattern than Western NDEs. Ian Stevenson suggested that these differences may not be simply cultural. There may be real differences between the experience of death in India and in the West, and these differences may depend on differences in the policy of higher-dimensional beings towards Indians and Westerners. I would suggest that cross-cultural differences in close-encounter experiences may depend on similar differences in policy. Thus, in traditional cultures, higher-dimensional beings may continue to relate to humans in accordance with ancient norms, but in modern high-tech societies they may adopt other modes of behavior in response to changing circumstances.

The smallpox lady may be an example of the traditional mode of interaction between humans and higher-dimensional beings. But whatever the right interpretation may be, here is the story:

> The first time was when I had smallpox, and there was no one at home. It was daytime, maybe noon, and I saw this lady, with long, long robes, long, long cloth. Abnormally long, so that if she tried to walk though our door she would stumble over it. She was at the height of that cabinet [a filing cabinet about five feet tall]. She had sharp features, long face, and curly hair.
>
> I had a lot of those smallpox sores, so I was feeling very disturbed. My mother was not there, and so I was worried. It was becoming too much for me. I was thinking, "My mother is not here," and feeling not cared for. That time I saw her at that height [five feet]. She was sitting, but there was nothing there, nothing to sit on. She was just sitting like this, with one leg over the other leg. She was looking down, and she was telling—it's not like telling. She was telling, but there was no sound there. It was not like my language spoken or anything, but she was telling, or you could say she conveyed somehow, "Don't worry, in two days you will be all right. Everything will be all right."
>
> Then she said that some children will be taken. Two blocks away from our house there is what we call the police line. Some policemen's residences are there. They're one-floor houses, in lines between the fencing. So she mentioned to me that "in the police line some children will be taken, but nothing will happen to you."
>
> At that time, I was worried about my sickness, and I didn't feel any fear or ask who is this person, why is she in this space. These things never came to my mind. I was just looking for someone to tell me "You will be all right." If I remember correctly, I would think this must have been the first time. I was quite small. How small was I? Four, five, like that. Maybe five. That was the first time.

I met this lady again, at least twice, maybe three times, but it was always when there was smallpox around in the city. Interestingly enough, in the police line two children died, and they were known to us too. And I recovered on the second day, even though so many smallpox sores were there on my body. During the evening of the second day I felt very thirsty, and I wanted to drink some water. They told me not to look in the mirror, because it looks horrible when you have this. But when I went to get some water, I looked at the mirror, and they had all become dry. This happened in two days, but they were expecting that I might have to be suffering for at least two more weeks, and they were bringing neem leaves and so on, to sooth the burning.

But I have seen that this person was always there at the time when there was smallpox—in midsummer then the smallpox breaks out. Now they say they have controlled it by vaccination. I doubt it, but they say that. Traditionally, people conduct a festival at that time because they say smallpox is an expansion of Durgā [the universal Mother Goddess]. She brings it, and if she is pleased then you won't suffer. But certainly, one hundred percent, I am sure this lady was not Durgā. She was not somebody that high. At the same time, she was not somebody from this planet, that is quite sure. She had an air of authority.

Another time I saw her as she was coming down the staircase in a close friend's house. There were four children there, and they all had smallpox. People used to ask my mother to come and read scriptures when there were such diseases, or when someone would die, or was on the death bed. There is a story about Durgā. My mother would go and read that as a religious ceremony, and she would make a feast. I would go with her. All four children had smallpox, and they were upstairs. Other children were told, "Do not go up because this is contagious. You can get it easily." They were my friends, so I wanted to go up and see them. But everyone was told not to go—strict instructions.

I just found a time when everybody was busy with the ceremony, and I walked up the stairs. As I went up, I saw her moving down the stairs, but not walking, not stepping down. And significantly she had this long cloth which would not be necessary just for covering the body—a very long, long cloth. It's like drapery, very beautiful, very good looking.

And because by this time I had grown up a little bit, I was studying her features. Back when I was suffering too, I didn't give any notice to this. I just saw there was somebody who told me that I would be all right. But this time I looked very clearly. My description is from the second or third time of seeing her. She wore white, pure white—not just white but a special, creamy white type of thing. It was white cloth.

Her hair was normal but maybe curly. It was not as black as a South Indian lady's hair, but it wasn't blond like here [in America]. One thing I noticed was that her waist was very, very thin. Later I studied in scriptures about four categories of women's bodies, as well as Apsarās and Gandharvas. She looked like a celestial, and her waist was very thin. It will be difficult to relate a thin waist like that with the size of her breasts. With such a big bosom and such a thin waist, somebody here would look like she was going to crash. A very thin waist, long, long thighs, and this cloth was worn with a *kacha* like a *brahmacārī dhoti* from the waist downwards. And this big cloth was at the top with a piece of cloth tight in the back, like you see in the sculptures. But her overcloth, the cloth that she was wearing, was really long.

Her face had very sharp features. She was very beautiful looking, and she didn't shock. When you see her you don't feel shocked; you feel like respecting some reverential person. You don't feel like you do when you see a goblin [Bhūta] or a ghost and there is shock. There is no shock. I have had some visions of some forms of Durgā, but that makes you feel like you are in front of a military officer. Durgā makes you feel like that, but she [the smallpox lady] doesn't make you feel like that. It was as if a student met the vice principal of the college walking on the sidewalk of the university—unofficial, but you know it is a very highly placed person.

When she was coming down the stairs she said, "Your friends are all right. Because you are so concerned about them, I went to see them, and they are all right." Then I developed a desire to have some contact with this person. The staircase is like this, and I am here. So I am standing there purposefully in her way. I think I was planning to say, "Why don't you come visit our house?" or "When can I see you?" or something like that. But one thing that happens in many of these incidents is that they look at you and make you "nonthinkable"—you won't be able to think too many things. You're so attracted to looking at them and appreciating the situation that before you can think of anything, they are gone. So they very comfortably do that.

And she was moving like this—floating. But her motion had nothing to do with the bending of the stair. It was on a different track. The stair made no difference to her, you could see that. But here is an interesting point. These things don't make any difference, but they use the stairs for going up. Why is that? I have an answer for that, but it doesn't come in here. They use the door. They don't have to go through the door, but they use the door. They use the stairs. They don't climb, but they use the stairs.

So she was coming like this, and then she just went through me in the sense that she was there and then she was not there. I looked behind me, and she was there with a very big smile on her face, like saying, "See, you're trying to stop me and ask me something, and I am already gone." A big smile on her face.

And she had a very big forehead. We have these four classes of human women's bodies. They're discussed in the scripture—our human female forms. But this being doesn't belong to any of those four. So she is not from here. But, of course, I did not understand all these things at that time. It is hard for me to go back and say exactly what I felt then, because these later understandings are coming up and confusing me now.

I didn't think this experience was very important at that time. Later, after having so many other experiences, this thing has become very important to me. She is a *kuladevata.* I found it much later. A particular family lineage is protected by that person. Because I belong to that particular family lineage, she took special interest in me. So because I was concerned about those boys, she visited there, even though they are not in our family line.

Maybe one more time I have seen her, but this time I have grown up. I have not reached that point in my life where I have taken to some nonbelieving. I haven't taken to that yet. That much I remember.

It was in a festival. There is a very ancient temple of Durgā in that city—she protects the city. They have a summer festival where they make a fountain in front of the temple hall. They put a lemon there, and the lemon goes up and down in the water. I would go there every day to see that, and I would stand there looking at it for a long time. "How did they get it to do this?"

They started at 4:00 p.m., and nobody was in the temple. Around 5:00 p.m., everyone would come. So I would go around the temple and come back. Because going around the temple will bestow some benediction or something, you know. So if you feel tired you try to catch two things together—trying to get a walk and get something good done. So I was going around, and there in the back of the temple there is the worship of *kanyās,* virgins.

So I was going around there, and I suddenly remembered this

lady—just like that. When I remembered, immediately, there she was under a banyan tree next to a big platform of cement. She was standing just in front of this. And she was in a very good, beautiful pose, blessing like this—like a dancing pose. And I was walking so fast toward that place that it was like being pulled. I came this close, almost, so that if she was breathing, in our terminology, it would be on my face.

She is very tall compared to our ladies in South India and even to Rajput ladies. Standing close to her, I didn't feel like I do when I stand close to normal people, like my sister. It was as if you stood next to a deity or a rose flower. But in the meantime I could see exactly the shape of the hand, and the breast, and the thighs, and everything. But they are still not like mine. So that was the first time I had a good understanding of an experience like that. They exist, these people. They have a form like us, but still it is not like our form.

Her skin had a nicer color than my hand. The skin is there, but I didn't feel like I would if I was standing next to a girl. It was not like that. It was almost as if I stood in front of a beautiful painting of Sarasvatī, or the deity form of Durgā. Because you know the deity is a person, and you don't see it so much like stone. You don't think this is a statue; you know this is Sarasvatī or Durgā.

But anyway, because I was that close, I saw she was breathing. And I had no wrong idea when I was looking at her. I was very respectful. She spread that type of atmosphere when she was there. You feel like going down and begging for some benediction or something like that.

Then she said they make a hole in the lemon, and when they put the fountain on they put it like this, and then it goes like that. So I laughed. I looked back to see if anybody was there, and there was nobody. So then she said that even if somebody was there they would not notice. She spoke this in my language, and there was sound. I could hear sound. This breathing was there. Her breathing was comparatively very slow—like a sick person would breath. But she had very exquisite features, very beautiful. And I also noted that she had a dot on the forehead, because I was very close.

I noted her lips. They moved, but they didn't synchronize with the words. So she was speaking something else. I have analyzed that she must have been speaking something else, and it gets to me so I am hearing in my language. So I thought that this sounds like dubbing in a movie.

And then she said, "You will be able to see all of us. You will see many of us." She reminded me that even as a very small child, whenever they took me to temples, when I would see Gaṇeśa [one of the principal Devas], I would loudly call him older brother. Everybody would call out "*Jaya* Gaṇeśa" or something, but I would say "*anna.*" *Anna* means "Oh, older brother." So she was mentioning that just like you called Gaṇeśa *anna,* so she said, "You have contact with us. And you're protected." And she said, "If you were not, you would have looked at me the same way that you would look at anybody else [i.e., lustfully]." And she said, "No, you're protected. We can protect you from that."

Then she said, "Actually as long as you have a desire . . ." I am using these words now, but it was not like that. Her words were more simple and not so philosophical. If you have a desire to enjoy, then we won't give you that protection. So she said, and this was very distinct, "Learn to see every woman as an expansion of Durgā." And she put her hand on my head. And I felt, "Oh my God, what is this?" It was not like what you feel if somebody touches you. Suddenly I felt cool. It was cool, but it was a very wonderful experience. It was like an experience rather than a touch.

She caressed me like this from the back, on the head, like you do to a young boy. And it really felt comfortable, like an affectionate mother. And somehow I had great respect for that person. You felt like it was a very respectable person in a higher position who comes to deal with somebody poor—like the queen came and shook hands with somebody. Then I asked, "Will I see you again, will I meet you again?" So she said, "Not unless it is necessary, if I am needed." I haven't seen her after that. That was the last time.

The Spear of Kārttikeya

The next story gives a further indication of Kannan's background. This story is quite distinct from typical reports of UFO close encounters, but it will not seem at all unusual to persons acquainted with accounts of Indian saints and mystics. The experience that Kannan reports could be classified as a "religious vision." Like the story of the smallpox lady and many UFO reports, it features intelligently directed phenomena that seem to emerge from another dimension.

In this case, however, the phenomena are explicitly connected with the traditional Vedic deity named Kārttikeya. In the Vedic literature,

Kārttikeya is the chief military general of the Devas. He is the son of Lord Śiva, and he was raised by virgins dwelling in the constellation Kṛttikā (the Pleiades). It is noteworthy that people in India are still reporting explicit experiences related to such Vedic deities.

One feature of this story is that Kannan seemed to have unusual knowledge about Kārttikeya—knowledge that he presumably acquired in a previous life. This ties in with the smallpox lady's statement that Kannan would have regular contact with higher beings and that he was somehow connected with them. As I have pointed out in previous chapters, many UFO contactees also claim to have a special relationship with higher beings, and some also claim that this dates back to a previous life.

Here is the story:

> Once I ran away from home and went to the Saṅgameśvara temple. It is a Śiva temple, and Kārttikeya is there. They have the Kārttikeya deity, and they have a peacock and a spear next to the peacock. Usually that's how it's done in Śaivite temple worship. They would have the deity's vehicle at the front [i.e., the peacock], and then the weapon of that particular deity.
>
> I went to the temple and did all the things you're supposed to do. I used to learn what things should be done in a Śiva temple without asking anybody. I always knew where to turn, where to sit, and where to stand. There is a whole ceremony for visiting Śiva's temple. Śiva's temple is a replica of Kailāsa, and Viṣṇu's temple is replica of Vaikuṇṭha. So the etiquette you follow as you enter into Vaikuṇṭha you do in the Viṣṇu temple. And if it is a Śaivite temple, you do everything exactly according to the custom in Kailāsa.
>
> So I used to do that very naturally. I used to tell my elders, "You're supposed to do this here and do that there. Why didn't you do this here?" In the beginning there were some objections, but later they used to take me if they wanted to go to some Śiva temple. "Take Kannan, and he will explain everything to you." They thought it is some special blessing on me.
>
> So I did the whole ceremony. It takes about 45 minutes to do all that. Then I came to Kārttikeya's place and sat down. I used to sit in a yoga pose, even when I was very small, and I sat there and kept looking at Kārttikeya. This deity has six heads, is sitting on a peacock, and has a spear. The temple worship was over, and the priest was going out. He went around me, but he didn't see me, and he went off. The temple was locked outside. And there was nobody there except

the gods and myself. So I just sat there. I did not sleep or anything.

The whole night was over. The next day the priest came and asked if I just sat there. There were ladies who came every day, morning and evening. They noticed, and they were kind of curious. What is this boy sitting here in this one spot? And I looked at them, but then I was looking at the spear again. So they started saying "*Sadhu,*" and soon there was a little crowd.

People left some offerings in front of me, some fruit. So the priest came and saw this, and he asked, "Where is your mother? What are you doing here? Why are you sitting like this?" He was trying to make the situation normal, but he was feeling nervous that I was sitting like that. So that evening, after *sandhya ārati,* he came to me and said, "You're going to sit here like this? Well if another saint is going to come, what can we do? When you feel hungry you take this." Then he said, "You must know, there is no place for passing [urinating, etc.] in this temple. I am locking, and I am going. I'll come and see you tomorrow morning." I made no response. I was just looking at the spear, so he went off.

When everything became silent outside, a spear came from inside that room. There is a stone spear here, but this spear was like light. It came from there, and then it stood right in the place where the stone spear was—just hovering there, going this way and that. It looked like light, but it was like gold—metallic. It is metal, but there is so much power in it that you just see the light. I felt that was what I was waiting for. I looked at it, and I was very happy. I folded my hands. During all this time a kind of locking was in my body. But that was completely opened up, and I felt completely normal.

There is a prayer to the spear in the Tamil language. It's actually a *mantra.* It has seed power in it. That sound came out. The sound was there—a very powerful voice of a hundred or two hundred people chanting. The prayer says that the spear in the hand of Skanda [Kārttikeya] gives protection. If one sees that as a fact it will become truth. By looking at the spear all the ghosts will go away, and this spear is the destroyer of all enemies of the Devas. Among the eight Lakṣmīs, it gives a feast to the Lakṣmī of courage. It killed Surapadma. Surapadma was a demon who got a benediction that only a five-year-old child can kill him. The prayer goes on like this about the glories of the spear. So I heard this like the sound of the ocean. It was like two hundred people chanting.

There was nobody in the temple, and I was not feeling any awkwardness, shock, or fear, or anything. I was feeling completely normal.

So much sound was coming, and I was chanting with that. So I chanted this verse, and the spear came like this, maybe this close. So I got up and paid my full obeisances. And then I got up and stood like this, and it was right there, for about two or three minutes. And then, just like that, it disappeared. Then I sat down again, and I took some fruit.

In the morning the priest came, and he looked at me and said, "Ho, so effulgent." Then the crowd started coming. It was a special day for the Saṅgameśvara temple, Tuesday, Śiva's day. So a bigger crowd was sitting there in front of me, and I was normal now. So one lady was asking me if her sick grandson will be all right. I took a fruit and gave it to her, and she went off. I was distributing all the fruits. Whatever anybody would ask, I would give a fruit.

One person asked me a question about a saint by the name Kumāra-gurupara. It is the name of a saint, and it is also the name of Kārttikeya. So he was asking me if Kumāra-gurupara was Kārttikeya, and I said no. That created a little stir among the people. Is this small boy answering questions like this? Then it was more questions on Kārttikeya. What are the different holy places of Kārttikeya? And then I would say this place is special for this, this place is special for that. And I start telling about Saṅgameśvara's glories, and then I spoke of the seven Śiva temples of the city. I mentioned many things that most people don't know.

While this was going on, one lady came to the temple who knows our family, and she informed them that I was there. She said, "Your son has become a big *swamijī* there. Everybody is listening to him." So then they had been looking for me for a few days already. That day my brother took a bicycle and came there. He just came straight into the temple to where I was and gave me a slap. "You're a rascal. Mother is crying." They were all coming to him and saying, "This *sadhu* has so much knowledge, so don't do like this." But my brother was not at all impressed, and he put me on the bicycle and took me back.

Encounter with a Jaladevata

In Indian tradition, a Jaladevata is a being who gives protection to people whose lives are endangered in a particular natural body of water, such as a lake or a stretch of a river. Here is a story of an encounter with a Jaladevata that took place recently in Māyāpura, near the town of Navadvīpa in West Bengal (about three hours' drive north of Calcutta). This happened near the end of June 1992. Māyāpura is an area of small villages and temples surrounded by miles of rice paddies. It is situated on a point of land bounded by a branch of the Ganges on one side and by the Jalangi River on the other.

The story was recounted by the wife of Kannan, who was in Māyāpura at the time and knew the people involved. Kannan translated as she told the story in her native language:

> Some boys from the *gurukula* school went swimming in the Ganges, and a five-year-old boy named Bhāgavat went with them. He doesn't know how to swim, but because all the boys went there, he also went with them. One boy took him on a bicycle, and the parents were just behind. So everyone jumped into the Ganges, and this boy also jumped, thinking, "That's what I'm supposed to do."
>
> The parents reached the riverbank about five or six minutes after the boys, because the boys were taking bicycles. They asked, "Where is Bhāgavat?" And all the boys were looking at each other, saying, "Oh, where is he?" Nobody knew. Then Dvaipāyana, one of the boys, showed the mother, "There is Bhāgavat there." And you could see only his finger sticking up above the surface of the water. The current there is very powerful, but they saw him. His hand was up, you could only see the finger, but he was staying in the same spot.
>
> He was not moving. The current is very fast, but he was not moving. So then, because the mother got all upset, Dvaipāyana jumped in, and he knows swimming. Near this side of the river there is a big current, but at a little distance there is a sand bank where the children who can swim will play. But the boy was on this side, the strong current side. So Dvaipāyana jumped in and went there and brought him. And the boy was not suffocating. He was just normal.
>
> When the mother asked what happened, the boy said that the currents were pulling him, and he was going to drown. Just then he saw a lady who lifted him up within the water. She was holding him within the water. She had a crown; she had earrings; she was dressed very nicely and looked very beautiful, and she was holding him. So for some time he felt the current was pulling him, but after that he was held in her hands within the water. That is what he said. He kept telling his mother, "That lady was very beautiful." And he was asking if that was Mother Ganges.

UFOs over Māyāpura

The last story from Kannan is a typical report of a UFO sighting in the nocturnal light category. This could have been a bolide, but that is perhaps ruled out by the fact that it was said to change speed from fast, to slow, and back to fast again. I include the story to show that typical UFO sightings are reported in India.

The story also shows how a person of native Indian culture naturally identifies such a phenomenon as a *vimāna.* A curious point is that Kannan used an example from the *Rāmāyaṇa* to argue that the UFO was able to expand and contract in size. He brought in this idea to explain how someone could be flying in something that looked so small.

A change in size was not directly observed in this sighting, but such changes are occasionally mentioned in UFO reports. For example, Betty Andreasson reported that she saw a UFO being shrunk by a factor of two or more, even though it was occupied by two human abductees and a number of UFO beings.[1] A possibly related observation was made by Steven Kilburn, an abductee interviewed by Budd Hopkins. Kilburn testified that he was brought into a UFO that looked much bigger on the inside than on the outside.[2]

Here is Kannan's story:

> I know this happened during the Desert Storm battle, because I had a class in the evening hearing the BBC news of the battle with the Bhakti Śāstri boys. We were following Saddam Hussein very meticulously—all his movements. So I would put the boys all in front of my house in the evening. We would lay out a mat, put out the lights, and then play the radio under the stars. But one time the battle news was not ready, and they were talking about some irrelevant topic like how Charlie Chaplin was taking a dance class, and the boys were making fun of it. They were all making a joke, in a very nonofficial mood, and it was about 8:00 p.m.
>
> We had these two huts facing each other, so we are sitting by this hut facing this side, and this is our living room hut. I was sitting there, and I just looked up and saw this very bright blue light over the other hut. It started from the Dhruva star, the Polestar, that can be seen behind our long building. It started from there, and from there it was moving very fast. Then when it came over our temple area, it went really slow. It had a tail in the back that started small and became bigger. And very clearly there was a gross object in front of it. It was not like a star, and it was not very high.

I was already seeing it, and one boy said, "What is that, Prabhu?" And then this other boy said, "What is that? What is that?" We all got up, and we were looking at it. There were six of us, five boys and me. My wife was in the kitchen cooking something, and she came out too because we were shouting, "Ho, what is that!" So it was at maybe one and a half palm tree's height from the roof of the hut, which is not very high. It was less than the height of the conch building [five stories]. It was very clear. Most of them were concentrating on the light. I looked in the front, and I saw that it was a clear object. It was not a star, and it was not very far in the sky. It was right there. And it was also like, moving around itself, but it was very slow as it came over our ISKCON area. Then it went up toward the Jalangi River, to the *gośāla* [cow barn] or maybe a little more. And then it took speed. It was like somebody who slows down to look at something.

It was very interesting, and the boys kept asking about it. I said, "Yeah, you know, people respect Māyāpura, so there must be somebody traveling in there." But if the whole thing looked so small at that height, then it must have been very small. So what could it be? Boys ask all kinds of questions. But anyway, to my understanding it was actually a *vimāna*. But for some reason some *vimānas* can become bigger or they can become smaller. As they go through certain areas they become small or big, according to the area. This is very clearly shown by Hanumān checking out the *Puṣpaka-vimāna* in Śrī Laṅkā. First it was a two-seater. Then it was becoming so big. And finally, when Rāma flew on that, it was bigger than a city. He took the whole army of Vānaras to Ayodhyā for His coronation. So it was bigger than a city.

It is only in our eyes that it looks that small. But actually I am sure that it was bigger than all of our four huts together. But they just made it smaller at that time, probably just for passing through this respectable area . . .

I distinctly remember that it was lower than the height of our conch building. It was so close, and you can't do anything about it. In Māyāpura, hawks fly much higher than that. It was an abnormal thing, but we were completely helpless just looking at it. When I described it to a Muslim gentleman, an old farmer, he said that in the sky over Māyāpura there are so many things like this. Things come, things go. People come, people go—so many things happen because this is Mahāprabhu's place. His land is just next to ours behind the *gośāla*. The language he used was, "It is not wonderful that such things are seen in the sky over Māyāpura."

Bibliography

Air Force, October 1994, "Air Force on Roswell," *MUFON UFO Journal,* No. 18, p. 13.

AFR, 14 Sept. 1959, "Air Force Regulation No. 200-2: Unidentified Flying Objects (UFO)," Department of the Air Force, U.S. Government Printing Office.

Agrawal, D. P., *et al.*, eds., 1985, "Climate and Geology of Kashmir, the Last 4 Million Years," in *Current Trends in Geology,* Vol. VI, New Delhi: Today & Tomorrow's Printers and Publishers.

Ahmad, S. Maqbul and Bano, Raja, 1984, *Historical Geography of Kashmir,* New Delhi: Ariana Publishing House.

Albertson, Maury and Shaw, Margaret, eds., May 22–25, 1992, *Proceedings of the International Symposium on UFO Research,* Fort Collins, Colorado: International Association for New Science.

Amnesty, March 1983, "Political Abuse of Psychiatry in the USSR," 322 8th Avenue, New York, N.Y.: Amnesty International USA.

Angelucci, Orfeo M., 1955, *The Secret of the Saucers,* Amherst, WI: Amherst Press.

Apte, Vaman S., 1965, *The Practical Sanskrit-English Dictionary,* Delhi: Motilal Banarsidass.

Augustine, St., 1952, *The City of God,* in Robert Hutchins, ed., *Great Books of the Western World,* Vol. 18, Chicago: Encyclopaedia Britannica.

Bailey, Alice A., 1951, *The Unfinished Autobiography,* New York: Lucis Pub. Co.

Bailey, Alice A., 1962, *A Treatise on Cosmic Fire,* New York: Lucis Pub. Co.

Bamzai, Prithivi, 1973, *A History of Kashmir,* New Delhi: Metropolitan Book Co.

Bearden, Thomas, 1979, "A Mind/Brain/Matter Model Consistent with Quantum Physics and UFO Phenomena," *1979 MUFON UFO Symposium Proceedings,* Seguin, Texas: MUFON, Inc., pp. 78–112.

Bentley, John, 1825, *Historical View of the Hindu Astronomy,* Osnabruck: Biblio Verlag, reprinted in 1970.

Berlitz, Charles and Moore, William, 1980, *The Roswell Incident,* London: Granada.

Bershad, Michael, 1988, *UFO,* Vol. 3, No. 2, pp. 32–33.

Bhagavad-gītā: see Bhaktivedanta Swami Prabhupāda, A. C., 1983, *Bhagavad-gītā As It Is,* Los Angeles: The Bhaktivedanta Book Trust.

Bhāg. Pur.: see Bhaktivedanta Swami Prabhupāda, A. C., 1982a, *Śrīmad-Bhāgavatam,* Los Angeles: The Bhaktivedanta Book Trust.

Bhaktisiddhānta Sarasvatī Goswami Ṭhākura, 1985, *Śrī Brahma-Saṁhitā,* Los Angeles: The Bhaktivedanta Book Trust.

Bhaktivedanta Swami Prabhupāda, A. C., 1982a, *Śrīmad Bhāgavatam,* Los Angeles: The Bhaktivedanta Book Trust. (References to commentaries in this work identify the volume and page number. Volumes are identified as Canto N, Part M, where N runs from 1 to 10, and M runs from 1 to 4. References to verses are given as *Bhāg. Pur.* x.y.z, where x is the canto, y is the chapter in that canto, and z is the verse in that chapter.)

Bhaktivedanta Swami Prabhupāda, A. C., 1982b, *The Nectar of Devotion,* Los Angeles: The Bhaktivedanta Book Trust.

Bhaktivedanta Swami Prabhupāda, A. C., 1983, *Bhagavad-gītā As It Is,* Los Angeles: The Bhaktivedanta Book Trust. (References to verses are given as *Bhagavad-gītā* x.y, where x the chapter, and y is the verse in that chapter.)

Bhaktivedanta Swami Prabhupāda, A. C., 1986, *Kṛṣṇa,* Los Angeles: The Bhaktivedanta Book Trust.

Bloecher, Ted; Clamar, Aphrodite; and Hopkins, Budd, 1985, "Summary Report on the Psychological Testing of Nine Individuals Reporting UFO Abduction Experiences," Fund for UFO Research, Box 277, Mt. Rainier, Md. 20712.

Blum, Howard, 1990, *Out There,* New York: Simon and Schuster.

Bowen, Charles, 1969a, "Few and Far Between" in Charles Bowen, ed., *The Humanoids,* Chicago: Henry Regnery Co., pp. 13–26.

Bowen, Charles, 1969b, "Interesting Comparisons" in Charles Bowen, ed., *The Humanoids,* Chicago: Henry Regnery Co., pp. 239–48.

Boylan, Richard J. and Boylan, Lee K., 1994, *Close Extraterrestrial Encounters,* Tigard, Oregon: Wild Flower Press.

Braude, Stephen E., 1991, *The Limits of Influence,* London: Routledge.

Bullard, Thomas E., 1987, *UFO Abductions: The Measure of a Mystery,* Washington, D.C.: Fund for UFO Research.

Carpenter, John, 1991, "Double Abduction Case: Correlation of Hypnosis Data," *Journal of UFO Studies,* n.s. 3, pp. 91–114.

Carpenter, John, 1992, "The Reality of the Abduction Phenomenon."

Carpenter, John, March 1993a, "Gerald Anderson: Disturbing Revelations," *MUFON UFO Journal,* No. 299, pp. 6–9.

Carpenter, John, April 1993b, "Reptilians and Other Unmentionables," *MUFON UFO Journal,* No. 300, pp. 10–11.

Cannon, Martin, undated, *The Controllers: A New Hypothesis of Alien Abductions.*

Chalker, Bill, 1987, "The UFO Mystery in Australia," *MUFON 1987 UFO* Symposium Proceedings, Seguin, Texas: MUFON, Inc., pp. 166–79.

Childress, David H., 1991, *Vimana Aircraft of Ancient India and Atlantis,*

Stelle, Illinois: Adventures Unlimited Press.
Condon, Edward U., 1969a, *Scientific Study of Unidentified Flying Objects,* D. S. Gillmor, ed., New York: E. P. Dutton & Co.
Condon, Edward U., Dec. 1969b, "UFOs I Have Loved and Lost," *Bulletin of the Atomic Scientists,* Vol. 15, No. 10.
Congress, April 5, 1966, "Unidentified Flying Objects: Hearing by Commit tee on Armed Services of the House of Representatives, Eighty-ninth Congress, Second Session," Washington: U.S. Government Printing Office.
Conroy, Ed, 1989, *Report on Communion,* New York: William Morrow and Co., Inc.
Corliss, William R., Nov.-Dec. 1990, "Science Frontiers No. 72," Glen Arm, Md.: The Sourcebook Project.
Council, April 5, 1985, "Scientific Status of Refreshing Recollection by the Use of Hypnosis," *Journal of the American Medical Association,* Vol. 253.
Creighton, Gordon, 1969a, "The Villa Santina Case," in Charles Bowen, ed., *The Humanoids,* Chicago: Henry Regnery Co., pp. 187–99.
Creighton, Gordon, 1969b, "The Amazing Case of Antonio Villas Boas," in Charles Bowen, ed., *The Humanoids,* Chicago: Henry Regnery Co., pp. 200–38.
Curtis, William, 1989, "Contactee: Firsthand," *UFO,* Vol. 4, No. 3, pp. 36–38.
Daly, Lawrence and Pacifico, J. Frank, December 1991, "Opening the Doors to the Past," *Champion Magazine,* pp. 43–47.
Dickinson, Terence, December 1974, "The Zeta Reticuli Incident," *Astronomy,* Vol. 2, No. 12, pp. 4–18.
Dikshitar, Ramachandra, 1944, *Aerial Warfare in Ancient India,* Madras: University of Madras.
Dobzhansky, Theodosius, 1972, "Darwinian Evolution and the Problem of Extraterrestrial Life," *Perspect. Biol. Med.,* Vol. 15, pp. 157–75.
Downing, Barry H., 1968, *The Bible and Flying Saucers,* Philadelphia: J. B. Lippincott.
Downing, Barry H., 1981, "Faith, Theory and UFOs," *1981 MUFON UFO* Symposium Proceedings, Seguin, Texas: MUFON, Inc., pp. 35–42.
Druffel, Ann and Rogo, D. Scott, 1988, *The Tujunga Canyon Contacts,* New York: Penguin Books.
Earley, George, 1990, *UFO,* Vol. 5, No. 4, p. 40.
Elliot, H. M., 1875, *The History of India as Told by Its Own Historians,* Vol. VI, London: Trubner and Co.
Elkins, Don and Rueckert, Carla, 1977, *Secrets of the UFO,* P.O. Box 5195, Louisville, Ky. 40205: L/L Research.
Elkins, Don; Rueckert, Carla; and McCarty, James A., 1984, *The Ra Material,* West Chester, PA: Whitford Press.

Evans-Wentz, W. Y., 1966, *The Fairy-Faith in Celtic Countries,* New York: University Books.

Farberman, Rhea, Nov. 11, 1994, "APA Council Addresses Controversy over Adult Memories of Childhood Sexual Abuse," American Psychological Association.

Fawcett, Lawrence and Greenwood, Barry J., 1984, *The UFO Cover-up,* formerly *Clear Intent,* New York: Prentice Hall Press.

Fiore, Edith, 1989, *Encounters,* New York: Ballantine Books.

Fiore, Edith, 1992, "Close Encounters of the Fourth Kind," *Proceedings of the International Symposium on UFO Research,* M. Albertson and M. Shaw, eds., Fort Collins, Colorado: International Association for New Science, pp. 41– 45.

Fowler, Raymond, 1979, *The Andreasson Affair,* Englewood Cliffs, New Jersey: Prentice Hall.

Fowler, Raymond, 1981, *Casebook of a UFO Investigator,* Englewood Cliffs, New Jersey: Prentice Hall.

Fowler, Raymond, 1990, *The Watchers,* New York: Bantam Books.

Fowler, Raymond, 1993, *The Allagash Abductions,* Tigard, Oregon: Wild Flower Press.

Friedman, Stanton and Moore, William L., 1981, "The Roswell Incident: Beginning of the Cosmic Watergate," *1981 MUFON UFO Symposium Proceedings,* Seguin, Texas: MUFON, Inc., pp. 132–53.

Friedman, Stanton, May, 1990, "Final Report on Operation Majestic 12," available from UFORI, P.O. Box 3584, Sta. B., Fredericton, New Brunswick, Canada.

Friedman, Stanton, 1991, "Update on Crashed Saucers in New Mexico," *MUFON 1991 International UFO Symposium Proceedings,* Seguin, Texas: MUFON, Inc.

Frye, Roland Mushat, 1983, *Is God a Creationist?,* New York: Charles Scribner's Sons.

Fuller, John G., 1966, *The Interrupted Journey,* New York: The Dial Press.

Fuller, Paul and Randles, Jenny, 1991, "Crop Circles: A Scientific Answer to the UFO Mystery?" in Terence Meaden, ed., *Circles from the Sky,* London: Souvenir Press.

Ganguli, Kisari Mohan, 1976, *The Mahabharata of Krishna-Dwaipayana Vyasa,* Vol. VI, *Drona Parva,* New Delhi: Munshiram Manoharlal (first published by Pratap Chandra Roy).

Garcia, Encarnacion, September 1993, "Caso Guarapiranga," *UFO,* ed. A. J. Gevaerd, No. 25, pp. 8–14.

Good, Timothy, 1988, *Above Top Secret,* New York: Quill, William Morrow.

Green, Robin, 1985, *Spherical Astronomy,* Cambridge: Cambridge Univ. Press.

Gresh, Bryan, October 1993, "Soviet UFO Secrets," *MUFON UFO Journal,*

No. 306, pp. 3–7.
Haines, Richard F., 1990, *Advanced Aerial Devices Reported During the* Korean War, Los Altos, California: LDA Press.
Haley, Leah A., 1993, *Lost was the Key,* Tuscaloosa, Alabama: Greenleaf Publications.
Hall, Richard H., ed., 1964, *The UFO Evidence,* Washington, D.C.: NICAP.
Hall, Richard H., 1988, *Uninvited Guests,* Santa Fe, New Mexico: Aurora Press.
Harder, James, 1992, "What We Might Learn from Extraterrestrial Contact," *Proceedings of the International Symposium on UFO Research,* M. Albertson and M. Shaw, eds., Fort Collins, Colorado: International Association for New Science, pp. 137– 47.
Hartland, Edwin S., 1891, *The Science of Fairy Tales,* London: Walter Scott.
Henry, Richard C., 1988, "UFOs and NASA," *Journal of Scientific Exploration,* Vol. 2, No. 2, pp. 93–142.
Hickson, Charles and Mendez, William, 1983, *UFO Contact at Pascagoula,* Tucson, Arizona: published by Wendelle C. Stevens.
Hind, Cynthia, 1981, "African Encounters: Case Investigations," *1981 MUFON UFO Symposium Proceedings,* Seguin, Texas: MUFON, Inc., pp. 80–91.
Hoagland, Hudson, Feb. 1969, "Beings from Outer Space—Corporeal and Spiritual," *Science,* Vol. 163, No. 3868.
Hopkins, Budd, 1981, *Missing Time,* New York: Ballantine Books.
Hopkins, Budd, 1987, *Intruders,* New York: Ballantine Books.
Hopkins, Budd, 1988, "Q&A: Budd Hopkins," *UFO,* Vol. 3, No. 2, pp. 17–21.
Howe, Linda Moulton, 1989a, *An Alien Harvest,* Littleton, Colorado: Linda Moulton Howe Productions.
Howe, Linda Moulton, 1989b, "An Alien Harvest," *MUFON 1989 International UFO Symposium Proceedings,* Seguin, Texas: MUFON, Inc., pp. 114–23.
Howe, Linda Moulton, 1992, "The UFO Jigsaw Puzzle," *Proceedings of the International Symposium on UFO Research*, M. Albertson and M. Shaw, eds., Fort Collins, Colorado: International Association for New Science, pp. 1–23.
Hridayānanda Goswami, 1992, *Mahābhārata,* Los Angeles: Bhaktivedanta Book Trust.
Huneeus, Antonio, 1991, *A Study Guide to UFOs, Psychic and Paranormal Phenomena in the U.S.S.R.,* Abelard Productions.
Hynek, J. Allen, 1972a, *The UFO Experience,* Chicago: Henry Regnery Co.
Hynek, J. Allen, 1972b, "Twenty-one Years of UFO Reports," in Carl Sagan and Thornton Page, eds., *UFO's—a Scientific Debate,* Ithaca and London: Cornell Univ. Press., pp. 37–51.
Hynek, J. Allen, 1976, "Swamp Gas Plus Ten—and Counting," *1976 MUFON Symposium Proceedings,* Seguin, Texas: MUFON Inc., pp. 76–83.

Hynek, J. Allen; Imbrogno, Philip J.; and Pratt, Bob, 1987, *Night Siege,* New York: Ballantine Books.

Jacobs, David M., 1988, "Post-Abduction Syndrome," *MUFON 1988 International UFO Symposium Proceedings,* Seguin, Texas: MUFON Inc., pp. 87–102.

Jacobs, David M., 1992, *Secret Life,* New York: Simon and Schuster.

Johnson, Donald A., 1989, "UFO Car Pursuits: Some New Patterns in Old Data," *MUFON 1989 International UFO Symposium Proceedings,* Seguin, Texas: MUFON, Inc., pp. 136–46.

Johnston, Francis, 1979, *Fatima,* Rockford, Ill.: Tan Books.

Josyer, G. R., trans., 1973, *Vymaanika-Shaastra Aeronautics,* by Maharshi Bharadwaaja, propounded by Subbaraya Sastry, Mysore, India: published by G. R. Josyer

Jung, Carl G., 1959, *Flying Saucers, a Modern Myth of Things Seen in the Skies,* New York: Harcourt Brace and Co.

Kafton-Minkel, Walter, 1989, *Subterranean Worlds,* Port Townsend, Washington: Loompanics Unlimited.

Kanjilal, Dileep Kumar, 1985, *Vimāna in Ancient India,* Calcutta: Sanskrit Pustak Bhandar.

Keene, M. Lamar, 1976, *The Psychic Mafia,* as told to Allen Spraggett, New York: St. Martin's Press.

Kinder, Gary, 1987, *Light Years,* New York: Pocket Books.

Kuettner, Joachim P., July 1971, "McDonald: A Last Respect," *Astronautics and Aeronautics,* p. 18.

Kumari, Ved, 1973, *The Nilamata Purana,* Vol. II, Srinagar: J & K Academy of Art, Culture and Languages.

Lanning, Kenneth, January 1992, *Investigator's Guide to Allegations of "Ritual" Child Abuse,* Federal Bureau of Investigation.

Lawson, Alvin H., 1977, "What Can We Learn from Hypnosis of Imaginary Abductions," *MUFON 1977 UFO Symposium Proceedings,* Seguin, Texas: MUFON, Inc., pp. 107+.

Layne, Meade, March-April 1953, *Round Robin,* Vol. 8, No. 6, San Diego: Borderland Sciences Research Associates.

Leslie, Desmond and Adamski, George, 1953, *The Flying Saucers Have Landed,* New York: The British Book Centre.

Lorenzen, Coral, 1969, "UFO Occupants in United States Reports" in Charles Bowen, ed., *The Humanoids,* Chicago: Henry Regnery Co., pp. 143–76.

Lorenzen, Coral and Lorenzen, Jim, 1976, *Encounters with UFO Occupants,* New York: Berkeley Publishing Co.

Maccabee, Bruce S., 1985, Foreword to Bloecher *et al.,* "Summary Report on the Psychological Testing of Nine Individuals Reporting UFO Abduc-

tion Experiences," Mt. Rainier, Md.: Fund for UFO Research.
Maccabee, Bruce S., May 1990a, "Recent UFO videotapes," *MUFON UFO Journal,* No. 265, pp. 3–7.
Maccabee, Bruce S., 1990b, "Report on the Analysis of the Gulf Breeze, Florida, UFO Videotape of December 28, 1987," Mt. Rainer, MD: Fund for UFO Research.
Maccabee, Bruce and Sainio, Jeffrey, December 1993, "Cruise-Missile UFO Disappears," *MUFON UFO Journal,* No. 308, pp. 3–7.
Mack, John, 1994, *Abduction,* New York: Charles Scribner's Sons.
Marchetti, Victor, May 1979, "How the CIA Views the UFO Phenomenon," *Second Look,* Vol. 1, No. 7, Washington, D.C., pp. 2–7.
Markowitz, William, 1980, "The Physics and Metaphysics of Unidentified Flying Objects," in D. Goldsmith, ed., *The Quest for Extraterrestrial Life,* Mill Valley, Calif.: Univ. Science Books, pp. 255–61.
McDonald, James E., April 22, 1967, "UFOs: Greatest Scientific Problem of our Times?", prepared from presentation at annual meeting of the American Society of Newspaper Editors, Washington, D.C.
McDonald, James E., July 1971, "Air Force Observations of an Unidentified Object in the South-Central U.S., July 17, 1957," *Astronautics and Aeronautics,* pp. 66–70.
Meaden, Terence, 1990, "Crop circles and the plasma vortex" in Ralph Noyes, ed., *The Crop Circle Enigma,* Bath, UK: Gateway Books, pp. 76–98.
Menzel, Donald H., and Boyd, Lyle G., 1963, *The World of Flying Saucers,* Cambridge: Harvard Univ. Press.
Misra, Janardan, 1988, *Veda and Bharat* (India), New Delhi: Ess Ess Publications
Moore, William, 1985, "Crashed Saucers: Evidence in Search of Proof," *MUFON 1985 UFO Symposium Proceedings,* Seguin, Texas: MUFON, Inc., pp. 131–79.
Moore, William and Friedman, Stanton, 1988, "MJ-12 and Phil Klass: What are the Facts?", *MUFON 1988 UFO Symposium Proceedings,* Seguin, Texas: MUFON, Inc.
Myers, F. W. H., 1961, *Human Personality and its Survival of Bodily Death,* New York: University Books, Inc.
NAS, Feb. 15, 1972, "Annual Report of the National Academy of Sciences, Fiscal year 1968–69."
Nathan, Kanishk, 1987, "UFOs and India: Ancient and Contemporary," *MUFON 1987 UFO Symposium Proceedings,* Seguin, Texas: MUFON, Inc., pp. 68–82.
O'Brien, Christian, 1985, *The Genius of the Few,* Wellingborough, Northamptonshire: Turnstone Press Limited.
Ofshe, Richard and Watters, Ethan, March/April 1993, "Making Monsters,"

Society, Vol. 30, No. 3, pp. 4–16.
Oppert, Gustav, 1967, *On the Weapons, Army Organization, and Political Maxims of the Ancient Hindus with Special Reference to Gunpowder and Firearms*, Ahmedabad: The New Order Book Co.
Page, Thornton, 1972, "Education and the UFO Phenomenon," in Carl Sagan and Thornton Page, eds., *UFO's—a Scientific Debate*, Ithaca and London: Cornell Univ. Press, pp. 3–10.
Pasricha, Satwant and Stevenson, Ian, 1986, "Near-Death Experiences in India," *The Journal of Nervous and Mental Disease*, The Williams & Wilkins Co., pp. 165–70.
Phillimore, J. S., 1912, *Philostratus in Honor of Apollonius of Tyana*, Vol. 1, Oxford: Clarendon Press.
Phillips, Ted R., 1981, "Close Encounters of the Second Kind: Physical Traces," *1981 MUFON UFO Symposium Proceedings*, Seguin, Texas: MUFON, Inc., pp. 93–129.
Pickthall, Marmaduke, 1976, *The Glorious Koran*, Albany: SUNY Press.
Pius XII, Pope, 1950, "Humani Generis," Papal Encyclical.
Powers, William T., 7 April 1967, "Analysis of UFO Reports," *Science*, Vol. 156, p. 11.
Prabhu, C. S. R., 1992, "A Preliminary Report on the Study and Investigation of some Ancient Scientific Shastras," Hyderabad, India.
Raghavan, V., 1956, "Yantras or Mechanical Contrivances in Ancient India," Transaction No. 10, Bangalore: The Indian Institute of Culture.
Randle, Kevin D. and Schmitt, Donald R., 1991, *UFO Crash at Roswell*, New York: Avon Books.
Randle, Kevin D. and Schmitt, Donald R., 1994, *The Truth About the UFO Crash at Roswell*, New York: M. Evans, 216 E. 49th St., Fax: (212) 486-4544.
Randles, Jenny, 1988, *Alien Abductions: The Mystery Solved*, New Brunswick, N.J.: Inner Light Publications.
Rhine, Louisa, 1961, *Hidden Channels of the Mind*, New York: William Stone Associates.
Ribera, Antonio, Oct.-Nov. 1966, "The Landing at Villares del Saz," *Flying Saucer Review*, pp. 28–30.
Rickard, Robert, 1990, "Clutching at Straws," in Ralph Noyes, ed., *The Crop Circle Enigma*, Bath, UK: Gateway Books, pp. 62–71.
Ring, Kenneth, 1985, *Heading Toward Omega*, New York: William Morrow, Quill.
Ring, Kenneth, 1992, *The Omega Project*, New York: William Morrow.
Roach, Franklin, 1972, "Astronomer's Views on UFO's," in Carl Sagan and Thornton Page, eds., *UFO's—a Scientific Debate*, Ithaca and London: Cornell Univ. Press., pp. 23–33.
Roll, William, 1977, "Poltergeists," in *Handbook of Parapsychology*, Benjamin Wolman, ed., New York: Van Nostrand, pp. 382–413.

Rosen, Steven, 1991, *The Lives of the Vaishnava Saints, Shrinivas Acharya, Narottam Das Thakur, Shyamananda Pandit,* Brooklyn, New York: Folk Books.

Sabom, Michael, 1982, *Recollections of Death: A Medical Investigation,* New York: Harper & Row.

Sagan, Carl, 1972, "UFO's: The Extraterrestrial and Other Hypotheses," in Carl Sagan and Thornton Page, eds., *UFO's—a Scientific Debate,* Ithaca and London: Cornell Univ. Press., pp. 265–75.

Salter, John R., 1989, *An Account of the Salter UFO Encounters of March 1988,* privately distributed.

Sanchez-Ocejo, Virgilio and Stevens, Wendelle C., 1982, *UFO Contact from Undersea,* Tucson, Arizona, P.O. Box 17206: UFO Photo Archives.

Sanderson, Ivan, 1970, *Invisible Residents,* New York: World Publishing Co.

Sastrin, Bapu Deva, trans., 1860, *Surya-siddhanta,* Calcutta: Baptist Mission Press, reprinted in *Bibliotheca Indica,* New Series No. 1, Hindu Astronomy I.

Satsvarūpa Dāsa Goswami, 1984, *A Handbook for Kṛṣṇa Consciousness,* Port Royal, Pa.: GN Press.

Saunders, David R., 1968, *UFOs? Yes!,* New York: The World Publishing Co.

Savadove, Larry, 1986, "Contact," a videotape produced by Spencer Young and James Margellos.

Schuessler, John F., 1985, "The Medical Evidence in UFO Cases," *MUFON 1985 UFO Symposium Proceedings,* Seguin, Texas: MUFON, Inc., pp. 79–86.

Schuessler, John F. and O'Herin, Edward, 1993, "Truck Driver Injured by UFO: The Eddie Doyle Webb Case," *MUFON 1993 International UFO Symposium Proceedings,* Seguin, Texas: MUFON, Inc., pp. 59–84.

Schwarz, Berthold E., 1988, *UFO Dynamics: Psychiatric and Psychic Aspects of the UFO Syndrome,* Moore Haven, Florida: Rainbow Books.

Scully, Frank, Oct. 12, 1949, "Scully's Scrapbook," *Variety,* pp. 61–62.

Scully, Frank, 1950, *Behind the Flying Saucers,* New York: Henry Holt and Co.

Seligman, Michael (producer), October 14, 1988, "UFO Cover-Up? Live," distributed by Lexington Broadcast Service.

Shastri, Hari Prasad, 1976, *The Ramayana of Valmiki,* London: Shanti Sadan. Three volumes.

Sitchin, Zecharia, 1976, *The 12th Planet,* New York: Avon.

Sitchin, Zecharia, 1980, *The Stairway to Heaven,* New York: St. Martin's Press.

Simpson, George Gaylord, 1964, *This View of Life,* New York: Harcourt, Brace, and World.

Śiva Purāṇa, 1991, *The Śiva Purāṇa,* Part II, Delhi: Motilal Banarsidass.

Slater, Elizabeth, June 30, 1983a, "Conclusions on Nine Psychologicals," Mt. Rainier, Md.: Fund for UFO Research.

Slater, Elizabeth, October 30, 1983b, "Addendum to Conclusions on Nine Psychologicals," Mt. Rainier, Md.: Fund for UFO Research.

Spanos, Nicholas P., Cross, Patricia A., Dickson, Kirby, and DuBreuil, Susan C., 1993a, "Close Encounters: An Examination of UFO Experiences," *Journal of Abnormal Psychology,* Vol. 102, No. 4, pp. 624–32.
Spanos, Nicholas P., Burgess, Cheryl A., and Burgess, Melissa F., 1993b, "Past-Life Identities, UFO Abductions, and Satanic Ritual Abuse," *The International Journal of Hypnosis,* Vol. 42, No. 4, pp. 433–45.
Stein, M. A., 1979, *Kalhaṇa's Rājataraṅgiṇī,* Delhi: Motilal Banarsidass, Vol. I.
Steinman, William S., 1986, *UFO Crash at Aztec,* Tucson, Arizona: UFO Photo Archives.
Stevens, Wendelle C. and Herrmann, William J., 1981, *UFO . . . Contact from Reticulum: A Report of the Investigation,* Tucson, Arizona: published by Wendelle C. Stevens.
Stevens, Wendelle C., 1982, *UFO . . . Contact from the Pleiades: A Preliminary Investigation Report,* Tucson, Arizona: UFO Photo Archives.
Stevens, Wendelle C. and Dong, Paul, 1983, *UFOs over Modern China,* Tucson, Arizona: UFO Photo Archives.
Stevens, Wendelle C. and Roberts, August, 1986, *UFO Photographs Around the World,* Tucson, Arizona: UFO Photo Archives.
Stevenson, Ian, 1971, "The Substantiality of Spontaneous Cases," in W. G. Roll, ed., *Proceedings of the Parapsychological Association,* Number 5, Duke Station, Durham, North Carolina.
Stevenson, Ian, 1977, "Reincarnation: Field Studies and Theoretical Issues," in Benjamin Wolman, ed., *Handbook of Parapsychology,* New York: Van Nostrand Reinhold Co., pp. 631–63.
Stevenson, Ian, January 1988, "Interview" by Meryle Secrest in *Omni,* Vol. 10, pp. 77– 80+.
Stevenson, Ian, January 1994, "A Case of Psychotherapist's Fallacy: Hypnotic Regression to "Previous Lives'," *American Journal of Clinical Hypnosis,* pp. 188–93.
Story, Ronald, ed., 1980, *The Encyclopedia of UFOs,* New York: Doubleday and Co.
Strieber, Whitley, 1988, *Transformation,* New York: Avon Books.
Taff, Barry, 1989, "Anatomy of an 'EBE,'" *UFO,* Vol. 4, No. 3, pp. 18–22.
Telano, Rolf, 1963, *The Flying Saucers,* Clarksburg, W. Va.: Saucerian Books.
Thompson, Richard, 1981, *Mechanistic and Nonmechanistic Science,* Los Angeles: Bhaktivedanta Book Trust.
Thompson, Richard, 1989, *Vedic Cosmography and Astronomy,* Los Angeles: Bhaktivedanta Book Trust.
Thompson, Richard, July/August 1991, "Astronomy and the Antiquity of Vedic Civilization," *Back to Godhead,* Vol. 25, No. 4, Los Angeles: Bhaktivedanta Book Trust.

Turner, Karla, 1992, *Into the Fringe,* New York: Berkeley Books.
Turner, Karla, 1993, Lecture at UFO EXPO/NY, November 1993, Alpha Omega Cassette Enterprises, 516 S. Oak Knoll #5, Pasadena, Calif. 91101.
Twiggs, Denise and Twiggs, Bert, 1992, *Secret Vows,* Tigard, Oregon: Wild Flower Press.
Vallee, Jacques, 1969a, *Passport to Magonia,* Chicago: Henry Regnery Co.
Vallee, Jacques, 1969b, "The Pattern Behind the UFO Landings," in Charles Bowen, ed., *The Humanoids,* Chicago: Henry Regnery Co., pp. 27–76.
Vallee, Jacques, 1979, *Messengers of Deception,* Berkeley, Calif.: And/Or Press.
Vallee, Jacques, 1988, *Dimensions: A Casebook of Alien Contact,* Chicago: Contemporary Books.
Vallee, Jacques, 1989, "Recent Field Investigations into Claims of UFO Related Injuries in Brazil," *MUFON 1989 International UFO Symposium Proceedings,* Seguin, Texas: MUFON, Inc., pp. 32–41.
Vallee, Jacques, 1990a, *Confrontations,* New York: Ballantine Books.
Vallee, Jacques, 1990b, "Five Arguments Against the Extraterrestrial Origin of Unidentified Flying Objects," *Journal of Scientific Exploration,* Vol. 4, No. 1, pp. 105–17.
Vallee, Jacques, 1991, *Revelations,* New York: Ballantine Books.
Vallee, Jacques, 1992, *UFO Chronicles of the Soviet Union,* New York: Ballantine Books.
Van Buitenen, J. A. B., trans., 1973, *The Mahabharata,* Book 1, Chicago: The Univ. of Chicago Press.
Van Buitenen, J. A. B., trans., 1975, *The Mahabharata,* Books 2 and 3, Chicago: The Univ. of Chicago Press.
Vasiliev, Leonid L., 1963, *Experiments in Distant Influence,* London: Wildwood House.
Velasco, Jean-Jacques, 1987, "Scientific Approach and Results of Studies into Unidentified Aerospace Phenomena in France," *MUFON 1987 International UFO Symposium Proceedings,* Seguin, Texas: MUFON, Inc., pp. 51–58.
Video 1, 1991, "UFO . . . Abduction, a True Story," videotape on the William Herrmann case from UFO Archives; distributed by Arcturus Films.
Video 2, 1990, "UFOs: The Best Evidence," a TV documentary narrated by George Knapp, Las Vegas, Nevada: KLAS-TV, Channel 8.
Video 3, 1991, "The UFO Report—Sightings," a TV documentary narrated by Tim White; supervising producer: Linda M. Howe.
Video 4, "Recollections of Roswell," a videotape with research by Don Berliner, Stanton T. Friedman, William L. Moore, Kevin Randle, and Don Schmitt, Mt. Rainer, Md.: Fund for UFO Research.
Walters, Ed and Walters, Francis, 1990, *The Gulf Breeze Sightings,* New York: William Morrow and Co., Inc.

Walters, Ed and Walters, Francis, 1994, *UFO Abductions in Gulf Breeze*, New York: Avon Books.

Walton, Travis, 1978, *The Walton Experience*, New York: Berkley Pub. Corp.

Webb, David F., 1985, "The Influence of Hypnosis in the Investigation of Abduction Cases," *MUFON 1985 UFO Symposium Proceedings*, Seguin, Texas: MUFON, Inc., pp. 88–95.

Webb, Walter N., 1988, "Encounter at Buff Ledge: a UFO Case History," *MUFON 1988 International UFO Symposium Proceedings*, Seguin, Texas: MUFON, Inc., pp. 20–36.

Wescott, Roger, April 21, 1991, personal communication.

Wescott, Roger, Dec. 16, 1992, personal communication.

Wheeler, John A., 1962, *Geometrodynamics*, New York: Academic Press.

Wilson, H. H., 1865, *The Viṣṇu Purāṇa*, Vol. II, London: Trubner & Co.

Wilson, H. H., 1989, *The Viṣṇu Purāṇa*, Vol. I, Jawahar Nager, Delhi: Nag Publishers.

Wright, Dan, February 1994a, "The Entities, Part I," *MUFON UFO Journal*, No. 310, pp. 3–7.

Wright, Dan, March 1994b, "The Entities, Part II," *MUFON UFO Journal*, No. 311, pp. 3–7.

Zeidman, Jennie, 1989, "The Mansfield Helicopter Case," *MUFON 1989 International UFO Symposium Proceedings*, Seguin, Texas: MUFON, Inc., pp. 12–30.

Zinsstag, Lou, 1990, *UFO . . . George Adamski, their Man on Earth*, Tucson, Arizona: UFO Photo Archives.

Notes

Introduction

1. Thompson, 1981.
2. Leslie and Adamski, 1953, pp. 85– 88.
3. Leslie and Adamski, 1953, pp. 95–100, 104.
4. Leslie and Adamski, 1953, p. 100.
5. Ganguli, 1976, p. 417.
6. Maccabee, 1990b, pp. 32–33.
7. Walters, 1990.
8. Earley, 1990, p. 40.
9. Kinder, 1987.
10. Stevens, 1982.
11. Amnesty, 1983, p. 9.
12. Vallee, 1991.
13. Downing, 1968, 1981.

Chapter 1

1. Howe, 1989a, p. 3.
2. Howe, 1989a, pp. 4–5.
3. Braude, 1991, p. 12.
4. Hoagland, 1969.
5. Condon, 1969b, pp. 7–8.
6. Markowitz, 1980, p. 258.
7. Markowitz, 1980, p. 258.
8. Markowitz, 1980, p. 258.
9. Sagan, 1972, pp. 267–71.
10. Sagan, 1972, p. 271.
11. Fowler, 1981, pp. 221–22.
12. See the *MUFON 1989 International UFO Symposium Proceedings.*
13. Hall, 1964, p. 53.
14. Hall, 1964, p. 3.
15. Condon, 1969a, p. 543.
16. Page, 1972, p. 9.
17. Page, 1972, pp. 9–10.
18. Roach, 1972, pp. 30–31.
19. Condon, 1969a, p. 516.
20. Condon, 1969a, p. 516.
21. Vallee, 1979, p. 7.
22. Hall, 1964, pp. 3–4.
23. Hall, 1964, pp. 2–3.
24. Condon, 1969a, pp. 839–40.
25. Markowitz, 1980, p. 255.
26. Markowitz, 1980, p. 255.
27. Powers, 1967.
28. Markowitz, 1980, p. 258.
29. Hynek, 1972b, pp. 47–48.
30. Hynek, 1972a, p. 157.
31. Fowler, 1979, p. 9.
32. Hynek, 1972a, p. 139.
33. Congress, 1966, p. 6046.
34. Congress, 1966, p. 6009.
35. Hynek, 1972a, pp. 185–86.
36. McDonald, 1967, p. 1.
37. McDonald, 1967, p. 6.
38. Hynek, 1972a.
39. Menzel and Boyd, 1963, p. 269.
40. McDonald, 1967, p. 8.
41. McDonald, 1971.
42. Kuettner, 1971, p. 18.
43. Henry, 1988, p. 122.
44. Fowler, 1981, pp. 224–25.
45. Congress, 1966, p. 5992.
46. Fowler, 1981, pp. 224–25.

Chapter 2

1. Hynek, 1972b, pp. 44–49.
2. Fowler, 1981, pp. 54, 153.
3. Fowler, 1981, p. 153.
4. Vallee, 1990b, p. 107.
5. Vallee, 1990b, pp. 108–11.
6. Fowler, 1981.

7. Fowler, 1981, pp. 107, 127, 129.
8. Randles, 1988, p. 53.
9. Cannon.
10. Cannon, p. 55.
11. Haley, 1993.
12. Turner, 1992, pp. 223–32.
13. Vallee, 1991, pp. 149–50.
14. Boylan, 1994, p. 50.
15. Stevenson, 1971, p. 93.
16. Hynek, 1972b, pp. 41–42.
17. Stevenson, 1971, pp. 100–1.
18. Stevenson, 1971, pp. 101–2.
19. Stevenson, 1971, p. 103.
20. Stevenson, 1971, p. 105.
21. Stevenson, 1971, p. 107.
22. Ofshe and Watters, 1993, p. 5.
23. Farberman, 1994, p. 2.
24. Daly and Pacifico, 1991, p. 45.
25. Spanos, *et al.,* 1993b.
26. Stevenson, 1994.
27. Stevenson, 1994, p. 192.
28. Lanning, 1992, p. 3.
29. Stevenson, 1971, p. 110.
30. Stevenson, 1971, p. 111.
31. Hopkins, 1981, pp. 23–28.
32. Hopkins, 1981, p. 28.
33. Hopkins, 1981, pp. 28–30.
34. Hopkins, 1981, pp. 32–33.
35. Hopkins, 1981, pp. 34–35.
36. Schwarz, 1988, p. 239.
37. Congress, 1966, p. 6018.
38. Congress, 1966, p. 6016.
39. Congress, 1966, p. 6039.
40. Vallee, 1969b, p. 28.
41. Fowler, 1981, p. 3.
42. Creighton, 1969a, pp. 187–99.
43. Congress, 1966, p. 6009.
44. Schwarz, 1988, pp. 69–77.
45. Schwarz, 1988, p. 78.
46. Schwarz, 1988, p. 80.
47. Schwarz, 1988, p. 83.
48. Ribera, 1966, pp. 28–29.
49. Lorenzen, 1969, p. 148.
50. Hind, 1981, p. 80.
51. Hind, 1981, p. 81.
52. Velasco, 1987, p. 56.
53. Velasco, 1987, pp. 56–57.
54. Phillips, 1981, pp. 93–129.
55. Lorenzen, 1976, pp. 406–15.
56. Condon, 1969a, p. 115.
57. Chalker, 1987, p. 176.
58. Fuller and Randles, 1991.
59. Johnson, 1989.
60. Johnson, 1989, p. 140.
61. Condon, 1969a, pp. 85–86.
62. Stevens and Roberts, 1986.
63. Stevens and Roberts, 1986, p. 38.
64. Stevens and Roberts, 1986, p. 39.
65. Stevens and Roberts, 1986, p. 34.
66. Maccabee, 1990a, p. 3.
67. Maccabee and Sainio, 1993, pp. 4–7.
68. Good, 1988, p. 292.
69. Good, 1988, pp. 292–93.

Chapter 3

1. Condon, 1969a, p. 503.
2. Condon, 1969a, p. 894.
3. Condon, 1969a, p. 506.
4. Condon, 1969a, pp. 902–3.
5. Condon, 1969a, pp. 903–4.
6. Hynek, 1972a, p. 174.
7. Hynek, 1972a, p. 175.
8. Condon, 1969a, p. 509.
9. Condon, 1969a, p. 509.
10. Condon, 1969a, p. 46.
11. Condon, 1969a, p. 511.
12. Fawcett and Greenwood, 1984, p. 125.
13. Condon, 1969a, p. 519.
14. Condon, 1969a, pp. 519–20.
15. Condon, 1969a, pp. 915–16.
16. Hall, 1964, pp. 19–22.
17. AFR, 1959, paragraph 9.
18. Fawcett and Greenwood, 1984, p. 165.
19. Fawcett and Greenwood, 1984, p. 166.

20. Hall, 1964, p. 84.
21. Hall, 1964, p. 84.
22. Condon, 1969a, p. 852.
23. AFR, 1959, paragraph 19.
24. Condon, 1969a, p. 175.
25. Condon, 1969a, p. 153.
26. Hall, 1964, p. 84.
27. Hall, 1964, p. 4.
28. Haines, 1990, pp. 28–29.
29. Condon, 1969a, p. 250.
30. Condon, 1969a, p. 256.
31. Condon, 1969a, p. 254.
32. Congress, 1966, p. 6073.
33. Hynek, 1972a, p. 70.
34. Condon, 1969a, p. 541.
35. Condon, 1969a, p. 543.
36. Condon, 1969a, p. 544.
37. Congress, 1966, pp. 6006–7.
38. Condon, 1969a, p. 547.
39. Lorenzen, 1969, p. 171.
40. Condon, 1969a, p. 528.
41. Saunders, 1968, p. 247.
42. Condon, 1969a, p. 1.
43. Condon, 1969a, p. 4.
44. Condon, 1969a, p. 5.
45. Condon, 1969a, pp. 5–6.
46. Hynek, 1976, p. 82.
47. Hynek, 1976, p. 83.
48. Condon, 1969a, pp. 266–67.
49. Condon, 1969a, p. 267.
50. Condon, 1969a, p. 268.
51. Condon, 1969a, pp. 72–73.
52. Fowler, 1981, p. 51.
53. NAS, 1972, p. 96.
54. Good, 1988.
55. Fawcett and Greenwood, 1984.
56. Huneeus, 1991.
57. Vallee, 1992.
58. Gresh, 1993.
59. Stevens and Dong, 1983.
60. Stevens and Dong, 1983, pp. 291–98.
61. Fowler, 1981, pp. 190–91.
62. Fowler, 1981, p. 186.
63. Fowler, 1981, pp. 187–88.
64. Good, 1988, pp. 497–500.
65. Good, 1988, pp. 498–99.
66. Corliss, 1990, p. 4.
67. This report will appear in the revised version of Albertson and Shaw, 1992.
68. Condon, 1969a, p. 722.
69. Berlitz and Moore, 1980.
70. Friedman and Moore, 1981.
71. Moore, 1985.
72. Randle and Schmitt, 1991.
73. Video 4.
74. Randle and Schmitt, 1991, pp. 130–31.
75. Randle and Schmitt, 1991, p. 50.
76. Randle and Schmitt, 1991, p. 110.
77. Vallee, 1991, p. 83.
78. Randle and Schmitt, 1991, p. 110.
79. Randle and Schmitt, 1991, p. 31.
80. Randle and Schmitt, 1991, p. 37.
81. Randle and Schmitt, 1991, p. 227.
82. Randle and Schmitt, 1991, p. 181.
83. Randle and Schmitt, 1991, opposite p. 177.
84. Randle and Schmitt, 1991, p. 93.
85. Randle and Schmitt, 1991, pp. 114–15.
86. Carpenter, 1993a.
87. Randle and Schmitt, 1994.
88. Air Force, 1994.
89. Blum, 1990, pp. 233–34.
90. Vallee, 1991, p. 74.
91. Good, 1988, pp. 544–51.
92. Friedman, 1990.
93. Moore and Friedman, 1988, p. 227.
94. Wescott, 1991.
95. Good, 1988, pp. 525–26.
96. Steinman, 1986, p. 337.
97. Steinman, 1986, p. 345.
98. Marchetti, 1979, p. 5.
99. Howe, 1989a, pp. 143–59.
100. Seligman, 1988.
101. Video 2, 1990.
102. Condon, 1969a, p. 522.
103. Condon, 1969a, p. 520.

Chapter 4

1. Hynek, 1972a, p. 142.
2. Webb, 1988.
3. Fuller, 1966.
4. Fuller, 1966, p. 260.
5. Fuller, 1966, p. 261.
6. Randles, 1988, p. 170.
7. Hopkins, 1981, p. 7.
8. Hopkins, 1981, p. 240.
9. Vallee, 1990b, p. 112.
10. Fowler, 1979, pp. 33–38.
11. Scully, 1950, p. 29.
12. Steinman, 1986.
13. Steinman, 1986, p. 291.
14. Scully, 1949, p. 61.
15. Fuller, 1966, p. 263.
16. Ring, 1992, p. 214.
17. Ring, 1992, p. 219.
18. Fowler, 1979, p. 35.
19. Moore, 1985, p. 152.
20. Zeidman, 1989, pp. 14–16.
21. Zeidman, 1989, pp. 17–18.
22. Fowler, 1990, p. 236.
23. Hopkins, 1981, pp. 212–13.
24. Creighton, 1969b, pp. 234–36.
25. Schuessler, 1985.
26. Hopkins, 1981, p. 140.
27. Fiore, 1992, p. 44.
28. Fiore, 1992, pp. 41, 44.
29. Randles, 1988, pp. 180–81.
30. Lorenzen, 1976, pp. 406–15.
31. Carpenter, 1993b, p. 11.
32. Bullard, 1987.
33. Randles, 1988, p. 180.
34. Taff, 1989.
35. Dobzhansky, 1972, p. 173.
36. Dobzhansky, 1972, p. 173.
37. Simpson, 1964, p. 259.
38. Wright, 1994a, pp. 6–7.
39. Wright, 1994b, pp. 4–5.
40. Simpson, 1964, p. 268.
41. Creighton, 1969b.
42. Hopkins, 1987, pp. 198–99, 215–20.
43. Hopkins, 1987.
44. Conroy, 1989, p. 152.
45. Conroy, 1989, p. 356.
46. Jacobs, 1992.
47. Fowler, 1990, pp. 20–30.
48. Randles, 1988, p. 187.
49. Randles, 1988, pp. 77–78.
50. Randles, 1988, p. 134.
51. Randles, 1988, pp. 139–40.
52. Randles, 1988, p. 141.
53. Randles, 1988, p. 146.
54. Stevens and Herrmann, 1981, pp. 136–37.
55. Vallee, 1990b, pp. 112–13.
56. Hopkins, 1981, p. 98.
57. Conroy, 1989, p. 152.
58. Hopkins, 1988, p. 20.
59. Stevens and Herrmann, 1981.
60. Sanchez-Ocejo and Stevens, 1982.
61. Salter, 1989.
62. Salter, 1989, p. 25.
63. Hopkins, 1981, p. 94.
64. Hopkins, 1981, p. 229.
65. Hopkins, 1987, p. 53.
66. Fuller, 1966, p. 238.
67. Druffel and Rogo, 1988, p. 73.
68. Randles, 1988, p. 60.
69. Randles, 1988, p. 61.
70. Randles, 1988, p. 170.
71. Council, 1985, p. 1.
72. Council, 1985, p. 3.
73. Council, 1985, pp. 3–4.
74. Fowler, 1993.
75. Carpenter, 1991.
76. Lawson, 1977.
77. Randles, 1988, pp. 202–3.
78. Maccabee, 1985, p. i.
79. Randles, 1988, pp. 179, 189–90.
80. Webb, 1985.
81. Bershad, 1988.
82. Hopkins, 1981, pp. 77–78.
83. Stevenson, 1977, p. 635.
84. Fiore, 1989, pp. 176–212.
85. Fiore, 1989, p. 177.
86. Bloecher *et al.*, 1985.

87. Slater, 1983a, p. 18.
88. Slater, 1983b, p. 33.
89. Slater, 1983b, p. 36.
90. Conroy, 1989, pp. 237–40.
91. Randles, 1988, p. 206.
92. Ring, 1992, p. 112.
93. Ring, 1992, p. 142.
94. Ring, 1992, p. 239.
95. Mack, 1994, p. 406.
96. Boylan, 1994, p. 50.
97. Spanos, *et al.*, 1993a.
98. Spanos, *et al.*, 1993a, p. 628.
99. Spanos, *et al.*, 1993a, p. 628.
100. Spanos, *et al.*, 1993a, p. 629.
101. Spanos, *et al.*, 1993a, p. 631.
102. Keene, 1976.
103. Braude, 1991.
104. Braude, 1991, pp. 128–29.
105. Braude, 1991, pp. 131–32.
106. Braude, 1991, p. 131.
107. Braude, 1991, pp. 137–38.
108. Braude, 1991, p. 138.
109. Schwarz, 1988, pp. 274–76.
110. Roll, 1977.
111. Schwarz, 1988, pp. 289–92.
112. Schwarz, 1988, p. 293.
113. Turner, 1992.
114. Turner, 1993.
115. Walters, 1994.
116. Fowler, 1993.
117. Vallee, 1988, pp. 173–76.
118. Vallee, 1988, pp. 176, 259.
119. Vallee, 1990a, p. 117.
120. Vallee, 1990a, p. 119.
121. Stevens and Herrmann, 1981, pp. 179–80.
122. Strieber, 1988.
123. Randles, 1988, p. 86.
124. Fowler, 1979, p. 189.
125. Fowler, 1979, p. 199.
126. Rhine, 1961.
127. Vallee, 1990b, p. 116.
128. Schwarz, 1988, pp. 517–18.
129. quoted in Randles, 1988, p. 220.
130. Bearden, 1979, p. 86.
131. quoted in Conroy, 1989,p. 287.
132. Randles, 1988, pp. 220–21.

Chapter 5

1. Fowler, 1981, p. 12.
2. Story, 1980, pp. 2–4.
3. Story, 1980, p. 52.
4. Story, 1980, pp. 145–46.
5. Story, 1980, pp. 227–29.
6. Story, 1980, pp. 380–81.
7. Jung, 1959, pp. 154–62.
8. Jung, 1959, pp. 146–47.
9. Zinsstag, 1990, p. 76.
10. Zinsstag, 1990, p. 92.
11. Zinsstag, 1990, pp. 73–74, 90.
12. Zinsstag, 1990, p. 80.
13. Zinsstag, 1990, p. 93.
14. Hickson and Mendez, 1983, p. 2.
15. Fowler, 1979, p. 99.
16. Stevens and Herrmann, 1981 and Video 1, 1991.
17. Stevens and Herrmann, 1981, pp. 36–49.
18. Stevens and Herrmann, 1981, p. 136.
19. Stevens and Herrmann, 1981, pp. 212–26.
20. Sanchez-Ocejo and Stevens, 1982.
21. Sanchez-Ocejo and Stevens, 1982, p. 28.
22. Sanchez-Ocejo and Stevens, 1982, p. 35.
23. Sanchez-Ocejo and Stevens, 1982, p. 49.
24. Sanchez-Ocejo and Stevens, 1982, pp. 115–16.
25. Sanchez-Ocejo and Stevens, 1982, p. 12
26. Sanchez-Ocejo and Stevens, 1982, pp. 67, 116–18.
27. Sanchez-Ocejo and Stevens, 1982, pp. 124–33.

28. Sanchez-Ocejo and Stevens, 1982, pp. 47, 54, 56, 67, 77.
29. Vallee, 1988, pp. 6–9.
30. Druffel and Rogo, 1988, pp. 118–28.
31. Druffel and Rogo, 1988, p. 120.
32. Stevens and Herrmann, 1981, pp. 140–41.
33. Lorenzen, 1976, pp. 101–11.
34. Dickinson, 1974.
35. Stevens and Herrmann, 1981, p. 232.
36. Stevens and Herrmann, 1981, p. 213.
37. Green, 1985, p. 504.
38. Lorenzen, 1976, p. 107.
39. Steinman, 1986, pp. 578–94.
40. Steinman, 1986, p. 590.
41. Stevens and Herrmann, 1981, pp. 139, 152.
42. Steinman, 1986, pp. 588–89.
43. Steinman, 1986, p. 583.
44. Steinman, 1986, p. 602.
45. Steinman, 1986, p. 105.
46. Moore, 1985, p. 144.
47. Fawcett and Greenwood, 1984, p. 170.
48. Moore, 1985, p. 145.
49. Stevens and Herrmann, 1981, p. 378.
50. Vallee, 1988, pp. 240–41.
51. Vallee, 1988, p. 240.
52. Stevens and Herrmann, 1981, p. 40.
53. Sanchez-Ocejo and Stevens, 1982, p. 128.
54. Carpenter, 1992, p. 21.
55. Hall, 1988, pp. 282–83.
56. O'Brien, 1985.
57. Sitchin, 1976.
58. Pius XII, 1950.
59. Howe, 1989a, p. 151.
60. Howe, 1989a, p. 256.
61. Elkins *et al.*, 1984, pp. 107–8, 174, 187.
62. Telano, 1963, p. 38.
63. Howe, 1992, p. 5.
64. Howe, 1992, p. 23.
65. Sitchin, 1976, pp. 312–61.
66. Fowler, 1990, p. 204.
67. Fowler, 1990, p. 49.
68. Fowler, 1979, pp. 96–99.
69. Strieber, 1988, p. 52.
70. Stevens and Herrmann, 1981, p. 284.
71. Elkins *et al.*, 1984, pp. 92–93, 161.
72. Layne, 1953, p. 7.
73. Howe, 1989a, p. 324.
74. Harder, 1992, p. 140.
75. Fowler, 1979, pp. 144–45.
76. Fowler, 1990, pp. 340–41.
77. Randles, 1988, p. 176.
78. Fowler, 1979, p. 187.
79. Salter, 1989, pp. 26, 29.
80. Hopkins, 1987, pp. 273–74.
81. Randles, 1988, p. 174.
82. Fowler, 1990, p. 213.
83. Fowler, 1990, p. 49.

Chapter 6

1. Van Buitenen, 1975, p. 308.
2. Sastrin, 1860, p. 11.
3. Thompson, 1989, p. 25.
4. Thompson, 1989, pp. 74–78, 85–88.
5. *Bhāg. Pur.* 10.75.13, 19, 22, 23, 26.
6. Bhaktivedanta, 1986, p. 649.
7. McDonald, 1971, p. 66.
8. *Bhāg. Pur.* 10.76.4–12.
9. Van Buitenen, 1975, pp. 264–65.
10. Bowen, 1969b, pp. 245–47.
11. Bhaktivedanta, 1986, p. 655.
12. Vallee, 1969a, p. 68.
13. Hickson and Mendez, 1983, pp. 152–55.
14. Van Buitenen, 1975, p. 202.
15. Fowler, 1979, pp. 174–75.
16. Bhaktivedanta, 1982a, Canto 4, Part 3, pp. 31–32.

17. *Bhāg. Pur.* 11.15.12.
18. Hridayānanda, 1992, Part 7, p. 271.
19. Stevens and Herrmann, 1981, pp. 177–80.
20. Randles, 1988, p. 184.
21. Hopkins, 1987, p. 109.
22. *Bhāg. Pur.* 10.62.16–21.
23. Druffel and Rogo, 1988, p. 50.
24. Fiore, 1989, p. 116.
25. Fiore, 1989, pp. 233–34.
26. Randles, 1988, pp. 84– 85.
27. Fowler, 1979, pp. 22–23.
28. Fowler, 1979, pp. 33–34.
29. *Bhāg. Pur.* 11.15.21.
30. Hopkins, 1987, pp. 229–30.
31. Fowler, 1993, p. 93.
32. Roll, 1977, p. 389.
33. Hridayānanda, 1992, Part 16, p. 484.
34. Hridayānanda, 1992, Part 16, p. 485.
35. Van Buitenen, 1975, p. 691.
36. Van Buitenen, 1975, p. 691.
37. Thompson, 1989, pp. 47–84.
38. Wilson, 1865, pp. 263–68.
39. Van Buitenen, 1975, p. 692.
40. Van Buitenen, 1975, p. 693.
41. Strieber, 1988, pp. 216–17.
42. Strieber, 1988, p. 218.
43. Strieber, 1988, p. 219.
44. Hridayānanda, 1992, Part 13, pp. 447–48.
45. Bhaktivedanta, 1982b, p. 12.
46. Myers, 1961, p. 142.
47. Vasiliev, 1963, p. 211.
48. Vasiliev, 1963, p. 213.
49. Vasiliev, 1963, p. 144.
50. Vallee, 1988, p. 192.
51. Boylan, 1994, p. 148.
52. Hopkins, 1981, pp. 178, 188–89.
53. Hopkins, 1981, pp. 195–96.
54. Strieber, 1988, p. 106.
55. Walters, 1990, pp. 256–58.
56. Hopkins, 1981, pp. 220–21.
57. See Randles, 1988.
58. Satsvarūpa, 1984, pp. 29–30.
59. Randles, 1988, p. 22.
60. Hopkins, 1981, pp. 82–83.
61. Fowler, 1979, p. 15.
62. Randles, 1988, p. 222.
63. Shastri, 1976, Vol. II, p. 95.
64. Van Buitenen, 1975, pp. 299–300.

Chapter 7

1. Raghavan, 1956, p. 21.
2. Raghavan, 1956, p. 8.
3. Phillimore, 1912, p. 88.
4. Elliot, 1875, pp. 477–78.
5. Oppert, 1967.
6. Raghavan, 1956, pp. 11–12.
7. Raghavan, 1956, p. 15.
8. Raghavan, 1956, p. 31.
9. Raghavan, 1956, pp. 29–30.
10. Leslie and Adamski, 1953, pp. 91–93.
11. Raghavan, 1956, p. 30.
12. Dikshitar, 1944, p. 281.
13. Sastrin, 1860, p. 90.
14. Kanjilal, 1985, p. 1.
15. Kanjilal, 1985.
16. Nathan, 1987.
17. Childress, 1991.
18. Nathan, 1987, p. 71.
19. Josyer, 1973, pp. vi–vii.
20. Kanjilal, 1985, pp. 2–3.
21. Prabhu, 1992.
22. Prabhu, 1992, p. 5.
23. Josyer, 1973, pp. 3–6.
24. Stevens and Herrmann, 1981, pp. 223–24.
25. Nathan, 1987, p. 70.
26. Nathan, 1987, p. 70.
27. *Bhāg. Pur.* 10.34.12–13.
28. *Bhāg. Pur.* 10.64.30.
29. *Bhāg. Pur.* 3.22.17.
30. *Bhāg. Pur.* 3.23.12–21.

31. *Bhāg. Pur.* 3.23.40–41.
32. *Bhāg. Pur.* 8.10.16–18.
33. *Bhāg. Pur.* 4.3.12.
34. Stevens, 1982, pp. 77–78.
35. *Śiva Purāṇa*, 1991, p. 807.
36. Van Buitenen, 1975, p. 549.
37. Van Buitenen, 1975, p. 550.
38. Sanderson, 1970.
39. Fowler, 1979.
40. Sanchez-Ocejo and Stevens, 1982.
41. Van Buitenen, 1975, p. 46.
42. Van Buitenen, 1975, p. 48.
43. Van Buitenen, 1975, p. 308.
44. Van Buitenen, 1975, p. 309.
45. Wilson, 1865, pp. 269–72.
46. Wilson, 1865, pp. 263–68.
47. *Bhāg. Pur.* 5.26.5.
48. Hridayānanda, 1992, Part 11, p. 396.
49. Hridayānanda, 1992, Part 19, p. 507.
50. Druffel and Rogo, 1988, pp. 46–47.
51. Curtis, 1989, p. 37.
52. *Bhāg. Pur.* 2.2.24.
53. Bhaktivedanta, 1982a, Canto 4, Part 2, p. 182.
54. *Bhāg. Pur.* 3.26.32–44.
55. Wheeler, 1962.
56. *Bhāg. Pur.* 3.26.34.
57. *Bhāg. Pur.* 4.12.19–20.
58. *Bhāg. Pur.* 4.12.34–35.

Chapter 8

1. Vallee, 1988, pp. 38–39.
2. *Bhāg. Pur.* 4.10.4–4.11.7.
3. Evans-Wentz, 1966, pp. 283–84.
4. Evans-Wentz, 1966, pp. 308–11, 314–15.
5. Wescott, 1992.
6. Evans-Wentz, 1966, p. 102.
7. Hartland, 1891, p. 101.
8. Jacobs, 1992, pp. 167–68.
9. Van Buitenen, 1973, p. 301.
10. *Bhagavad-gītā* 7.10, 14.4.
11. *Bhāg. Pur.* 4.14.35.
12. *Bhāg. Pur.* 4.14.43–44.
13. Conroy, 1989, p. 363.
14. Augustine, 1952, pp. 416–17.
15. Evans-Wentz, 1966, pp. 112–13.
16. Mack, 1994, p. 417.
17. Haley, 1993, p. 84.
18. Twiggs, 1992.
19. Jacobs, 1992, p. 206.
20. Boylan, 1993, p. 179.
21. Turner, 1992, p. 172.
22. Walters, 1994, p. 95.
23. Evans-Wentz, 1966, pp. 346–47.
24. Hartland, 1891, pp. 162–63.
25. Kafton-Minkel, 1989, p. 191.
26. *Bhāg. Pur.* 5.24.9–10.
27. *Bhāg. Pur.* 5.24.16.
28. *Bhāg. Pur.* 9.3.28–32.
29. *Bhāg. Pur.* 10.13.40.
30. Sastrin, 1860, p. 2.
31. Randles, 1988, p. 162.
32. Mack, 1994, p. 404.
33. Ring, 1992, p. 221.
34. Fowler, 1979, pp. 65–86.
35. Fowler, 1979, pp. 33–37.
36. Rickard, 1990, p. 63.
37. Vallee, 1988, p. 19.
38. Vallee, 1988, p. 16.
39. Stein, 1979, pp. 35–40.
40. Kumari, 1973.
41. Bamzai, 1973, pp. 52–53 and Ahmad and Bano, 1984, pp. 33–35.
42. Agrawal, 1985, p. 67.
43. Agrawal, 1985, p. 7.
44. Johnston, 1979, p. 46.
45. *Bhāg. Pur.* 10.33.3–4.
46. Johnston, 1979, p. 46.
47. Johnston, 1979, p. 47.
48. Johnston, 1979, p. 62.
49. Johnston, 1979, p. 63.
50. Johnston, 1979, p. 55.
51. Johnston, 1979, p. 65.

52. Meaden, 1990, p. 88.
53. Johnston, 1979, pp. 27–28.
54. Johnston, 1979, p. 24.
55. Stevens and Roberts, 1986, pp. 14–22.
56. Vallee, 1988, p. 196.
57. Johnston, 1979, p. 51.
58. Johnston, 1979, p. 25.
59. Vallee, 1988, p. 202.
60. Vallee, 1988, p. 214.
61. Pickthall, 1976, pp. xxi–xxii.

Chapter 9

1. Schwarz, 1988, pp. 196–207.
2. Schwarz, 1988, pp. 196, 210.
3. Schwarz, 1988, p. 210.
4. *Bhāg. Pur.* 10.66.32–33.
5. *Bhāg. Pur.* 3.26.1–72.
6. *Bhagavad-gītā* 14.6–8.
7. *Bhāg. Pur.* 5.24.4–6.
8. *Bhāg. Pur.* 10.6.27–29.
9. Howe, 1989a, pp. 104–5.
10. Howe, 1989a, p. 23.
11. Howe, 1989a, p. 4.
12. Howe, 1989a, p. 4.
13. Good, 1988, pp. 303–5.
14. Fowler, 1990, p. 16.
15. Conroy, 1989, pp. 376–77.
16. Fowler, 1990, p. xii.
17. Howe, 1989a, p. 2.
18. Howe, 1989a, p. 16.
19. Howe, 1989a, p. 17.
20. Howe, 1989a, p. 21.
21. Howe, 1989a, p. 22.
22. Howe, 1989b, p. 119.
23. Conroy, 1989, pp. 177–79, and Howe, 1989a, pp. 301–39.
24. Howe, 1989a, pp. 341–73.
25. Steinman, 1986, pp. 581, 589.
26. Blum, 1990, pp. 230, 233.
27. Blum, 1990, p. 230.
28. Howe, 1989a, p. 341.
29. Fowler, 1990, pp. 9–10.
30. Steinman, 1986, p. 421.
31. Sanchez-Ocejo and Stevens, 1982, p. 115.
32. Howe, 1989a, p. 120.
33. Howe, 1989a, pp. 116–20.
34. Vallee, 1988, p. 20.
35. Vallee, 1988, p. 21.
36. Hridayānanda, 1992, Part 13, pp. 449–50.
37. Randles, 1988, p. 160.
38. Garcia, 1993.
39. Vallee, 1989, pp. 32–41.
40. Schuessler and O'Herin, 1993, p. 66.
41. Schuessler and O'Herin, 1993, p. 75.
42. Randles, 1988, p. 175.
43. Walton, 1978, pp. 23–78.
44. Haines, 1990, pp. 19–20.
45. Haines, 1990, p. 20.
46. Walton, 1978, p. 28.
47. Schwarz, 1988, p. 242.
48. Schwarz, 1988, p. 242.
49. Schwarz, 1988, p. 244.
50. Schwarz, 1988, p. 250.
51. Schwarz, 1988, p. 245.
52. *Bhāg. Pur.* 7.10.54–55.
53. *Bhāg. Pur.* 2.7.37.
54. Van Buitenen, 1975, p. 420.
55. Shastri, 1976, Vol. I, pp. 38–39.
56. Shastri, 1976, Vol. I, p. 39.
57. Elkins *et al.,* 1984, p. 25.
58. *Bhāg. Pur.* 5.19.21.
59. Thompson, 1989, pp. 53, 56–57, 65.
60. Evans-Wentz, 1966, pp. 46–47.
61. Hridayānanda, 1992, Part 6, pp. 154–55.
62. Hridayānanda, 1992, Part 6, p. 155.
63. Hridayānanda, 1992, Part 6, p. 156.

Chapter 10

1. Sabom, 1982, p. 76.
2. For example, see Ring, 1985, and Sabom, 1982.
3. Sabom, 1982, p. 91.
4. Fowler, 1990, p. 174.
5. Fowler, 1990, p. 176.
6. Strieber, 1988, pp. 201–4.
7. Fowler, 1990, p. 346.
8. Druffel and Rogo, 1988, p. 70.
9. Druffel and Rogo, 1988, pp. 72–78.
10. Fowler, 1990, p. 10.
11. Fowler, 1990, p. 11.
12. Fowler, 1990, p. 144.
13. Fowler, 1990, p. 150.
14. Fowler, 1990, p. 151.
15. Randles, 1988, pp. 218–19.
16. Jacobs, 1988, pp. 91–92.
17. Jacobs, 1988, p. 99.
18. Randles, 1988, p. 222.
19. Hopkins, 1981.
20. Hopkins, 1981, p. 130.
21. Fowler, 1990, pp. 324–25.
22. Fowler, 1990, p. 270.
23. Fowler, 1990, p. 233.
24. Fowler, 1990, p. 234.
25. Fowler, 1990, p. 235.
26. Fowler, 1990, p. 232.
27. Myers, 1961, pp. 138–40.
28. Hopkins, 1987, pp. 49–51.
29. Fuller, 1966.
30. Pasricha and Stevenson, 1986, p. 166.
31. Stevenson, 1988, p. 116.
32. Pasricha and Stevenson, 1986, p. 167.
33. *Bhāg. Pur.* 6.1–2.
34. Druffel and Rogo, 1988, pp. 36–37, 86–87.
35. Strieber, 1988, p. 241.
36. Strieber, 1988, p. 210.
37. Strieber, 1988, p. 214.
38. Strieber, 1988, pp. 152–58.
39. Strieber, 1988, p. 109.
40. Howe, 1989a, pp. 154–55.
41. Zeidman, 1989, pp. 14–18.
42. Fowler, 1990, pp. 189–90.
43. Turner, 1993.
44. Turner, 1993.
45. *Bhāg. Pur.* 11.15.23
46. Hridayānanda, 1992, Part 11, pp. 409–10.
47. Van Buitenen, 1973, pp. 298–99.
48. Rosen, 1991, pp. 84–85.
49. Rosen, 1991, pp. 63–64.
50. Rosen, 1991, pp. 119–39.

Chapter 11

1. Fowler, 1979, pp. 143–44.
2. Fowler, 1979, p. 121.
3. Fowler, 1979, p. 141.
4. Howe, 1989a, pp. 118–19.
5. Howe, 1989a, p. 119.
6. Elkins and Rueckert, 1977.
7. Elkins and Rueckert, 1977, p. 82.
8. Elkins and Rueckert, 1977, p. 30.
9. Elkins and Rueckert, 1977, p. 20.
10. Sanchez-Ocejo and Stevens, 1982, pp. 146–47.
11. Sanchez-Ocejo and Stevens, 1982, p. 151.
12. Sanchez-Ocejo and Stevens, 1982, pp. 154–55.
13. Sanchez-Ocejo and Stevens, 1982, p. 133.
14. Randles, 1988, p. 171.
15. Bowen, 1969a, pp. 17–18.
16. Bhaktisiddhānta, 1985, p. 59.
17. Vallee, 1988, p. 40.
18. Vallee, 1988, p. 40.
19. Hynek *et al.*, 1987, p. 158.
20. Seligman, 1988.
21. Howe, 1989a, p. 151.
22. Bailey, 1951, pp. 162–63.

23. Bailey, 1962.
24. Bailey, 1951, pp. 88–89.
25. Bailey, 1951.
26. Bailey, 1962, p. 907.
27. Bailey, 1962, pp. 809, 910.
28. Bailey, 1962, p. 948.
29. Bailey, 1962, p. 594.
30. Bailey, 1951, p. 256.
31. Bailey, 1962, p. 678.
32. Sanchez-Ocejo and Stevens, 1982, p. 80.
33. Sanchez-Ocejo and Stevens, 1982, pp. 81–82.
34. Sanchez-Ocejo and Stevens, 1982, p. 81.
35. Sanchez-Ocejo and Stevens, 1982, p. 84.
36. Sanchez-Ocejo and Stevens, 1982, pp. 71–72.
37. Sanchez-Ocejo and Stevens, 1982, p. 73.
38. Bailey, 1962, p. 242.
49. Fowler, 1990, p. 11.
40. Fowler, 1990, p. 145.
41. Johnston, 1979, p. 28.
42. Angelucci, 1955, p. 32.
43. Angelucci, 1955, p. 34.
44. Elkins *et al.*, 1984, pp. 70–71, 90.
45. Elkins *et al.*, 1984, p. 78.
46. Elkins and Rueckert, 1977, p. 49.
47. Elkins and Rueckert, 1977, p. 50.
48. Elkins and Rueckert, 1977, p. 50.
49. *Bhāg. Pur.* 10.89.51–56.
50. Bhaktivedanta, 1982b.
51. Elkins *et al.*, 1984, p. 79.
52. *Bhāg. Pur.* 1.1.1.

Appendix 1

1. Frye, 1983, p. 15.
2. Apte, 1965, p. 1018.
3. See Thompson, 1989.
4. Bentley, 1825, p. xxvii.
5. Misra, 1988, p. 10.
6. Misra, 1988, p. 12.
7. Wilson, 1989, p. vi.

Appendix 2

1. Fowler, 1990, pp. 81–82.
2. Hopkins, 1981, p. 68.

Index

Made in the USA
Columbia, SC
12 March 2018